AF614756

METHODS IN MOLECULAR BIOLOGY™

Series Editor
John M. Walker
School of Life Sciences
University of Hertfordshire
Hatfield, Hertfordshire, AL10 9AB, UK

For further volumes:
http://www.springer.com/series/7651

Retinal Development

Methods and Protocols

Edited by

Shu-Zhen Wang

Department of Ophthalmology, University of Alabama at Birmingham, Birmingham, AL, USA

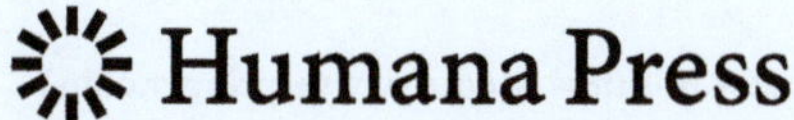

Editor
Shu-Zhen Wang
Department of Ophthalmology
University of Alabama at Birmingham
Birmingham, AL, USA

Please note that additional material for this book can be downloaded from http://extras.springer.com

ISSN 1064-3745 ISSN 1940-6029 (electronic)
ISBN 978-1-61779-847-4 ISBN 978-1-61779-848-1 (eBook)
DOI 10.1007/978-1-61779-848-1
Springer New York Dordrecht Heidelberg London

Library of Congress Control Number: 2012938950

© Springer Science+Business Media, LLC 2012
All rights reserved. This work may not be translated or copied in whole or in part without the written permission of the publisher (Humana Press, c/o Springer Science+Business Media, LLC, 233 Spring Street, New York, NY 10013, USA), except for brief excerpts in connection with reviews or scholarly analysis. Use in connection with any form of information storage and retrieval, electronic adaptation, computer software, or by similar or dissimilar methodology now known or hereafter developed is forbidden.
The use in this publication of trade names, trademarks, service marks, and similar terms, even if they are not identified as such, is not to be taken as an expression of opinion as to whether or not they are subject to proprietary rights.

Printed on acid-free paper

Humana Press is part of Springer Science+Business Media (www.springer.com)

Preface

In recent years, there have been major advances in the concepts and methodologies used in the study of retinal development at both cellular and molecular levels. These advanced methodologies have allowed and will continue to allow researchers to gain new insights into the molecular mechanisms underlying retinal development. Additionally, the retina, being part of the central nervous system (CNS) and accessible for direct, in vivo experimental manipulations, has historically served and continues to serve as an "eye" to "seeing" the molecular and cellular events underpinning CNS development.

In this volume, a group of distinguished researchers offer insightful and detailed protocols for a wide range of experiments that fall into six theme categories. Theme I describes several methodologies for manipulating gene expression in vivo, including conditional gene inactivation (knockout), generation of transgenic Xenopus, and electroporation of embryonic chick and adult mouse eyes. These methodologies are instrumental in uncovering the molecular mechanisms of retinal development, while some may be adapted to investigating potential therapeutic reagents. Theme II focuses on techniques for tracing cell fates with modernized classic blastomere manipulation in Xenopus and with Cre-based technique in mouse and in zebrafish. Theme III covers several protocols of in vitro systems, which have become increasingly popular in biological and biomedical laboratories. Theme IV presents protocols to study retinal regeneration and stem cell-based replacement, two research areas with heightened interests due to their therapeutic implications. Theme V centers on ERG (function) recording and noninvasive imaging, which are likely needed for future analyses of retinal development. Theme VI is devoted to other emerging, cutting-edge methodologies, including laser microdissection for studying miRNA and DNA methylation, 3C (chromosomal conformation capture), Exome-seq, and RAN-seq. These emerging methodologies will empower investigators in the retinal development field and other fields, such as epigenetics and gene discovery. As such, this volume provides methodologies crucial to the success of increasingly more complex and often challenging investigations in the fields of retinal development and other biological and biomedical research.

As in other volumes of the "Methods in Molecular Biology" series, each chapter in this volume contains a "Notes" section, in which expert researchers offer key insightful instructions, and yet not in excessive details, to assist users of this volume to successfully execute specific experiments.

This volume is designed as a reference manual for scientists with various levels of experience, from those who wish to set foot in the field of retinal development to those who wish to enrich the battery of techniques used in their research. It also provides a framework of methodologies that can be modified and applied to studies of the development of nonretinal tissues, the pathological processes of certain retinal degenerative diseases, and the development of gene- and cell-based therapies.

Birmingham, AL, USA *Shu-Zhen Wang*

Contents

Part VI Emerging Methodologies

Contributors

TERI L. BELECKY-ADAMS • *Department of Biology & Center for Regenerative Biology and Medicine, Indiana University-Purdue University Indianapolis, Indianapolis, IN, USA*

LENNART BERGLIN • *Department of Ophthalmology, Emory University, Atlanta, GA, USA*

CHRISTIANA J. BERNAL • *Department of Ophthalmology, Emory University, Atlanta, GA, USA*

JEFFREY H. BOATRIGHT • *Department of Ophthalmology, Emory University, Atlanta, GA, USA*

MATTHEW J. BROOKS • *Neurobiology Neurodegeneration and Repair Laboratory, National Eye Institute, National Institutes of Health, Bethesda, MD, USA*

JING CHEN • *Department of Ophthalmology and Visual Sciences, University of Michigan, Ann Arbor, MI, USA*

SHIMING CHEN • *Department of Ophthalmology and Visual Sciences, Washington University School of Medicine, St. Louis, MO, USA; Department of Developmental Biology, Washington University School of Medicine, St. Louis, MO, USA*

MICAH A. CHRENEK • *Department of Ophthalmology, Emory University, Atlanta, GA, USA*

MOLLY E. CLARK • *Departments of Vision Sciences & Optometry, University of Alabama at Birmingham, Birmingham, AL, USA*

QIAN DING • *University of Rochester Eye Institute, University of Rochester, Rochester, NY, USA*

HEITHEM M. EL-HODIRI • *Center for Molecular and Human Genetics, The Research Institute at Nationwide Children's Hospital, Columbus, OH, USA*

LIN GAN • *Department of Neurobiology and Anatomy, Center for Neural Development and Disease, University of Rochester Eye Institute, University of Rochester, Rochester, NY, USA*

DANIEL GOLDMAN • *Department of Biological Chemistry, Molecular & Behavioral Neuroscience Institute, University of Michigan, Ann Arbor, MI, USA*

PENNY GOODMAN • *Department of Ophthalmology, Emory University, Atlanta, GA, USA*

ALECIA K. GROSS • *Department of Vision Sciences, Evelyn F. McKnight Brain Institute University of Alabama at Birmingham, Birmingham, AL, USA; Department of Cell Biology, University of Alabama at Birmingham, Birmingham, AL, USA*

JR. LASZLO HACKLER • *Avidin Ltd, Szeged, Hungary*

MOHAMMAD HAERI • *Department of Neuroscience & Physiology, SUNY Upstate Medical University, Syracuse, NY, USA*

PETER HITCHCOCK • *Department of Ophthalmology and Visual Sciences, University of Michigan, Ann Arbor, MI, USA*
YI-WEN HSIEH • *Department of Developmental Biology, Cincinnati Children's Research Foundation, Cincinnati, OH, USA*
SCOTT R. HUDSON • *Department of Biology & Center for Regenerative Biology and Medicine, Indiana University-Purdue University Indianapolis, Indianapolis, IN, USA*
ANU JAYABALU • *Department of Biological Structure, Institute for Stem Cells and Regenerative Medicine, University of Washington, Seattle, WA, USA*
KANGXIN JIN • *Department of Pediatrics, Center for Advanced Biotechnology and Medicine, Graduate Program in Molecular Genetics, Microbiology and Immunology, UMDNJ-Robert Wood Johnson Medical School, Piscataway, NJ, USA*
MIKE O. KARL • *Department of Biological Structure, Institute for Stem Cells and Regenerative Medicine, University of Washington, Seattle, WA, USA*
LISA E. KELLY • *Center for Molecular and Human Genetics, The Research Institute at Nationwide Children's Hospital, Columbus, OH, USA*
BARRY E. KNOX • *Department of Neuroscience & Physiology, SUNY Upstate Medical University, Syracuse, NY, USA*
TIMOTHY W. KRAFT • *Department of Vision Sciences, University of Alabama at Birmingham, Birmingham, AL, USA*
DEEPAK A. LAMBA • *Department of Biological Structure, Institute for Stem Cells and Regenerative Medicine, University of Washington, Seattle, WA, USA*
YI-CHAO LI • *Department of Biomedical Engineering, University of Alabama at Birmingham, Birmingham, AL, USA*
JING LUO • *Department of Ophthalmology and Visual Sciences, University of Michigan, Ann Arbor, MI, USA*
WENXIN MA • *National Eye Institute, National Institutes of Health, Bethesda, MD, USA*
TOMOHIRO MASUDA • *Wilmer Eye Institute, Johns Hopkins University School of Medicine, Baltimore, MD, USA*
SHANNATH L. MERBS • *Wilmer Eye Institute, Johns Hopkins University School of Medicine, Baltimore, MD, USA*
SALLY A. MOODY • *Department of Anatomy and Regenerative Biology, The George Washington University School of Medicine and Health Sciences, Washington, DC, USA*
JACOB NELLISSERY • *Neurobiology Neurodegeneration and Repair Laboratory, National Eye Institute, National Institutes of Health, Bethesda, MD, USA*
JOHN M. NICKERSON • *Department of Ophthalmology, Emory University, Atlanta, GA, USA*
VERITY F. OLIVER • *Wilmer Eye Institute, Johns Hopkins University School of Medicine, Baltimore, MD, USA*
YI PAN • *Center for Molecular and Human Genetics, The Research Institute at Nationwide Children's Hospital, Columbus, OH, USA*

GUANG-HUA PENG • *Department of Ophthalmology and Visual Sciences, Washington University School of Medicine, St. Louis, MO, USA*
RINKI RATNA PRIYA • *Neurobiology Neurodegeneration and Repair Laboratory, National Eye Institute, National Institutes of Health, Bethesda, MD, USA*
ZHAO QIN • *Department of Molecular, Cellular, and Developmental Biology, University of Michigan College of Literature, Science, and the Arts, Ann Arbor, MI, USA; Developmental Genetics Program, Skirball Institute of Biomolecular Medicine, New York University School of Medicine, New York, NY, USA*
HARSHA KARUR RAJASIMHA • *Neurobiology Neurodegeneration and Repair Laboratory, National Eye Institute, National Institutes of Health, Bethesda, MD, USA*
RAJESH RAMACHANDRAN • *Department of Biological Chemistry, Molecular & Behavioral Neuroscience Institute, University of Michigan, Ann Arbor, MI, USA*
PAMELA A. RAYMOND • *Department of Molecular, Cellular, and Developmental Biology, University of Michigan College of Literature, Science, and the Arts, Ann Arbor, MI, USA*
T. MICHAEL REDMOND • *National Eye Institute, National Institutes of Health, Bethesda, MD, USA*
THOMAS A. REH • *Department of Biological Structure, Institute for Stem Cells and Regenerative Medicine, University of Washington, Seattle, WA, USA*
AARON REIFLER • *Department of Biological Chemistry, Molecular & Behavioral Neuroscience Institute, University of Michigan, Ann Arbor, MI, USA*
JEROME E. ROGER • *Neurobiology Neurodegeneration and Repair Laboratory, National Eye Institute, National Institutes of Health, Bethesda, MD, USA*
JOSHUA SAMMONS • *Department of Cell Biology, University of Alabama at Birmingham, Birmingham, AL, USA*
ANAND SWAROOP • *Neurobiology Neurodegeneration and Repair Laboratory, National Eye Institute, National Institutes of Health, Bethesda, MD, USA*
SCOTT TAYLOR • *Department of Ophthalmology and Visual Sciences, University of Michigan, Ann Arbor, MI, USA*
SARIKA TIWARI • *Department of Biology & Center for Regenerative Biology and Medicine, Indiana University-Purdue University Indianapolis, Indianapolis, IN, USA*
ANNA LA TORRE • *Department of Biological Structure, Institute for Stem Cells and Regenerative Medicine, University of Washington, Seattle, WA, USA*
JIN WAN • *Department of Biological Chemistry, Molecular & Behavioral Neuroscience Institute, University of Michigan, Ann Arbor, MI, USA*
SHU-ZHEN WANG • *Department of Ophthalmology, University of Alabama at Birmingham, Birmingham, AL, USA*
MENGQING XIANG • *Department of Pediatrics, Center for Advanced Biotechnology and Medicine, Graduate Program in Molecular Genetics, Microbiology and Immunology, UMDNJ-Robert Wood Johnson Medical School, Piscataway, NJ, USA*
RUN-TAO YAN • *Department of Ophthalmology, University of Alabama at Birmingham, Birmingham, AL, USA*

XIAN-JIE YANG • *Department of Ophthalmology, Molecular Biology Institute, Jules Stein Eye Institute, David Geffen School of Medicine, University of California, Los Angeles, CA, USA*
XIN-CHENG YAO • *Department of Biomedical Engineering, University of Alabama at Birmingham, Birmingham, AL, USA; Department of Vision Sciences, University of Alabama at Birmingham, Birmingham, AL, USA*
DONALD J. ZACK • *Wilmer Eye Institute and Departments of Molecular Biology and Genetics, Neuroscience, and Institute of Genetic Medicine, Johns Hopkins University School of Medicine, Baltimore, MD, USA*

Part I

Manipulation of Gene Expression In Vivo

Chapter 1

Conditional Control of Gene Expression in the Mouse Retina

Qian Ding and Lin Gan

Abstract

Conditional knockout is a powerful research tool for specific deletion of target genes, especially for the genes that are widely expressed and developmentally regulated. The development of the retina involves multiple intrinsic and extrinsic factors, many are required for embryonic development or expressed in multiple tissue or cell types. To study their roles in a spatial- or temporal-specific fashion, Cre/*loxP*-based gene-targeting approach has been utilized successfully. This chapter describes the methodology of conditional knockout approach in studying the development of the retina, using LIM homeobox gene *Isl1* as an example. It provides details on targeting vector design and construction, introducing the vector into embryonic stem (ES) cell, screening ES cell for the recombination events, injecting ES cells, and generating chimeric and null mice. It also discusses the current issues in the use of Cre/*loxP*-based gene-targeting approach.

Key words: Mouse, Gene targeting, Conditional knockout, Retinal development, Cre recombinase, ES cell

1. Introduction

Gene targeting is a powerful tool in studying *in vivo* function of mammalian genes. It allows researchers to generate a variety of mutations at specific murine genomic loci and investigate the gene function in a physiological context during development and adult stages. However, many genes have roles in multiple tissues or organs and are required for normal embryogenesis and survival. Their disruption in the germline often causes early embryonic lethality, preventing analysis at later stages (1). Moreover, deletion of gene in multiple cells and tissues where its function is required also renders problems of interpreting the function of the gene in a system or cell type as the phenotype of germline deletion due to cell-autonomous mechanisms rather than a combination of effects

Shu-Zhen Wang (ed.), *Retinal Development: Methods and Protocols*, Methods in Molecular Biology, vol. 884,
DOI 10.1007/978-1-61779-848-1_1, © Springer Science+Business Media, LLC 2012

in more than one tissue or cell type (2). To bypass these limitations, the site-specific recombination system is utilized to restrict the gene deletion at specific stages as well as in cell type- or tissue-specific fashion. Cre/*loxP* recombination is one of the most widely used and best refined system to generate conditional knockout mice to date (3).

Cre recombinase (Cre) is a 38-kDa protein from P1 bacteriophage that recognizes and catalyzes homologous recombination between two 34-bp *loxP* (locus of crossover in P1) sites, resulting in an excision of a fragment of DNA flanked by *loxP* sites (Fig. 1) (4). The Cre recombinase can be expressed under the control of cell- or tissue-specific promoters. Cre mice are generated by either transgenic or gene-targeting (knock-in) strategy. *loxP* sites are introduced

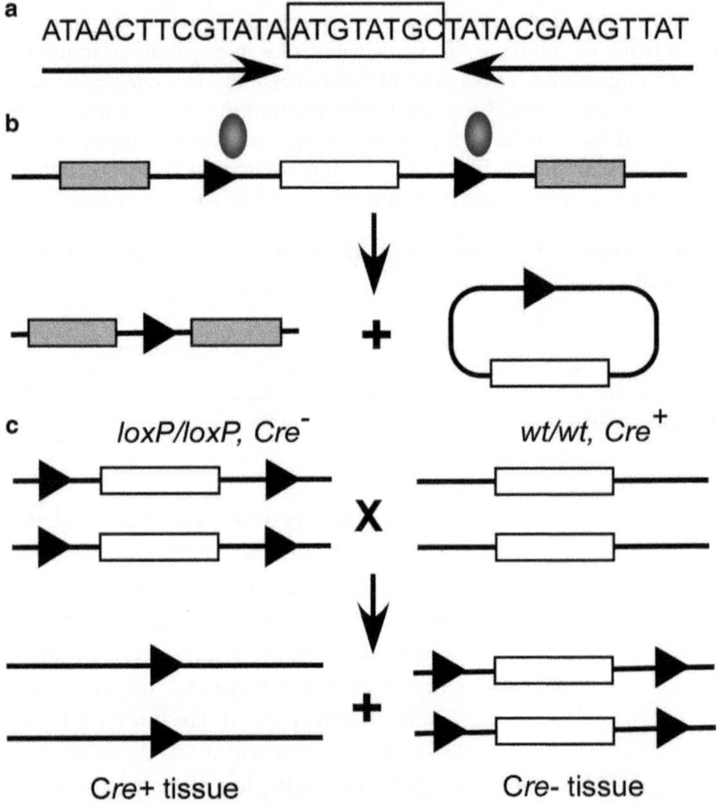

Fig. 1. Schematic diagram of tissue-specific controlling of gene expression by Cre recombinase. (**a**) The 34-bp *loxP* site consists of two inverse repeats (*arrow*) flanking an 8-bp core sequence (*box*) which confers the directionality of the site. (**b**) Cre recombinase (*gray*)- mediated gene excision between two *loxP* sites. (**c**) Generation of conditional knockout mouse by breeding a mouse with homozygous floxed gene (*left*) with a mouse carrying tissue-specific Cre (*right*). After two generations, the offspring carrying both homologous floxed gene and *Cre* have tissue-specific disruption of the targeted gene in Cre-expressing tissue. *Checked box*, homologue arms; *open box*, targeted gene; *black triangle*, *loxP* sites.

to the genome at desired sites by homologous recombination with targeting vector in embryonic stem (ES) cells. Then the *loxP* carrying mice generated from the ES cells are bred with Cre mice. The offspring carrying both *loxP* and Cre alleles will undergo the excision of DNA fragment from their genome in specific cells or tissues where Cre is expressed. In recent years, as more and more regulators in mouse retinal development have been identified, many Cre mouse lines have been established to exert the deleter function in the developing or the mature retina.

The mouse retina develops from optic vesicle (OV), a protrusion of the neuroepithelium of the diencephalon at embryonic day 8. The distal OV invaginates to form optic cup with the inner layer developing into neural retina (5). Birthdating analysis revealed that the retinal progenitor cells pass through a series of competent stages and generate different retinal cell types in a phylogenetically conserved order, with retinal ganglion cells, amacrine cells, horizontal cells and cone cells born before birth, and rod, bipolar, and Müller cells born after birth (6). Choosing a suitable mouse line with specific expression of Cre in retinal cell types of interest is necessary to conditionally inactivate a gene and elucidate its role of in the neural retina. This chapter will use the *Isl1* conditional knockout mouse as an example to illustrate the use of the Cre/*loxP* gene-targeting strategy in retinas. We will describe the general approach to the generation of conditional knockout mice for the study of gene function in the developing mouse retina and address current issues in gene targeting.

2. Materials

2.1. ES Cells and Mouse Lines

1. W4 ES cells.
2. C57BL/6J mice.
3. ROSA26-FLPe mice.

2.2. ES Cell Culture and Growth Condition (W4 ES Cells)

1. Dulbecco's minimal essential medium (DMEM) with high glucose, 15% fetal bovine serum (FBS) (heat activated at 56°C, 30 min), 2 mM glutamine (from 100× stock), 0.1 mM nonessential amino acids (from 100× stock), 50 μg/ml penicillin and streptomycin (aliquoted and stored at –20°C), and 1,000 U/ml LIF, 1 μM β-mercaptoethanol (aliquoted and stored at –20°C).
2. 0.1% Trypsin in PBS (store at 4°C).
3. 0.1% Gelatin in water (sterilize by autoclaving, store at 4°C).
4. Tissue culture plates with fibroblast feeder cells.
5. 2× Freezing medium: 60% DMEM, 20% FBS, and 20% DMSO.
6. Cell culture facility equipped with laminar flow cabinet, humidified incubator (5% CO_2:95% air, 37°C).

2.3. Electroporation

1. Electroporation apparatus.
2. Electroporation cuvettes.
3. Growing ES cells.
4. ES cell medium.
5. Gelatinized plates.
6. 200 mg/ml G418 in PBS.
7. 0.05% Trypsin in saline/EDTA.

2.4. Southern Blot

1. ES cell lysis buffer: 10 mM Tris–HCL, pH 7.5, 10 mM EDTA, 10 mM NaCl, 0.5% sarcosyl, and 1 mg/ml Proteinase K.
2. NaCl–ethanol solution: 660 μl of 5 M NaCl mixed with 50 ml 100% ethanol.
3. Restriction enzyme (RE) digestion cocktail: RE buffer, 10 mM spermidine, 0.2 mg/ml BSA, and 40 U RE each.
4. 0.8% Agarose. 1× TAE gel running buffer with ethedium bromide (EtBr).
5. Denature buffer: 0.1 M HCl.
6. Transfer buffer: 0.4 M NaOH.
7. Washing buffer: 2× SSC, 0.1% SDS in water.
8. Phosphorimaging system.

2.5. Microinjection

1. Blastocysts from pregnant female mice.
2. Microinjection medium: Hepes-buffered DMEM, 5% FBS, aliquoted and stored at −20°C.

2.6. Embryo Transfer

1. Embryo transfer hosts (female mice at 2.5 days of pseudopregnancy).
2. Anesthetics (avertin): dissolve 1 g of tribromoethanol in 80 ml of distilled water by gentle warming (40°C), add 1 ml of tertiobutyl alcohol, and mix well. Store at 4°C in dark.
3. Sterile surgical instruments.
4. Embryo transfer pipette.
5. Mouth pipette.
6. Anesthesia.
7. 70% Ethanol.
8. Wound clips.

2.7. Genotyping

1. Tail digestion buffer: 10 mM Tris–HCL (pH 8.0), 25 mM EDTA (pH 8.0), 100 mM NaCl, 1% SDS, and 0.2 mg/ml proteinase K.
2. Tail restriction digestion cocktail for Southern: 5–10 μg tail DNA, 3 μl of 10× RE buffer, 0.3 μl of 0.1 M spermidine, 0.3 μl of 2 mg/ml BSA, 40 U RE, and add H_2O to 30 μl.

3. PCR mixture: 1 μg DNA as template, 1× PCR buffer, 200 μM each dNTP, 1 U Taq polymerase in a 25 μl volume, 10 pmol each oligonucleotide primer 5′-GGTGCTTAGCGGT GATTTCCTC and 5′-GCACTTTGGGATGGTAATTGGAG to detect a 452-bp product of wild-type *Isl1* allele and a 512-bp product of *Isl1*^{CKO} allele.

3. Methods

3.1. Vector Design and Construction

The detailed information of a targeting vector is not within the scope of this chapter. In short, a targeting vector is composed of sequences homologous to those in the desired genomic integration site, consisting of upstream homologous region, the upstream *loxP* site, the region to be disrupted, the downstream *loxP* site, and the downstream homologous region, positive (to select the clones with vector DNA incorporated) and negative (to kill the clones with the whole vector integrated) selection markers, and plasmid backbone (7) (see Note 1). *loxP* sites are introduced into a targeting vector so that they flank the genomic region to be deleted. It is crucial that these insertions will not interfere with the normal expression of the gene. Therefore, *loxP* sites are often placed in introns without disrupting the splicing, and the mouse carrying the targeted gene is phenotypically wild type (see Note 2) (Fig. 1).

Here, we describe the generation of *Isl1* conditional knockout mice. ISL1 is one of the founding members of the LIM-homeodomain transcription factor family. Conventional *Isl1* knockout mice do not survive beyond E11.5. To assess the role of ISL1 in the retina development in the mid- to late gestation stages, an *Isl1* conditional knockout (*Isl1*^{CKO}) allele was generated (8). Among the four exons, Exon 2, which encodes the first LIM domain, was selected to be disrupted. To construct the targeting vector, we inserted a neomycin-resistance gene cassette flanked by FRT sites along with two *loxP* sites, one at the 5′ end of Exon 2 and the other at the 3′ end of Exon 2 (Fig. 2).

3.2. Manipulation of Mouse ES Cells

1. Gelatinize plates: add 5 ml of gelatin solution to each 10-cm plate and leave the plates at room temperature for 20 min. Aspirate off the gelatin and allow to air-dry.
2. Make feeder plates: start mouse fibroblast cell (STO cell) culture in a 10-cm plate with 15 ml of medium, add 6–10 ml of mitomycin C to the medium, and incubate the plate at 37°C for 2–3 h. Rinse with PBS twice, trypsinize and plate STO cells onto the gelatinized plates at a density of 4×10^6/10-cm plate.
3. Thaw an ES cell vial and plate on the feeder plate at a density of 2×10^6/10-cm plate (see Notes 3 and 4).
4. Change medium daily.

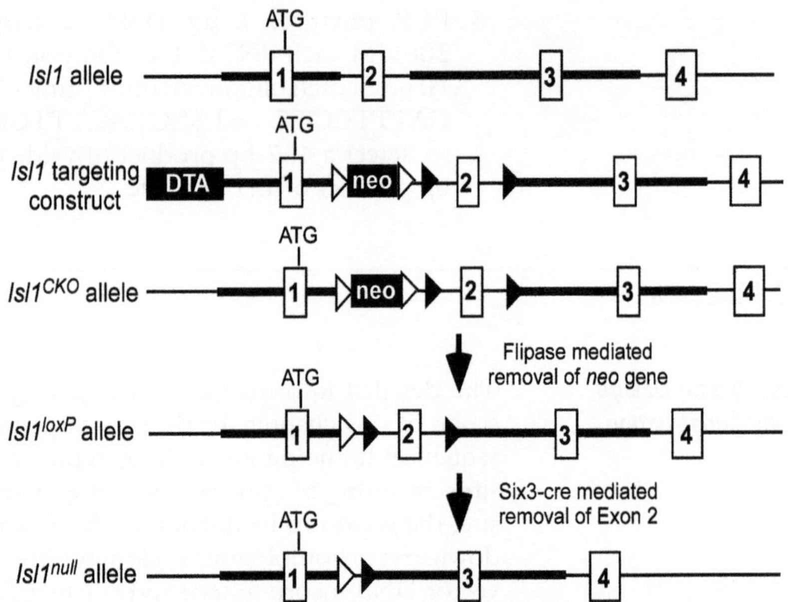

Fig. 2. Generation of *Isl1* conditional knockout allele. *Isl1* genomic structure and restriction enzyme map is shown at the *top*. *White boxes* are exons. *Thick bars* are the sequences used to generate homologous arms in the targeting vector; *open arrowhead*, FRT site for flipase recognition; *solid arrowhead*, *loxP* site for Cre recombinase recognition. *Neo* PGK-neomycin-resistance gene, *DTA* diphtheria toxin gene for negative selection of ES cells.

3.3. Electroporation of ES Cells

1. Linearize 200–300 μg of *Isl1* conditional knockout targeting vector DNA purified by CsCl gradient centrifugation method by *Not* I digestion, followed by extraction with phenol/chloroform, precipitation with two volumes of ethanol, washing with 70% ethanol. Resuspend the DNA in sterile TE or PBS at a concentration of 1 μg/μl.
2. Feed ES cells 3 h before harvesting. After collecting ES cells, wash the cell pellet twice with PBS. Resuspend ES cells in PBS to make the concentration of about $2 \times 10^7/0.75$ ml.
3. Add 20 μg of DNA to each 0.75 ml of cells and electroporate the mixture at 0.25 kV, 500 μF. After electroporation, place the mixture on ice for 10 min and plate out the cells on five plates seeded with mitomycin C-treated STO cells.
4. From the following day, feed the cells daily with ES medium supplemented with G418 at 200 μg/ml. Colonies should be visible 6 days after electroporation and picked up at 7–9 days after electroporation.

3.4. Screening for Targeted ES Cells by Southern Blot Analysis

1. Prepare 96-well plates with feeder cells.
2. Wash the ES cell-containing plates twice with PBS. Leave cells in PBS during picking. Pick the ES cell clones with tips under a dissecting microscope and transfer each colony into one well

of 96-well plate containing 35 μl of 0.05% trypsin at room temperature. Digest the colonies for 10–20 min in an incubator, add 80 μl of ES medium to neutralize the trypsin, mix well by pipetting, and transfer the cells to two or three 96-well plates with feeder cells. Change medium after 6 h and daily thereafter.

3. When the colonies in 96-well plates have reached at least 80% confluence, freeze the clones in one set of plates with freezing medium and keep the plates in −70°C freezer. Wash the cells in the other sets twice with PBS and add 50 μl of ES cell lysis buffer and incubate overnight at 55–60°C in a humid atmosphere.
4. Cool the plates on ice, then add 150 μl of NaCl–ethanol mix to each well to precipitate the DNA. Let the plate stand on the bench for 1 h. Discard the solution by inverting the plate, and add 150 μl of 70% ethanol to wash the wells. Invert the plates to discard the ethanol and briefly blot on paper towels. Repeat the washing step once. Air-dry the plates.
5. Make digestion cocktail and add 30 μl to each well. Incubate the plate in 37°C overnight in humid atmosphere.
6. Next day, add loading buffer, mix well, run the samples in 0.8% agarose gel. Treat the gel with 0.1 N HCl for 15–30 min with shaking, vacuum transfer the DNA with 0.4 M NaOH for 4 h to overnight to Hybond-N+ membrane. After transferring, briefly rinse the filter in Tris–HCl, pH 7.4, add Rapid Hyb and probe, and hybridize at 65°C in Hybaid Oven for overnight. Wash the filter with 2× SSC–0.1% SDS twice and image by phosphorimaging system.

3.5. Blastocyst Injection

1. Harvest E3.5 blastocysts from pregnant females.
2. Feed ES cells in growth phase for 1 h. Trypsinize the ES cells to single cells and resuspended in microinjection medium at 4°C.
3. Pick up a blastocyst with a micropipette and place it in the center of the field under the microscope.
4. Pick up 6–8 small round ES cells in the injection pipette. Place the injection pipette in the same focal plane as the blastocyst.
5. Apply force to pop the injection pipette through the trophectoderm wall.
6. Expel ES cells into the blastocoel cavity and remove the pipette.
7. Repeat this procedure until all the embryos are injected with ES cells and transfer the blastocysts to the culture medium for a brief period of culture in a 5% CO_2:95% air in 37°C incubator.

3.6. Embryo Transfer

1. Mate female mice with sterile males during oestrus to produce pseudopregnant mice, which serve as embryo transfer hosts on the third day of pseudopregnancy.

2. Weigh and anesthetize the recipient female with avertin.
3. Take up a column of air, medium, a small air bubble, 10–15 embryos to be transferred, and finally a small air bubble at the tip.
4. Clean the back of the mice with 70% ethanol and make a 1-cm transverse incision at the level of the first lumbar vertebra. Slide the incision to one side and cut the peritoneum wall. The uterus is exteriorized and the reproductive tract is pulled out with forceps.
5. Insert the tip of the transfer pipette into the lumen and expel the embryos with the air bubble as a marker showing the expulsion of the embryos.
6. Gently return the reproductive tract to the body cavity and repeat on the other side.
7. Close the skin incisions with a small wound clip.

3.7. Determination of Chimerism

1. A convenient way to determine the contribution of targeted ES cell in the offspring is to monitor the coat color. For example, W4 ES cells derived from agouti 129S6 mice are injected into blastocysts from albino C57BL/6J-*Tyr*$^{c\text{-}2J}$ mice to produce mouse chimeras with agouti/white coat color. Generally the degree of coat color chimerism represents the degree of contribution of germline.
2. ES cell lines commonly used are often derived from male embryos and the resulting chimeras are expected to be males.
3. Choose the highly chimeric animals to breed to mice of appropriate genetic background to obtain F1 heterozygotes for the *Isl1*^{CKO} allele (see Note 5). Besides producing F1 heterozygotes, this mating also confirms the germline transmission in the chimera by the genetic markers such as coat color and eye color of the offspring. The pups from W4 ES cell-derived gametes will be observed with black eye at birth and the agouti phenotype at around day 7. Half of the progenies with germline transmission are expected to be heterozygote and are determined by Southern blot or PCR genotyping methods.

3.8. Southern Blot and PCR Genotyping

1. Collect ≤1 cm of mouse tail into a 1.5-ml tube (day 8–10 pups are ideal).
2. Add 600 μl of digestion solution to the tail and incubate at 55°C with shaking for ≥4 h (best over night).
3. Let the tubes cool to room temperature. Extract with 500 μl phenol/chloroform (1:1) and transfer about 400 μl top layer with a wide-bore tip to a fresh tube.
4. Add 1 ml of ethanol, mix and precipitate the DNA for 1–2 min at room temperature, and wash DNA once with 70% ethanol.

5. Dry the DNA and dissolve in 50 μl of TE. The DNA concentration should be about 1 μg/μl.
6. Southern blot of tail DNA samples was performed as described in Subheading 3.2.
7. PCR genotyping was carried out using the following conditions: denaturation at 95°C for 5 min, followed by 35 cycles of 95°C for 30 s, 55–70°C for 30 s, and 72°C for 30 s.

3.9. Tissue-Specific Deletion

1. Cross the heterozygous mice for the floxed gene to ROSA26-FLPe mice to remove FRT-flanked neomycin-resistance gene.
2. Breed the mice without *neo* to *Six3-Cre* mouse line. *Six3-Cre* mice express Cre recombinase in the eye field and the ventral forebrain starting at E9–E9.5 and have been used successfully as an effective retina-specific deleter (9). In our experiments, Cre recombinase-mediated deletion of the *loxP*-flanked *Isl1* Exon 2 resulted in a null mutation via a reading frame shift (see Notes 6–10).
3. After one more generation, the progenies carrying homozygous $Isl1^{loxP/loxP}$ and *Six3-Cre* have *Isl1* deleted in a tissue-specific fashion (Fig. 3a). Usually, several breeding schemes can be developed, depending on the viability and fertility of each genotype. We incorporated a lacZ knock-in allele, an *Isl1* null mutant, into the breeding scheme. Thus, only one floxed gene needs to be excised by Cre recombinase to produce *Isl1*-null cells. Moreover, the knock-in lacZ reporter gene is used to trace *Isl1*-expressing cells (Fig. 3b) (see Note 11).

4. Notes

1. The construction of targeting vector is very important for the targeting efficiency (10). Usually longer homology arms increase targeting frequency (11, 12). The vector should be constructed from a DNA library that originates from the isogenic mouse strain that the ES cell line is derived. Base-pair mismatches could strongly affect the efficiency of intrachromosomal recombination in mammalian cells (13, 14). Besides, the targeting efficiency also has locus-specific variations depending on additional parameters, such as transcriptional activity, chromosomal location of the target genes, and spatial aspects of the nucleus (10).
2. When the *loxP* sites are inserted into the genomic loci, the selection of the region to be floxed is very important for both deletion and recombination efficiency. Usually the further the two *loxP* sites are apart, the less frequently the recombination

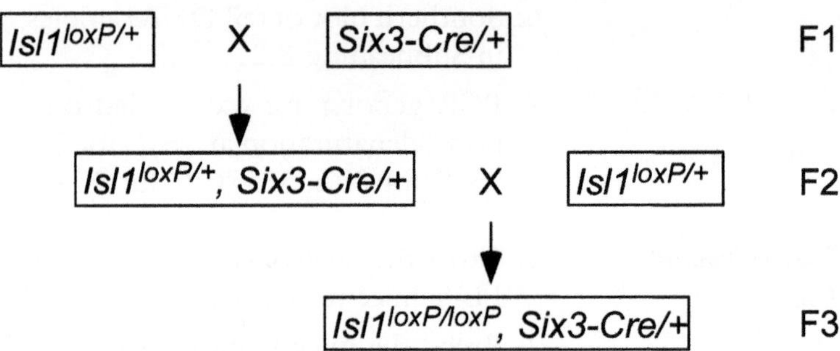

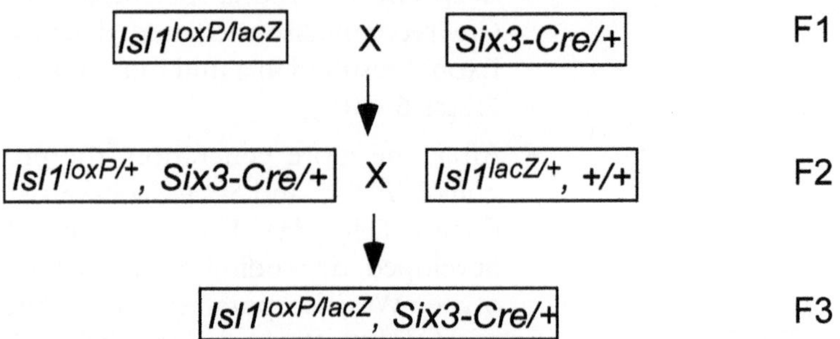

Fig. 3. Breeding scheme of *Isl1* conditional knockout. (**a**) Heterozygous breeding scheme. After two generations, F3 mice carrying homozygous for the floxed *Isl1* and *Six3-Cre* locus are present as *Isl1* conditional knockout. (**b**) Incorporation of lacZ allele into the breeding scheme. F3 mice harboring a null allele and a floxed allele undergoing excisional recombination by *Six3-Cre* recombinase.

event happens (15). Meanwhile the floxed sequence is also needed to be vital for the expression of the target gene. So theoretically, the segment with minimal length, which can lead to the disruption of the gene expression, should be selected (10). After the conditional knockout mice are generated, the deletion efficiency should be examined by in situ hybridization at mRNA level or immunohistochemistry at protein level.

3. The protocols listed in Subheading 3 provide only the guidelines for ES cell culture and necessary adjustments should be made daily according to ES cell growth. ES cell medium should be as fresh as possible and only medium less than 2 weeks old is used. Old leftover medium can be saved for STO cell culture.
4. ES cells should be split every 3 days or less. Growing ES cells undisturbed for 3 days could result in the differentiation.
5. ES cells from the 129 mouse strains were most widely used; however, different genetic background may affect how close

the model is to a desired purpose. In addition, a uniform genetic background allows the precise comparison of the mutant phenotypes to the control. Backcrossing is the straightforward way and is applicable to any inbred strain (16, 17). Usually the heterozygotes confirmed by Southern blot are crossed with mice of an inbred strain, and the heterozygous progenies from this mating are then backcrossed to the mice of the inbred stain again. Usually after seven generations of backcrossing, 99% of loci not linked to the mutant allele will be homozygous (18).

6. Cre mice can be generated by both transgenesis or through a knock-in strategy. Transgenic mouse is easier and faster to obtain, but due to the poorly defined promoter elements used to drive Cre expression, the specificity of the transgene expression is not very reliable (19). The use of knock-in approach by homologous recombination can ensure more faithful and regulated expression, but if the expression level from the promoter is low, the deletion will not be efficient enough. In the case of *Isl1* conditional knockout retina, *Six3-Cre* can delete *Isl1* in more than 90% of the cells from onset of *Isl1* expression, while we did not see the deletion until late embryonic stage by *Math5-Cre* mice ((8); unpublished observations by Ling Pan, University of Rochester).

7. Before the application of a newly generated *Cre* line, the spatial and temporal pattern of the recombination should be elucidated first. Breeding the *Cre* mouse with a reporter line such as the Z/EG mice (20). By observing the activation of reporter, we can monitor the Cre-mediated recombination events.

8. Besides crossing with *Cre* mice in retina-specific knockout, Cre recombinase can also be delivered by injecting the Cre-expressing virus directly into the vitreous cavity (21).

9. The targeted gene in conditional knockout mice can also be inactivated by inducible Cre. This induction is mediated by a ligand-binding reaction (22). Cre is expressed as a fusion protein with a mutated ligand-binding domain of the estrogen receptor and can be specifically activated by inducer, tamoxifen, a synthetic estrogen analog. Thus the gene inactivation depends on the temporal course of inducer administration.

10. For each conditional mouse line, several *Cre* lines can be chosen (Table 1). Even targeting the same cell type, different *Cre* lines may have different performance and onset time. In addition, using *Cre* lines for specific cell type can help to identify the cell-autonomous role of the targeted gene.

11. According to our breeding strategy showing in figure 3, 12.5% of the offspring will be conditional knockout mice. This percentage is under the assumption that the targeted gene and the *Cre* loci are not linked.

Table 1
Mouse lines expressing Cre recombinase in the developing retina

Promoter	Targeted cell types	References
Pax6	Distal neural retina	(23)
Six3	Neural retina	(9)
Chx10	Neural retina	(24)
Math5	RGC, amacrine, and horizontal cell	(25)
Thy1.2	RGC and neural retina	(26)
mRho	Rod bipolar cells and rod photoreceptor	(27)
Nes	Muller glia	(28)
Foxg1	Muller glia	(29)
Pcp2	Bipolar cells	(30)
ChAT	Starburst amacrine cell	(31)
Ptf1a	Amacrine and horizontal cell	(32)

Pax6 paired box gene 6, *Six3* six/sine oculis subclass of homeobox gene, *Math5* murine atonal homolog 5, *Thy1*.2 thymus cell antigen 1.2, *mRho* mouse rhodopsin, *Foxg1* Forkhead box G1, *Pcp2* purkinje cell protein 2, *ChAT* choline acetyl transferase, *Ptf1a* pancrease-specific transcription factor

References

1. Ferrara N, Carver-Moore K, Chen H, Dowd M, Lu L, O'Shea KS, Powell-Braxton L, Hillan KJ, Moore MW (1996) Heterozygous embryonic lethality induced by targeted inactivation of the VEGF gene. Nature 380:439–442
2. Lewandoski M (2001) Conditional control of gene expression in the mouse. Nat Rev Genet 2:743–755
3. Fukushige S, Sauer B (1992) Genomic targeting with a positive-selection lox integration vector allows highly reproducible gene expression in mammalian cells. Proc Natl Acad Sci USA 89:7905–7909
4. Sternberg N, Hamilton D (1981) Bacteriophage P1 site-specific recombination. I. Recombination between loxP sites. J Mol Biol 150:467–486
5. Ashery-Padan R, Gruss P (2001) Pax6 lights-up the way for eye development. Curr Opin Cell Biol 13:706–714
6. Cepko CL, Austin CP, Yang X, Alexiades M, Ezzeddine D (1996) Cell fate determination in the vertebrate retina. Proc Natl Acad Sci USA 93:589–595
7. Beglopoulos V, Shen J (2004) Gene-targeting technologies for the study of neurological disorders. Neuromolecular Med 6:13–30
8. Pan L, Deng M, Xie X, Gan L (2008) ISL1 and BRN3B co-regulate the differentiation of murine retinal ganglion cells. Development 135:1981–1990
9. Furuta Y, Lagutin O, Hogan BL, Oliver GC (2000) Retina- and ventral forebrain-specific Cre recombinase activity in transgenic mice. Genesis 26:130–132
10. Glaser S, Anastassiadis K, Stewart AF (2005) Current issues in mouse genome engineering. Nat Genet 37:1187–1193
11. Thomas KR, Deng C, Capecchi MR (1992) High-fidelity gene targeting in embryonic stem cells by using sequence replacement vectors. Mol Cell Biol 12:2919–2923
12. Hasty P, Rivera-Perez J, Bradley A (1991) The length of homology required for gene targeting in embryonic stem cells. Mol Cell Biol 11:5586–5591
13. te Riele H, Maandag ER, Berns A (1992) Highly efficient gene targeting in embryonic

stem cells through homologous recombination with isogenic DNA constructs. Proc Natl Acad Sci USA 89:5128–5132

14. Yang Y, Seed B (2003) Site-specific gene targeting in mouse embryonic stem cells with intact bacterial artificial chromosomes. Nat Biotechnol 21:447–451
15. Ringrose L, Chabanis S, Angrand PO, Woodroofe C, Stewart AF (1999) Quantitative comparison of DNA looping in vitro and in vivo: chromatin increases effective DNA flexibility at short distances. EMBO J 18:6630–6641
16. Simpson EM, Linder CC, Sargent EE, Davisson MT, Mobraaten LE, Sharp JJ (1997) Genetic variation among 129 substrains and its importance for targeted mutagenesis in mice. Nat Genet 16:19–27
17. Green EL (1966) Biology of the laboratory mouse. McGraw-Hill, New York, p 11
18. Papaioannou V, Johnson R (2000) Production of chimeras by blastocyst and morula injection of targeted ES cells. Gene targeting. Oxford University Press, New York, USA, pp 101–175
19. Le YZ (2011) Conditional gene targeting: dissecting the cellular mechanisms of retinal degenerations. J Ophthalmol 2011:806783
20. Novak A, Guo C, Yang W, Nagy A, Lobe CG (2000) Z/EG, a double reporter mouse line that expresses enhanced green fluorescent protein upon Cre-mediated excision. Genesis 28:147–155
21. Koike C, Nishida A, Ueno S, Saito H, Sanuki R, Sato S, Furukawa A, Aizawa S, Matsuo I, Suzuki N, Kondo M, Furukawa T (2007) Functional roles of Otx2 transcription factor in postnatal mouse retinal development. Mol Cell Biol 27:8318–8329
22. Feil R, Brocard J, Mascrez B, LeMeur M, Metzger D, Chambon P (1996) Ligand-activated site-specific recombination in mice. Proc Natl Acad Sci USA 93:10887–10890
23. Marquardt T, Ashery-Padan R, Andrejewski N, Scardigli R, Guillemot F, Gruss P (2001) Pax6 is required for the multipotent state of retinal progenitor cells. Cell 105:43–55
24. Rowan S, Cepko CL (2004) Genetic analysis of the homeodomain transcription factor Chx10 in the retina using a novel multifunctional BAC transgenic mouse reporter. Dev Biol 271:388–402
25. Yang Z, Ding K, Pan L, Deng M, Gan L (2003) Math5 determines the competence state of retinal ganglion cell progenitors. Dev Biol 264:240–254
26. Campsall KD, Mazerolle CJ, De Repentingy Y, Kothary R, Wallace VA (2002) Characterization of transgene expression and Cre recombinase activity in a panel of Thy-1 promoter-Cre transgenic mice. Dev Dyn 224:135–143
27. Le YZ, Zheng L, Zheng W, Ash JD, Agbaga MP, Zhu M, Anderson RE (2006) Mouse opsin promoter-directed Cre recombinase expression in transgenic mice. Mol Vis 12: 389–398
28. Zimmerman L, Lendahl U, Cunningham M, McKay R, Parr B, Gavin B, Mann J, Vassileva G, McMahon A (1994) Independent regulatory elements in the nestin gene direct transgene expression to neural stem cells or muscle precursors. Neuron 12:11–24
29. Kersigo J, D'Angelo A, Gray BD, Soukup GA, Fritzsch B (2011) The role of sensory organs and the forebrain for the development of the craniofacial shape as revealed by Foxg1-cre-mediated microRNA loss. Genesis 49:326–341
30. Zhang XM, Chen BY, Ng AH, Tanner JA, Tay D, So KF, Rachel RA, Copeland NG, Jenkins NA, Huang JD (2005) Transgenic mice expressing Cre-recombinase specifically in retinal rod bipolar neurons. Invest Ophthalmol Vis Sci 46:3515–3520
31. Ivanova E, Hwang GS, Pan ZH (2010) Characterization of transgenic mouse lines expressing Cre recombinase in the retina. Neuroscience 165:233–243
32. Nakhai H, Sel S, Favor J, Mendoza-Torres L, Paulsen F, Duncker GI, Schmid RM (2007) *Ptf1a* is essential for the differentiation of GABAergic and glycinergic amacrine cells and horizontal cells in the mouse retina. Development 134:1151–1160

Chapter 2

Generation of Transgenic *Xenopus* Using Restriction Enzyme-Mediated Integration

Mohammad Haeri and Barry E. Knox

Abstract

Transgenesis, the process of incorporating an exogenous gene (transgene) into an organism's genome, is a widely used tool to develop models of human diseases and to study the function and/or regulation of genes. Generating transgenic *Xenopus* is rapid and involves simple in vitro manipulations, taking advantage of the large size of the amphibian egg and external embryonic development. Restriction enzyme-mediated integration (REMI) has a number of advantages for transgenesis compared to other methods used to produce transgenic *Xenopus*, including relative efficiency, higher transgene expression levels, fewer genetic chimera in founder transgenic animals, and near-complete germ-line transgene transmission. This chapter explains the REMI method for generating transgenic *Xenopus laevis* tadpoles, including improvements developed to enable studies in the mature retina.

Key words: Transgenesis, Transgene, Reporter Gene, Cell-specific promoter, Gene expression, *Xenopus laevis*, Amphibians, Gene regulation, REMI

1. Introduction

Xenopus is a primary animal model in use for decades to understand vertebrate development (1), nuclear reprogramming (2), and metamorphosis (3, 4). In addition, *Xenopus* has been an extremely important system for elucidating the cell (5, 6), molecular (1) and circadian (7–10) biology of the retina, and is emerging in utilization for investigating retinal diseases (11–18) and regenerative mechanisms (19). The study of promoters and gene regulation are other applications of transgenesis technology (20–23). During the last two decades, a number of different transgenesis methods have been described; they include those mediated by restriction enzymes (24, 25), I-SceI meganuclease (26–28), transposons (29–31), phi-C31-integrase (32–34), and DNA injection (1).

Shu-Zhen Wang (ed.), *Retinal Development: Methods and Protocols*, Methods in Molecular Biology, vol. 884, DOI 10.1007/978-1-61779-848-1_2, © Springer Science+Business Media, LLC 2012

The advantages and drawbacks of the various transgenic *Xenopus* methods have recently been reviewed (35). Typically, restriction enzyme-mediated integration (REMI) is able to produce transgenic embryos easily without a large investment in expensive equipment or time-consuming training in specialized techniques. When a fluorescent reporter transgene is used, then embryos can be sorted early, e.g. neurulation, if a suitable promoter is used. Typically, between 20 and 50 transgenic tadpoles with uniform (not genetic chimera) expression patterns can be generated in a single day. However, the integration site of the transgene is random and copy number is variable between different primary transgenic animals (36). Natural mating of F_0 animals can be performed to create transgenic lines, often with less mosaicism and a high-level expression pattern than the original animals.

Overview. REMI transgenesis in *Xenopus* requires five steps:

1. *Preparation of interphase egg extract* (37). Eggs from hormonally stimulated females are collected and a cytoplasmic fraction is prepared after they have progressed in vitro into interphase. The extract is used to initiate decondensation of sperm chromatin and swelling of sperm nuclei, which is visible under a microscope. The egg extract is prepared in advance and stored at –80°C for at least 6 months.
2. *Preparation of sperm nuclei* (37). Intact sperm are isolated from whole minced testes of fully mature male frogs. Nuclei are prepared by hypotonic treatment of sperm and permeabilized with a mild detergent (lyso-PC) allowing egg extract, plasmid DNA, and restriction enzymes access to sperm chromatin in the REMI reaction. The sperm nuclei can be prepared in advance and stored at –80°C for at least 1 year.
3. *Restriction enzyme-mediated integration reaction* (25, 38). Permeabilized sperm nuclei are mixed with egg extract, linearized plasmid containing the transgene cassette, and restriction enzyme. The partially decondensed chromatin is rendered accessible to restriction enzyme generating breaks that can anneal with linearized plasmid. DNA ligase and repair activities in the egg cytoplasm (39) link the plasmid DNA into chromatin and closes restriction enzyme breaks. The reaction is carried out for several minutes before nuclear transplantation.
4. *Microinjection of sperm nuclei* (2). After the REMI reaction, sperm nuclei are transplanted into eggs. The glass needle must be chosen large enough to minimize damage to the fragile decondensed sperm nucleus while small enough to minimize trauma to the egg. A successful transplantation will generate a transgenic embryo expressing the gene of interest in a significant fraction of the injected eggs. The majority of integrations occur during the REMI and before the fertilization of the egg, thus

generating non-chimeric embryos. Although it is rare, we have observed transgenic tadpoles expressing the transgene unilaterally.

5. *Selection of transgenic embryos.* The final step is to select fertilized eggs and to sort out well-developing embryos at gastrulation, neurulation, etc. Cleaving transplanted embryos should be carefully transferred into fresh media with large-bore collecting pipettes. Often, a small tissue extrusion (bleb) on an otherwise normally developing embryo is observed on the second day after injection (dpi). These blebs are outpouching cells from the puncture hole in the vitelline membrane created during the injection. Although these blebs can be manually removed, it is preferable to allow them to fall off spontaneously, which usually occurs after neurulation. Eggs that receive damaged sperm nucleus, no nucleus, or more than one nucleus will exhibit aberrant cleavage patterns (multiple cleavage planes, partial or unilateral cleavage, pseudocleavage, and incomplete or shallow furrow). It is imperative to continuously sort normally developing embryos from the maldeveloped ones because the latter are easily infected and will compromise the healthy embryos. Typically, up to 50% of embryos are lost at each major developmental stage (i.e. gastrulation, neurulation, and feeding larvae). Approximately 30% of the tadpoles are transgene positive by genotyping a week after nuclear transplantation. Selection of tadpoles by transgene fluorescence will depend upon the promoter, level of expression, mosaicism, and stability of the particular gene chosen for study.

The number of integration sites as well as the number of transgenes per site (concatemers) may vary between one and more than ten, with some influence exerted by the amount of DNA and the restriction enzyme used in the REMI reaction. The exact transgene number and sites of integration can be determined by Southern blotting (36) while the transgene expression level can be accurately measured by any number of quantitative methods such as real-time PCR. Since transgenes are typically integrated as concatemers, the addition of more than one type of linearized plasmid with compatible ends will produce transgenic tadpoles expressing multiple different plasmids with a high frequency (over 90%).

Critical elements for high-yield transgenesis are: (1) high-quality eggs (e.g. have a clear maturation spot, even animal pole coloration and easily distinguishable equatorial border); (2) high-quality sperm nuclei (e.g. well permeabilized and concentrated at $5–8 \times 10^7$ nuclei/ml); (3) high-quality egg extract (e.g. creates an easily observable swelling of the sperm within a few minutes at room temperature); (4) highly purified linearized DNA (preferably prepared immediately before use).

2. Materials

2.1. Trangenesis

2.1.1. Hormones

1. Pregnant mare serum gonadotropin (PMSG): store PMSG aliquots of 100 U/100 μl dH_2O at –20°C. On the day of injection, dilute one aliquot of PMSG in 400 μl sterile MilliQ dH_2O for each frog (see Note 1).
2. Human chorionic gonadotropin (HCG): store HCG as powder in 500–700-U aliquots at 4°C. Dissolve 500–700 U of HCG in 500 μl of diluent provided by the manufacturer on the day of injection. Do not store dissolved HCG (see Note 2).

2.1.2. Stock Solutions

1. 10× MMR: 1 M NaCl, 20 mM KCl, 20 mM $CaCl_2(2H_2O)$, 10 mM $MgCl_2(6H_2O)$, and 50 mM HEPES(–Na). Adjust pH to 7.8 with 10 N NaOH. Bring volume up to 3 L with MilliQ dH_2O. Dispense into 1-L bottles, autoclave, and store at 4°C.
2. 8× Egg laying solution: 0.88 M NaCl, 16 mM KCl, 4.8 mM Na_2HPO_4, 0.25 M Tris base, 16 mM $NaHCO_3$, 16 mM $MgSO_4$ anhydrous, and 25 ml acetic acid. Adjust pH to 7.6 if necessary with 10 N NaOH or acetic acid. Bring volume up to 4 L with MilliQ dH_2O. Pour into a container and store at 4°C.
3. 2% Cysteine solution in 1× MMR: dissolve L-cysteine hydrochloride monohydrate (Sigma C7880–500G) in 1× MMR (2%). Adjust pH to 7.9 with sodium hydroxide pellets. Bring volume up to 400 ml with 1× MMR.
4. 100 mM $MgCl_2$: dissolve $MgCl_2$ in MilliQ dH_2O. Bring volume up to 100 ml with MilliQ dH_2O. Filter-sterilize and store at room temperature.
5. 6% Ficoll, 0.4× MMR solution: dissolve Ficoll in MilliQ dH_2O, add 10× MMR to reach a final concentration of 0.4× MMR. Bring volume up to 2 L with MilliQ dH_2O. Filter-sterilize and store at –20°C.
6. 6% Ficoll, 0.1× MMR solution: dissolve Ficoll in MilliQ dH_2O, add 10× MMR to reach a final concentration of 0.1× MMR. Bring volume up to 2 L with MilliQ dH_2O. Filter-sterilize and store at –20°C.
7. 0.1× MMR/10 μg/ml gentamicin.
8. 0.1× MMR, prepared with sterile MilliQ dH_2O.
9. Interphase egg extract aliquot (made and stored at –80°C).
10. Sperm dilution buffer: 250 mM sucrose, 75 mM KCl, 0.5 mM spermidine, and 0.2 mM spermine. Dissolve sucrose in MilliQ dH_2O. Add in other reagents. Bring volume up to 45 ml. Adjust pH to 7.3–7.5. Bring volume up to 50 ml with MilliQ dH_2O. Filter-sterilize, dispense into 1-ml aliquots and store at –20°C.
11. Linearized plasmid DNA (linearized, purified, adjusted to a concentration of 200 ng/μl, and stored at –20°C). Any restriction

enzyme that does not cut frequently might be suitable for linearization of the plasmid. Our first choice is *Xho*I, followed by *Nhe*I, *Not*I, *Sal*I, *Age*I, *Bam*HI, *Eco*RI, and *Apa*LI.

12. Restriction enzyme, 0.5 U/μl, diluted in compatible buffer. We usually use *Xho*I (New England Biolab) and dilute the enzyme in its 10× compatible buffer (NEB2). The restriction enzyme should be the same used for the linearization of the plasmid.

2.1.3. Other Materials

1. Plastic mesh-coated plates: Cut pieces of nylon mesh 600 μm (http://www.smallparts.com) and attach to bottom of 90-mm plate using chloroform (perform under hood). Glue the edge of the plastic mesh with melting plastic glue. Plates can be reused for many rounds of injections (Fig. 1).
2. Transplantation needles: flame-polish the ends of capillary pipettes to prevent clogging of the needle by small pieces of glass during the injection. If the glass pipettes are not silanized, prepare a silanization chamber; place a pack of glass capillary tubes (1.2 mm OD and 0.69 ID, Warner Instruments) in a

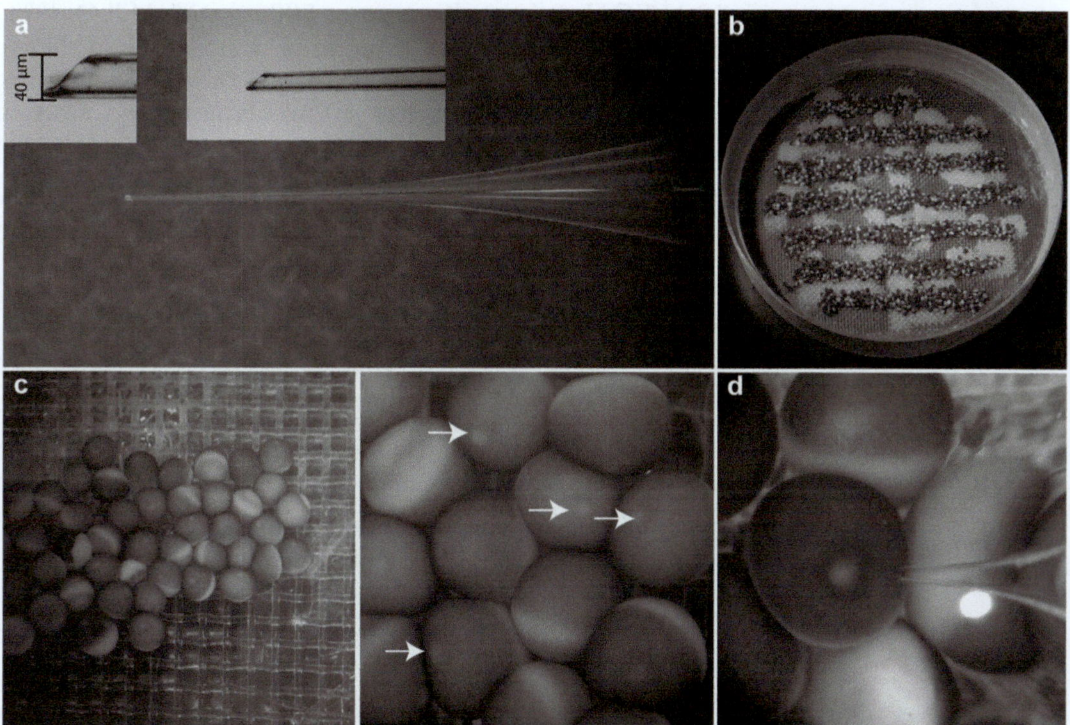

Fig. 1. Injection of eggs with sperm nuclei. (**a**) Injection needles with beveled tip. Glass needles are tapered and broken to create a 40–60-μm diameter beveled opening. (**b**) Eggs are dispersed in nylon mesh-covered plates along parallel lines with 4–6 eggs width on the mesh. A total of ~1,500 eggs are placed in each plate. (**c**) Healthy eggs should have clear maturation spots (*arrows*) and non-mottled appearance. The image on the *right* is at a higher magnification. (**d**) Eggs are injected around the maturation spot on the animal pole with continuous flow from the needle. The tip of the needle should approach the injection site and poke the membrane with a quick jabbing motion, penetrating only beyond the membrane as shown.

small beaker 1/3-filled with silane (*N,N*-dimethyltrimethylsilylamine, Sigma) in a dessicator overnight and bake them at 180°C for 2 h. Using a capillary glass puller, pull capillary glasses to make needles and clip the tip to create a 40–60-μm diameter opening. The clipping of the tip should be performed carefully with forceps under the dissecting microscope with a measuring grid to generate beveled tips. The sharp tip will make the injection of the eggs fast and easy and cause less damage to the membrane (Fig. 1a).

3. Transplantation unit: prepare a micromanipulator with a glass pipette holder, an oil-filled infusion pump (Harvard Apparatus), a glass syringe, and plastic tubing. Assemble the transplantation unit, fill the syringe with mineral oil and remove any bubbles in the syringe or along the plastic tubing (2.4 mm OD and 0.78 ID, Fisher). Set the infusion rate of the pump to 40 μl/h.

2.2. Interphase Egg Extract

2.2.1. Preparations

1. PMSG (see above).
2. HCG (see above).

2.2.2. Stock Solutions

1 M HEPES: dissolve HEPES in MilliQ dH_2O. Adjust pH to 8.2 with 10 N KOH. Bring volume up to 500 ml with MilliQ dH_2O and mix thoroughly until the solution is clear. If pH of diluted (10 mM) HEPES is not 7.7 (due to the pH of dH_2O), adjust to 7.7. Filter-sterilize, dispense into 50-ml aliquots, and store at –20°C (see Note 3).

1. 1.5 M Sucrose: dissolve sucrose in MilliQ dH_2O. Bring volume up to 100 ml with MilliQ dH_2O. Filter-sterilize, dispense into 10-ml aliquots, and store at –20°C.
2. 0.5 M EGTA: dissolve EGTA in dH_2O. Adjust pH to 7.7 with 10 N KOH. Adjust volume to 100 ml with MilliQ dH_2O. Filter-sterilize, dispense into 10 ml aliquots, and store at –20°C.
3. 1 M $CaCl_2$: dissolve $CaCl_2$ in MilliQ dH_2O. Bring volume up to 100 ml with MilliQ dH_2O. Filter-sterilize and store at room temperature.
4. 1 M $MgCl_2$: dissolve $MgCl_2$ in MilliQ dH_2O. Bring volume up to 100 ml with MilliQ dH_2O. Filter-sterilize and store at room temperature.
5. Protease inhibitors: chymostatin, leupeptin, hemisulfate, and pepestatin.
   (a) Prepare a protease inhibitor solution for each protease inhibitor; dissolve 50 mg of the protease inhibitor in 5 ml of DMSO, for a concentration of 10 mg/ml.
   (b) Dispense into 50-μl aliquots. Store at –20°C. We purchase the following protease inhibitors: Chymostatin (Sigma, C7268), leupeptin, hemisulfate (EMD, 108975), and Pepestatin A (MPbio, #19536825) (see Note 4).

6. 20× Extract buffer (XB) salt stock: 2 M KCl, 20 mM $MgCl_2$, and 2 mM $CaCl_2$. Prepare 1 L with MilliQ dH_2O. Filter sterilize, dispense into 500-ml aliquots, and store at –20°C.
7. Energy mix: 150 mM creatine phosphate, 20 mM ATP, and 20 mM $MgCl_2$. Prepare 10 ml with MilliQ dH_2O. Filter-sterilize (syringe filter), dispense into 500-μl aliquots, and store at –20°C.

2.2.3. Buffers

1. XB pH 7.7: 1× (1:20 dilutions of 20×) XB salts, 50 mM sucrose, and 50 mM HEPES (pH 8.2). Prepare 1 L with MilliQ dH_2O.
2. 2% Cysteine solution in 1× MMR: dissolve L-cysteine hydrochloride monohydrate (Sigma C7880-500G) in 1× MMR (2%). Adjust pH to 7.9 with sodium hydroxide pellets. Bring volume up to 400 ml with 1× MMR (see Note 5).
3. CSF-XB: 1× (1:20 dilutions of 20×) XB salts, 1 mM $MgCl_2$, 10 mM HEPES pH 8.2, 50 mM sucrose, and 5 mM EGTA. Prepare 500 ml with MilliQ dH_2O (see Note 6).
4. 10× MMR: 1 M NaCl, 20 mM KCl, 20 mM $CaCl_2(2H_2O)$, 10 mM $MgCl_2(6H_2O)$, and 50 mM HEPES(–Na). Adjust pH to 7.8 with 10 N NaOH. Prepare 3 L with MilliQ dH_2O. Dispense into 1-L bottles, autoclave, and store at 4°C.
5. 8× Egg laying solution: 0.88 M NaCl, 16 mM KCl, 4.8 mM Na_2HPO_4, 0.25 M Tris base, 16 mM $NaHCO_3$, 16 mM $MgSO_4$ anhydrous, and 2.5% (v/v) acetic acid. Adjust pH to 7.6 if necessary with 10 N NaOH or acetic acid. Prepare 4 L with MilliQ dH_2O. Pour into a container (carboy) and store at 4°C.
6. Versilube F-50.
7. Ten adult female frogs (*X. laevis*).
8. Other materials: 50-ml conical tubes, 14-ml round-bottom tubes (LPS, L285991), long cotton swabs, JS-13.1 swinging bucket rotor, 1- and 2-ml syringes, 18-gauge needles, SW 55Ti swinging bucket rotor, thin wall polyallomer 5 ml Beckman ultracentrifuge tubes (Beckman #326819), 0.6 ml sterile, low adhesion microfuge tubes (silanized, LPS), cut 100 μl tips, liquid nitrogen (crushed dry ice as alternative).

2.3. Sperm Nuclei

2.3.1. Buffer and Reagent Preparation

1. HCG (see above).

2.3.2. Stock Solutions

1. 1 M HEPES: dissolve HEPES in MilliQ dH_2O. Adjust pH to 8.2 with 10 N KOH. Bring volume up to 500 ml with MilliQ dH_2O and mix thoroughly until the solution is clear. Filter-sterilize, dispense into 50-ml aliquots, and store at –20°C (see Note 7).

2. 0.5 M EDTA pH 8.0: adjust pH with 10 N NaOH. Bring volume up to 100 ml with MilliQ dH_2O. Filter-sterilize and store at room temp.
3. 1.5 M KCl: dissolve KCl in MilliQ dH_2O. Bring volume up to 100 ml with MilliQ dH_2O. Filter-sterilize and store at room temperature.
4. 100 mM DTT (Sigma D0632): dissolve completely in MilliQ dH_2O. Bring volume up to 50 ml with MilliQ dH_2O. Filter-sterilize and store at –20°C.
5. 1% Tricaine (Sigma A5040): dissolve tricaine (ethyl 3-aminobenzoate methanesulfonate salt) in MilliQ dH_2O. Bring volume up to 500 ml with MilliQ dH_2O. Dispense into 500-ml bottles and store at –20°C.
6. 10 mM Spermidine trichloride (Sigma S2501): dissolve spermidine trichloride completely in MilliQ dH_2O. Bring volume up to 50 ml with MilliQ dH_2O. Filter-sterilize and store at –20°C.
7. 10 mM Spermine tetrahydrochloride (Sigma S1141): dissolve spermine completely in MilliQ dH_2O. Bring volume up to 50 ml with MilliQ dH_2O. Filter-sterilize and store at –20°C.
8. 10 mg/ml Leupeptin, hemisulfate (EMD, 108975): dissolve leupeptin in DMSO. Bring volume up to 5 ml with DMSO. Dispense into 50-μl aliquots and store at –20°C.
9. 0.3 M PMSF (Sigma P7626): dissolve PMSF completely in ethanol. Bring volume up to 5 ml with ethanol. Dispense into 100-μl aliquots and store at –20°C (see Note 8).
10. Lysolecithin (Sigma TypeI-L4129): dissolve lysolecithin completely in MilliQ dH_2O. Bring volume up to 5 ml with MilliQ dH_2O. Dispense into 100-μl aliquots and store at –20°C.
11. 10% Bovine serum albumin, FractionV (Sigma A7906): dissolve BSA in MilliQ dH_2O. Bring volume up to 50 ml with MilliQ dH_2O and store at –20°C.

2.3.3. Buffers

1. 2× Nuclear preparation buffer (NPB): 500 mM sucrose, 30 mM HEPES pH 8.2, 2 mM EDTA pH 8.0, 1.0 mM Spermidine trichloride, 0.4 mM Spermine tetrachloride, and 2 mM DTT. First dissolve sucrose in ~1,500 ml MilliQ dH_2O. Add the remaining stock solutions. Bring volume up to 2 L with MilliQ dH_2O. Filter-sterilize, dispense into 1 L bottles, and store at –20°C.
2. 1× NPB, 3% (w/v) BSA, protease inhibitor buffer: 1× NPB, 3% BSA, 10 μg/ml leupeptin, 0.3 mM PMSF. Prepare 50 ml with MilliQ dH_2O in a 50-ml conical tube and keep on ice.
3. 1× NPB, 0.3% (w/v) BSA: 1× NPB, 0.3% BSA. Bring volume up to 20 ml with MilliQ dH_2O. Prepare in a 50-ml conical tube and keep on ice.

4. Storage buffer: 1× NPB, 0.3% (w/v) BSA, 30% glycerol. Bring volume up to 10 ml with MilliQ dH_2O. Prepare in a 50-ml conical tube and keep on ice.
5. Sperm dilution buffer: 250 mM sucrose, 75 mM KCl, 0.5 mM spermidine, and 0.2 mM spermine. Dissolve sucrose in MilliQ dH_2O. Add in other reagents. Bring volume up to 45 ml. Adjust pH to 7.3–7.5. Bring volume up to 50 ml with MilliQ dH_2O. Filter-sterilize, dispense into 1-ml aliquots, and store at -20°C.
6. Five to ten sexually mature male frogs (*X. laevis*).
7. Other materials: JS-13 rotor, 1× MMR, 50-ml conical tubes, 14-ml round-bottom tubes (LPS, L285991), Petri plates, cut 1,000-μl tips, cut 100-μl tips, sterile cheesecloth, small glass funnel, fine #5 forceps, dissecting tools, pulverized dry ice (on a tray or in a bucket), 500 μl sterile low adhesion (silanized) microfuge tubes, pipettors, timer, and two ice buckets.

Notes and Cautions

- Tricaine is a carcinogen. Wear gloves throughout the procedure.
- All buffers and reagents must remain on ice throughout the procedure.
- Always use cut tips. Sperm nuclei are sheared if they pass through an uncut tip.

3. Methods

3.1. Trangenesis

3.1.1. Procedure

1. Prime 4–6 female frogs (*Xenopus laevis*) by injecting 100 U of PMSG into the dorsal lymph sac (Fig. 2a) 4–5 days prior to the HCG injection. Incubate the injected frogs in frog water at 16–18°C (see Note 9).
2. The evening before the day of the transgenesis procedure, inject each female frog with 500–700 U of HCG and place two frogs per tank containing frog water. Place the frogs in the 16–18°C incubator overnight. Change the water in the tank with fresh frog water after the injection. Use a temperature and (12 h/12 h) light/dark-controlled incubator and set the light onset at 7:00 am (see Note 10).
3. On the day of the procedure remove frogs from the incubator and place each frog in a tank filled with 1× egg laying solution (at frog-room-temperature, 20°C) (see Note 11).
4. Gently collect eggs using a collecting pipette and transfer eggs from each tank into one 50-ml conical tube. There should be a maximum of 15 ml of eggs in each tube. If there are more

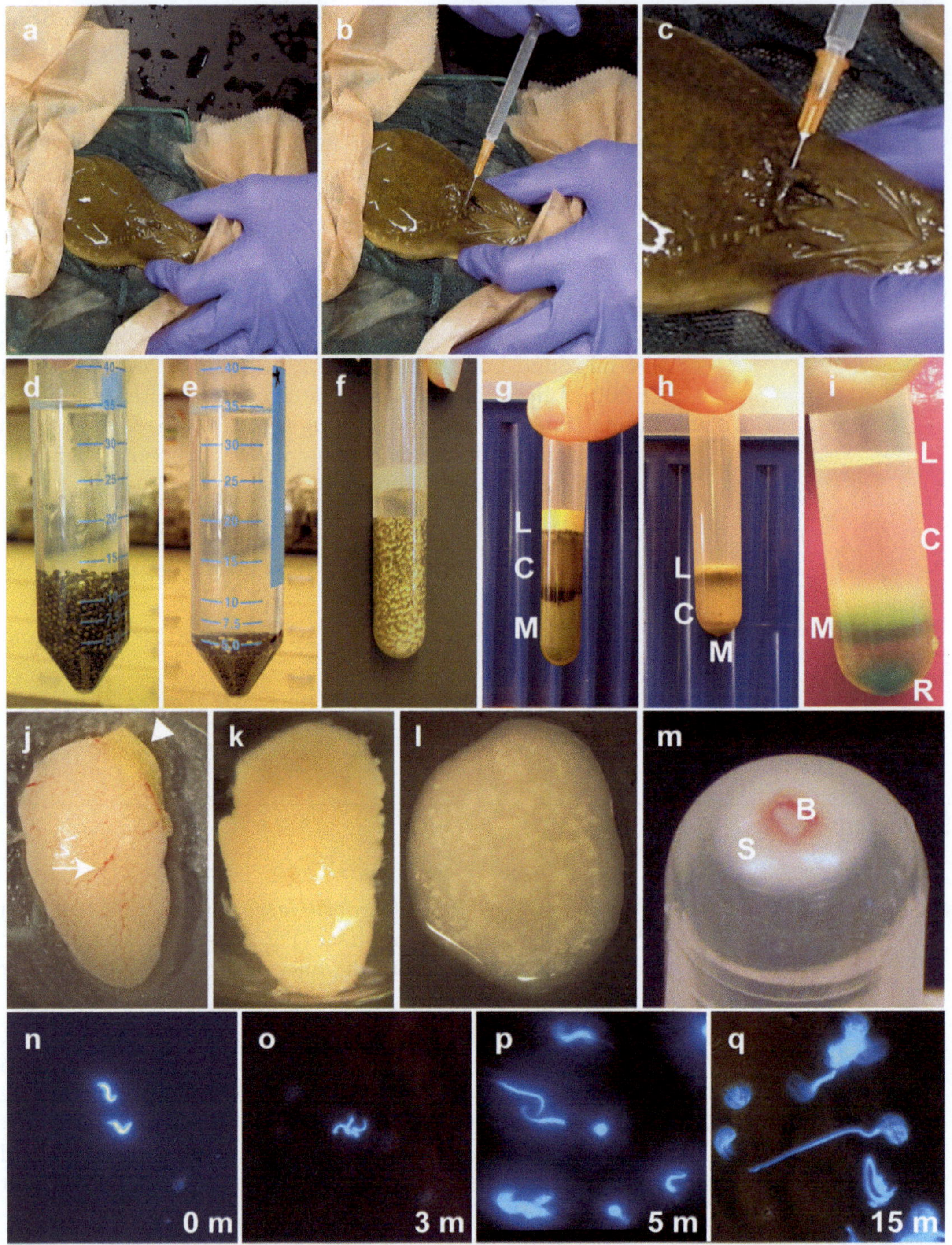

Fig. 2. Egg extract and sperm nuclei preparation. (**a–c**) Injection of females with hormones into the posterior lymph sac. Frogs are covered (including the eyes) with wet paper towel and injected under the skin into the dorsal lymph sac. (**d**) Washed eggs are settle to the bottom of the tube. (**e**) Eggs are treated with cysteine to remove jelly coat. After treatment, eggs are compact at the bottom of the tube. (**f–i**) Various stages in the egg extract preparation (see Subheading 3 for details). (**f**) After the first spin, eggs are packed but not broken and versilube replacing CSF-XB between the eggs. (**g**) Crushed eggs separated into

eggs collected from one tank, divide them into two or more tubes (see Note 12).

5. Decant as much liquid as possible. Wash eggs in 1× MMR three or more times, until the solution is clear.
6. Remove as much liquid as possible using a collecting pipette (see Note 13).
7. Fill up the conical tubes with 2% cysteine/1× MMR solution (to 45 ml line) (Fig. 2b). Put the cap on and gently invert tubes over and over to strip off the jelly coat from the eggs. Observe the eggs as they fall to the bottom of the tube. As their jelly coats are released, the eggs become more compact at the bottom of the conical tube. For instance, if the initial volume of eggs was 15 ml, the volume will drop to ~7 ml after proper dejellying (Fig. 2b). This process should take approximately 3–5 min depending on the thickness of the jelly coat and the freshness of the 2% cysteine/1× MMR solution. As soon as the eggs become compact and their jelly coats are released, immediately pour off the cysteine and wash eggs extensively with 1× MMR buffer to remove the cysteine (wash with 35 ml of 1× MMR buffer five times or more, until the solution is clear). Frequently, dead eggs (appearing white) will be found on top of the healthy eggs. Remove these dead eggs from the top using a collecting pipette (see Note 14).
8. Pour off the 1× MMR and remove the residual MMR by tilting the tube and removing the solution with a collecting pipette.
9. Add sufficient amount of room-temperature 6% Ficoll/0.4× MMR (~5 ml) to cover the treated eggs. The hyperosmolar 6% Ficoll/0.4× MMR solution protects treated eggs, which have lost their protective jelly coat and will be subjected to damage to their membrane during the injection.
10. Add 10 ml room-temperature 6% Ficoll/0.4× MMR to mesh-covered plates and transfer the eggs onto plates (Fig. 2.1b, c): first swirl the conical tubes to make the eggs float in the solution, then collect eggs in a plastic pipette, and dispense them

Fig. 2. (continued) three layers: *a thick yellow lipid layer on top* (**L**); the desired cytoplasmic layer which appears *gray* (**C**); the bottom layer appears gray/black and consists of unbroken eggs and membranes (**M**). (**h**) After the second cytoplasmic spin, the lipid and cytoplasmic layers are observed and the residual membranes pellet to the bottom. (**i**) After ultracentrifugation, the cytoplasm separates into 4 layers. Top layer: the *yellow lipid layer* (**L**); second layer: the *clear golden cytosol* (**C**); third layer: *membranes and mitochondria* (**M**); bottom layer: *glycogen and ribosomes* (**R**). Preparation of sperm nuclei. (**j**) Testes are removed and rinsed to remove blood and lipid. (**k**) Peripheral blood vessels and fat are manually stripped away. (**l**) Testes are macerated with fine forceps to release sperm. Sperm are treated with lysolecithin to release nuclei, which are collected by centrifugation. (**m**) The sperm nuclear pellet (**S**) often has a ring of red blood cells (**B**) at the bottom of the tube, which should be avoided. (**n–q**), In order to calibrate the egg extract for nuclear decondensation activity, sperm nuclei (stained with DAPI) are incubated for various times and amounts of egg extract. Swelling is easily observed in the microscope. Extensive swelling, such as that shown here in the 5 and 15 min samples should be avoided, and the extract diluted or time adjusted in the REMI reaction as necessary.

along parallel lines with 4–6 eggs width on the plastic mesh. Let the eggs settle down and stick to the mesh (5–10 min) before moving them to the dissecting microscope (this is the optimal time to perform the REMI reaction).

3.1.2. REMI Reaction

1. In a 1.5-ml silanized tube, first add 5 μl of linearized plasmid and subsequently add 4 μl of stock sperm nuclei using a cut pipette tip. Mix gently.
2. Incubate the mixed linearized plasmid and sperm nuclei 5 min at room temperature.
3. Add the following to the mixed linearized plasmid and sperm nuclei:
   (a) 5 μl Egg extract.
   (b) 1 μl of a 1:40 dilution of the restriction enzyme (0.5 U/μl) in its 10× buffer (see Note 15).
   (c) 2 μl of $MgCl_2$ (100 mM).
   (d) Bring the volume up to 32 μl with sperm dilution buffer (15 μl).
4. Warm up the tube containing the reaction between two fingers and incubate for 3 min.
5. Dilute the reaction with the approximate amount of sperm dilution buffer (already acclimatized to room temperature) needed to deliver one sperm nuclei per second based on the rate of fluid injection in the injector (we usually start with a sperm nuclei count of 5×10^5/ml, set the flow rate of the injector (Harvard Apparatus) to 40 μl/h and do a 1:50 dilution of the REMI reaction with the sperm dilution buffer).

3.1.3. Injection

1. Mix the REMI solution by flicking the tube. Using a pipette tip fitted with a piece of tygon tubing, draw in 40 μl of the REMI reaction from the middle portion of the 1.5-ml microfuge tube (the bottom or the surface of the tube might contain some unwanted artifacts that clog the needle). Avoid drawing in air bubbles.
2. Attach a clean and cut needle to the pipette tip and push the fluid through gently and continuously while keeping the tip of the needle upright. Looking at the rising level of the solution in the glass pipette, keep the pressure constant until small fluid drops come out of the tip of the needle. Remove the needle from the tube while the pipettor button is still pushed all the way to prevent adding air bubbles to the needle or drawing the fluid back into the pipette.
3. Attach the filled needle to the rubber tube connected to the automated pump (Harvard Apparatus), which is already running. Check the tip of the needle under the dissecting microscope

before starting injections. The tip of the needle and the fluid path should be free of blocking artifacts from the REMI or from pieces of glass. The fluid coming out of the tip should be visible due to its refraction index difference from the 6% Ficoll/0.4× MMR. If debris is seen within the glass needle, it should be changed even if the fluid is moving out of the tip. Artifacts tend to stick to sperm and this unpredictably reduces the number of sperm delivered to eggs. When the path is clear and the fluid is running, place the needle on the manipulator and tighten it without breaking the glass needle.

4. Inject eggs around the maturation spot on their animal pole while the fluid is running (Fig. 1e). Approach the tip of the needle to the injection site and poke the membrane with a quick jabbing motion, penetrating only beyond the membrane. As soon as the tip is inside the egg, draw back the needle quickly and move to the next egg. The injection time for each egg should take less than a second, being inside the egg approximately half a second (see Note 16).
5. When the injection is over, gently shake the plate to release the eggs from the mesh. Swirl the plate gently to bring the eggs to the center of the plate and pour the injected eggs into a new 50-ml plate. Store in the 18°C incubator for 3–6 h before sorting fertilized eggs into 6% Ficoll/0.1× MMR.
6. Three to six hours after injection, the eggs will have reached the 4-cell stage. Sort out fertilized eggs that have evenly dividing cells. Place them in a new plate containing 6% Ficoll/0.1× MMR. Incubate fertilized eggs at 18°C overnight (10–16 h) (see Note 17).
7. After 10–16 h, sort out good embryos into plates filled with sterilized 0.1× MMR/10 μg/ml gentamicin (prepare the solution using sterilized MilliQ water). Sort the embryos once again in the afternoon and change the plate filled with sterilized 0.1× MMR/10 μg/ml gentamicin. Store in the 18°C incubator overnight.
8. Sort out good embryos into new plates filled with sterilized 0.1× MMR and continue daily water changes with 0.1× MMR until day 7, when they are ready for sorting under the dissecting microscope and UV illumination.

3.2. Interphase Egg Extract Preparation

1. Prime ten female frogs (*X. laevis*) 3–5 days prior to HCG injection by injecting 100 U of PMSG into the dorsal lymph sac (Fig. 2a). Maintain at room temperature.
2. The evening before the extract preparation, inject each frog with 500–700 U of HCG (Chorulon) and place two frogs per tank containing frog water. Place the frogs in the 15–18°C incubator overnight (see Note 18).

3. On the day of the procedure remove frogs from the incubator and place each frog in a tank filled with 1× egg laying solution (at frog-room-temperature, 20°C) (see Note 19).
4. Gently collect eggs from each tank using a collecting pipette and place them into 50-ml conical tubes. Examine a sample of eggs from each tank under the dissecting microscope to determine if they are healthy (i.e. they have a good jelly coat and a clear maturation spot; in contrast, unhealthy eggs are large and white, overly speckled or stringy). There should be a total of 100 ml or more eggs collected into several tubes. Pull out any dead (white) or unevenly pigmented eggs (see Note 20).
5. Divide eggs between several 50-ml conical tubes to have approximately 20 ml of eggs in each tube. Pour off as much liquid as possible. Wash eggs with 1× MMR three times or more, until the solution is clear. Remove as much liquid as possible using a collecting pipette.
6. Fill up the conical tube with 2% cysteine solution (to the 45 ml line). Put the cap on and gently invert tubes repeatedly to remove the jelly coat from the eggs. Observe the eggs as they fall to the bottom of the tube. As their jelly coats are released, the eggs become more compact at the bottom of the conical tube. For instance, if the initial volume of eggs was 20 ml, the volume will drop to ~10 ml after proper dejellying (Fig. 2b). This process should take about 3–5 min depending on the thickness of the jelly coat and the freshness of the 2% cysteine solution. As soon as the eggs become compact and their jelly coats are released, pour off the cysteine and wash eggs with XB buffer to remove the cysteine (wash with 35 ml of XB buffer five times or more, until the solution is clear). Use a transfer pipette to remove dead eggs from the mix (dead eggs are typically white, larger than healthy ones, and tend to appear on the top).
7. Wash twice with 25 ml of protease inhibitor buffer in CSF-XB (10 μg/ml, a 1:1,000 dilution) (see Note 21).
8. Transfer collected eggs into 14-ml clear tubes (JS 13.1 rotor holds six tubes), using a wide-bore transfer pipette. Transfer equal amounts of eggs to each tube.
9. Remove as much CSF-XB+ protease inhibitor buffer as possible from each tube. Add 1 ml Versilube F-50.
10. Spin the tubes containing the dejellied eggs and Versilube in the JS-13.1 rotor until the rotor reaches 150 × g (1,000 rpm), followed by 30 s at 600 × g (2,000 rpm). Eggs will be packed after this spin but not broken (Fig. 2c). The Versilube will have replaced the CSF-XB between the eggs. An inverted meniscus between the Versilube and displaced CSF-XB should be clearly visible. Remove the excess CSF-XB and Versilube and balance the tubes (by transferring the eggs with a transfer pipette).

11. Spin the tubes for 10 min at 16,000×*g* (10,000 rpm) at 4°C in the JS-13.1 swinging bucket rotor to crush the eggs. The crushed eggs will separate into three layers (Fig. 2d):

Top layer	A thick yellow lipid layer which can gently be removed with a cotton swab (see Note 22)
Middle layer	The desired cytoplasmic layer which appears gray; occasionally there is a second darker cytoplasmic layer
Bottom layer	A black layer consisting of unbroken eggs and membranes

12. Collect the cytoplasmic layer by inserting an 18-gauge needle attached to a 2-ml syringe down the side of the tube, allowing the tip of the needle to be visible against the wall of the tube. Aiming for the middle layer, withdraw the cytoplasm slowly into the syringe and avoid mixing the extracted solution with other layers. Transfer cytoplasm to a fresh 14-ml tube on ice. Combine the cytoplasm from the two 14-ml tubes if necessary. Estimate the approximate volume of the cytoplasm in the tube and record it.
13. Add protease inhibitors, leupeptin and PepstatinA, to each tube for a final (10 μg/ml) (do a 1:1,000 dilution of stock solution) (see Note 23).
14. Spin the cytoplasm in the JS-13.1 swinging bucket rotor at 16,000×*g* (10,000 rpm) for 10 min for further separation of the cytoplasm (Fig. 2e).
15. Collect the cytoplasm as before. Expect to collect approximately 1.0 ml of cytoplasm from the initial 40 ml of eggs.
16. Measure the volume of the cytoplasm and add 1/20th volume of Energy Mix solution. Transfer the cytoplasm to thin-walled, polyallomer, 5-ml Beckman ultracentrifuge tubes (Beckman # 326819).

    Add 1 M $CaCl_2$ to each tube for a final concentration of 0.4 mM (1 μl/2.5 ml of cytosol) and incubate at room temperature for 15 min (see Note 24).

    Spin the cytoplasm in the Beckman SW55Ti swinging bucket rotor at 270,000×*g* (47,000 rpm) for 3 h at 4°C.
17. The cytoplasm will separate into four layers (Fig. 2f):

Top layer	The yellow lipid layer. This should be minimal at this step
Second layer	The cytosol, a clear golden color
Third layer	Membranes and mitochondria
Bottom	Glycogen and ribosomes

Gently remove the lipid layer with a cotton swab. Extract the cytosolic layer by inserting an 18-gauge needle attached to a 1-ml syringe down the side of the tube into the cytosol. Pull on the syringe with even, gentle pressure so that the layers do not mix.

18. Transfer the cytoplasmic layer to a fresh ultracentrifuge tube and spin again at 270,000 × g (47,000 rpm) for 45 min at 4°C (see Note 25).
19. Carefully remove the cytosolic layer with an 18-gauge needle attached to a 1-ml syringe. If the procedure is performed properly, the layers should not be as evident as before and the cytosol should have a clear golden color. If the cytosolic layer is cloudy, spin in a microfuge for 15–30 min at 4°C.
20. A typical yield for this prep, for an initial 100 ml of eggs, is 1–2 ml of high-speed cytosol.
21. Dispense into 20-μl aliquots on ice.
22. Flash-freeze in liquid nitrogen and store at –80°C.

3.3. Sperm Nuclei Preparation

1. Choose six mature males showing a distinguishable dark patch on their hand. Inject 100 U of HCG into the dorsal lymph sac 2–5 days before the procedure (see Fig. 2). Incubate the frogs at frog-room-temperature.
2. To isolate testes, anesthetize the sexually mature males in cold 1% tricaine (ethyl-3-aminobenzoate methanesulfonate salt) for 10–15 min. Verify that they are properly anesthesized by pinching their toes with tweezers or by flipping them onto their back (see Note 26).
3. Rinse one male at a time with water. Decapitate and pith the male using a paper clip. Open the abdomen by cutting the skin in the midline, followed by the abdominal wall. The addition of two transverse cuts at the base of the abdominal wall (hypogastric area) facilitates the isolation of testes. Isolate both testes from the frog using forceps and a pair of dissecting scissors. Avoid collecting the fat pad attached to the upper pole of the testes during isolation. Avoid damaging the testes.
4. Rinse the testes quickly with ice-cold 1× MMR and store in a 50-ml conical tube filled with ice-cold 1× NPB. When all testes are collected in the 50-ml conical tube, invert the tube several times. Wash the testes two or more times with ice-cold 1× NPB. Fill the 50-ml conical tube with fresh ice-cold 1× NPB and maintain it on ice. Invert the tube every 5 min.
5. Place one testis at a time in a Petri dish filled with ~5 ml ice-cold NPB, mounted on a cold surface to keep the solution cold. Under the dissecting microscope, use #5 forceps to clean

up the testes from any trace of blood found in the network of capillary/veins surrounding the testes. Rinse testes with ice-cold 1× NPB (see Note 27).

6. Prepare a new Petri dish (mounted on a cold surface) with a large drop (1 ml) of ice-cold 1× NPB. Place the rinsed testes in the 1× NPB. Using a pair of sharp #5 forceps, meticulously mince the testis into small pieces from one side to the other. Mince the testis until they reach a pudding-like texture and clumps are no longer visible to the naked eye.
7. Add 1 ml ice-cold 1× NPB to the lacerated testes. Collect the solution with a cut P1000 pipette and place it into a prechilled conical tube filled with 10 ml of ice-cold 1× NPB. If residual tissues remain on the plate, mince them again to reach the pudding-like texture and collect the macerated testes as described. If connective tissues remain on the plate, remove them and collect the macerated testes by adding additional ice-cold 1× NPB. Add all macerated testes to the same 50-ml conical tube.
8. Once all macerated testes have been collected, homogenize the solution by pipetting up and down several times using a collecting pipette.
9. Using a funnel and eight layers of cheesecloth presoaked in 1× NPB, filter the sperm solution into a prechilled 50-ml conical tube (see Note 28).
10. Divide the filtered sperm solution into 15-ml round-bottom polypropylene centrifuge tubes with tops.

3.3.1. Centrifugation

1. Spin the sperm solution using a prechilled JS13 rotor at 3,000×*g* (~4,500 rpm) for 10 min at 4°C.
2. Pour off the supernatant. Add 8 ml of ice-cold 1× NPB to each tube and repeat the centrifugation at 3,000×*g* (~4,500 rpm) for 10 min at 4°C.
3. Pour off the supernatant and dispose of the remnant wash solution by inverting the tube on Kim wipe tissues. There will be a noticeable ring of red blood cells appearing pink at the bottom of tube. Add 500 μl of 1× NPB to the pellet of each tube and resuspend sperm using a cut P1000 pipette. Avoid collecting the pink ring of blood cells (Fig. 2g). It is preferable to leave some sperm than to collect some blood.

3.3.2. Detergent Treatment

1. Incubate sperm solution on bench for 5–10 min to allow it to accommodate to room temperature.
2. Divide the sperm solution into two sets of tubes. Add 50 μl of lysolecithin to one set, and 100 μl of lysolecithin to the other (see Note 29).

3. Mix the solution gently by flicking the tubes. Incubate at room temperature for exactly 10 min.
4. Dilute each milliliter of the sperm solution with 10 ml of a solution of ice-cold 1× NPB/3% BSA/10 μg/ml leupeptin/0.3 mM PMSF.
5. Spin down the diluted sperm solution at 3,000×*g* (4,500 rpm) for 10 min at 4°C. Pour off the supernatant and add 11 ml of ice-cold 1× NPB/0.3% BSA buffer and gently resuspend the pellet.
6. Spin down the sperm solution again at 3,000×*g* (4,500 rpm) for 10 min at 4°C. Pour off the supernatant and discard any remaining solution by inverting the tube on Kim wipe tissues.
7. Resuspend the pellet in 300–500 μl of ice-cold 1× NPB/0.3% BSA/30% glycerol storage buffer for each frog used (i.e. if 6 frogs were used, use 3 ml).
8. It is recommended to determine the sperm nuclei count using a hemacytometer. Make a 1:100 dilution of sperm nuclei with sperm dilution buffer and load each side of the hemacytometer grid with 10–20 μl of the diluted sperm nuclei. Perform the sperm count for both the lysolecithin preparations. The sperm nuclei concentration should be in the range of $5–8 \times 10^7$ sperm nuclei/ml. For higher concentrations a second dilution is needed on the day of transgenics; for lower concentrations the second REMI dilution step should be corrected accordingly (see Subheading 3.1.2).
9. Dispense into 20-μl aliquots into prechilled (on dry ice) silanized 0.6-ml Eppendorf tubes sitting on a tray of pulverized dry ice. Swirl the sperm solution while dispensing the solution to prevent precipitation of sperm. Store aliquots at −80°C.
10. To verify the success of the prep in permeabilizing the membrane, perform the egg extract/sperm extract assay (Fig. 2h). Make a 1:50 dilution of sperm nuclei with sperm dilution buffer. Make a 1:1 solution of diluted sperm and egg extract (addition of 20 μl of diluted sperm with 20 μl of egg extract is recommended) and mix with a cut pipette tip. Place two drops of the mixture on a glass slide and place a coverslip. Incubate the slide at room temperature in a humidified chamber and observe the slide under the microscope every 5 min. This step must be performed for both lysolecithin sperm treatments. Clear elongation of sperm should be observed after 30 min (see Note 30).

4. Notes

1. ProSpec, HOR272 (http://www.prospecbio.com).
2. Intervet, Inc. ch-475-1, NADA#140-92T, chorionic gonadotropin, 10,000 U/vial.
3. Adjusting the pH requires approximately 30 ml KOH.
4. Each stock solution contains only one protease inhibitor.
5. Use within 1 h of making. If the pH is below 7.8, the solution will not be effective in dejellying eggs.
6. The final concentration of $MgCl_2$ will be 2 mM (($MgCl_2$) = 2 mM, with the $MgCl_2$ in XB salts).
7. If the pH of diluted (10 mM) HEPES is not 7.7 (due to pH of dH_2O), adjust the 1 M stock pH to 7.7.
8. TOXIC material, wear gloves!
9. During this period feed the frogs two times, one of which should be 2 days before the HCG injection. Change the frog water once during this period.
10. The interval between the injection time and the initiation of egg laying is approximately 14 h after HCG injection when frogs are kept at 16°C. A proper timing of injection is required to maximize the number of eggs collected.
11. The amount of 1× egg laying solution in the tank should be sufficient to cover the frog.
12. Look at a sample of eggs from each tank under the microscope to determine if they are healthy (Fig. 1, i.e. they have a good jelly coat and a clear maturation spot; in contrast, unhealthy eggs are large and white, overly speckled or stringy). If the eggs look healthy, collect them and avoid combining them. It is recommended not to combine eggs from different frogs because their jelly coats have different thicknesses, and therefore require different lengths of time for the dejellying step. Also, keep track of eggs collected from each tank from the beginning until the end in order to know which frogs are producing more and better fertilized eggs. Pull out any dead (white) or unevenly pigmented eggs.
13. Tilting the tube helps removing the solution using a collecting pipette. Residual MMR will dilute the 2% cysteine solution and decrease its strength and ability to remove the jelly coat from the eggs.
14. During the cysteine treatment, observe the jelly coats being stripped off the eggs by placing the tube between the eyes and the path of light from the window. As the tubes are continuously inverted, more jelly coats will be floating in the

solution and the eggs will become more compact at the bottom of the tube. Immediately pour off the cysteine solution and wash treated eggs. The cysteine treatment is a critical step. Overtreatment with cysteine makes the egg membrane vulnerable and weak and very soft during the injection. These eggs break easily even after successful fertilization. On the other hand, undertreatment with cysteine makes the injection extremely difficult and reduces the overall number of fertilized eggs.

15. We usually use XhoI (New England Biolab) for linearization and the corresponding buffer (NEB 2, New England Biolab).
16. To start the injection, center the first few eggs of the first row under the dissecting microscope. When the eggs are in the center of the viewfield, orient the needle toward those eggs, followed by fine manipulations of the tip toward the targeted eggs. Inject the eggs gently around the maturation spot in the animal pole. Eggs that are too firm or too soft to inject are not good indicators of a great yield. This again emphasizes the importance of the treatment with the 2% cysteine/1× MMR solution. Never stop the pump from running when you are filling or changing the needle or leaving the stage momentarily. Sperm quickly precipitate when the flow is interrupted. Be consistent and finish the injection of a full plate in 30–40 min, depending on the number of eggs on the plate.
17. Unevenly divided eggs and those with a very superficial division line typically will not survive. Do not keep injected eggs in 6% Ficoll/0.4× MMR longer than 6 h since this increases the number of mushroom embryos. The 4-cell stage is the optimal time for sorting. To increase the yield of fertilized eggs, perform another round of sorting at the blastula stage. If hesitant about the fairness of some fertilized egg, place them in a separate plate, because dying eggs will eventually compromise the healthy ones. Limit the number of sorted fertilized eggs to 50 per plate.
18. An adult female should weigh approximately 110 g. Adjust the amount of HCG for larger frogs.
19. The frogs are separated in a tank in case one or more of the frogs are producing unhealthy eggs. Unhealthy eggs (more than 10% white or stringy eggs) should not be used in the prep. For the following steps make all solutions before starting the extract prep. Once the procedure is started, carry out the steps without any delays. Optimally, the high speed spin should begin 45–60 min after dejellying the eggs.
20. It is recommended not to combine eggs from different frogs, because their jelly coats have different thicknesses, and therefore need different amounts of time for dejellying.

21. Make a 1:1,000 dilution of protease inhibitors using CSF-XB buffer as the diluent (add 1 μl of 10 mg/ml protease inhibitor stock solutions to every 1 ml of CSF-XB for a final concentration of 10 μg/ml of each protease inhibitor). Make enough buffer to wash each tube twice with 25 ml of this buffer. Keep the buffer on ice.

22. Removing the lipid layer facilitates the extraction of the cytoplasmic layer.

23. For each milliliter of cytoplasmic extract, add 1 μl of each stock protease inhibitor solution (10 mg/ml). Balance tubes carefully.

24. This treatment inactivates the CSF and pushes the extract into interphase. Balance the tubes perfectly! The tubes for this type of rotor must be filled to within 3–4 mm from the top. This allows enough space for the bucket cap and prevents the tubes from collapsing, thus losing some of the sample and damaging the centrifuge. If all the cytoplasmic extract fits into one tube, combine tubes. On the other hand, if there is not enough cytoplasmic extract to fill one tube, fill the tube with mineral oil.

25. Tubes must be perfectly balanced! Again, the tubes need to be filled to within 3–4 mm from the top.

26. Tricaine is a carcinogen. Wearing gloves is essential at all times during this procedure.

27. Testes are covered with a capsule and many vessels containing blood. Remove the adipose tissue and blood from the testes by cutting the tiny vessels and pushing the blood out by gently touching them with tweezers (sharp # 5 forceps). This is a crucial step since blood cells generate small clots and clog injection needles.

28. Do not squeeze the cheesecloth. This will carry over some connective tissue and cellular clumps; instead, add 5–10 ml of ice-cold 1× NPB to wash off sperm left in the cheesecloth.

29. This is the critical point in the procedure; the sperm nuclear membrane must be permeabilized in order for the restriction enzyme and egg extract to reach their targets. The amount of detergent needed to dissolve the nuclear membrane is dependent on the concentration of the sperm solution. To avoid counting the density of sperm nuclei at this step, it is possible to divide the tubes into two and treat both sets of tubes as two different preparations, one of which will be treated with 50 μl and the other with 100 μl of lysolecithin. One of the two treatments will obtain the sperm count suitable for producing transgenic tadpoles. When the sufficient amount of detergent needed is obtained repeatedly over several sperm extracts, treat all tubes with that amount.

30. If elongated sperms are observed only around the edge of the coverslip, the amount of solution is too small for the size of the coverslip; the coverslip should float on the solution. No change in the size of sperm after 30 min indicates poorly permeabilized membrane.

Acknowledgments

We thank Annabelle Pellerin and Maria Goralski for help in preparation of this manuscript. We acknowledge present and former lab and Center for Vision Research members who have helped develop this procedure over many years. This work was supported by the National Institutes of Health Grants EY-11256 and EY-12975 (B.E.K.), Research to Prevent Blindness (Unrestricted Grant to SUNY UMU Department of Ophthalmology), Fight for Sight (FFS) and Lions of CNY.

References

1. Kay BK, Peng HB (eds) (1991) *Xenopus laevis*: Practical uses in cell and molecular biology, vol. 36, Academic Press, San Diego
2. Gurdon J (2009) Nuclear reprogramming in eggs. Nat Med 15:1141–1144
3. Marsh-Armstrong N, Cai L, Brown DD (2004) Thyroid hormone controls the development of connections between the spinal cord and limbs during *Xenopus laevis* metamorphosis. Proc Natl Acad Sci USA 101:165–170
4. Furlow JD, Neff ES (2006) A developmental switch induced by thyroid hormone: *Xenopus laevis* metamorphosis. Trends Endocrinol Metab 17:40–47
5. Besharse JC (1986) Photosensitive membrane turnover: differentiated membrane domains and cell-cell interaction. In: Adler R, Farber DB (eds) The Retina: Part 1: pp. 297–362. Academic Press, New York
6. Gabriel RE (2000) Special issue: neurobiology of the anuran retina. Microsc Res Tech 50:325–424
7. Anderson FE, Green CB (2000) Symphony of rhythms in the *Xenopus laevis* retina. Microsc Res Tech 50:360–372
8. Hayasaka N, LaRue SI, Green CB (2010) Differential contribution of rod and cone circadian clocks in driving retinal melatonin rhythms in *Xenopus*. PLoS One 5:e15599
9. Hayasaka N, LaRue SI, Green CB (2002) In vivo disruption of *Xenopus* CLOCK in the retinal photoreceptor cells abolishes circadian melatonin rhythmicity without affecting its production levels. J Neurosci 22:1600–1607
10. Liu X, Green CB (2001) A novel promoter element, photoreceptor conserved element II, directs photoreceptor-specific expression of nocturnin in *Xenopus laevis*. J Biol Chem 276: 15146–15154
11. Baker SA, Haeri M, Yoo P, Gospe SM 3rd, Skiba NP, Knox BE, Arshavsky VY (2008) The outer segment serves as a default destination for the trafficking of membrane proteins in photoreceptors. J Cell Biol 183:485–498
12. Calvert PD, Schiesser WE, Pugh EN Jr (2010) Diffusion of a soluble protein, photoactivatable GFP, through a sensory cilium. J Gen Physiol 135:173–196
13. Choi RY, Engbretson GA, Solessio EC, Jones GA, Coughlin A, Aleksic I, Zuber ME (2011) Cone degeneration following rod ablation in a reversible model of retinal degeneration. Invest Ophthalmol Vis Sci 52:364–373
14. Iakhine R, Chorna-Ornan I, Zars T, Elia N, Cheng Y, Selinger Z, Minke B, Hyde DR (2004) Novel dominant rhodopsin mutation triggers two mechanisms of retinal degeneration and photoreceptor desensitization. J Neurosci 24:2516–2526
15. Knox BE, Schlueter C, Sanger BM, Green CB, Besharse JC (1998) Transgene expression in *Xenopus* rods. FEBS Lett 423:117–121
16. Muradov H, Boyd KK, Haeri M, Kerov V, Knox BE, Artemyev NO (2009)

Characterization of human cone phosphodiesterase-6 ectopically expressed in *Xenopus laevis* rods. J Biol Chem 284:32662–32669

17. Haeri M, Knox BE (2012) Rhodopsin mutant P23H destabilizes rod photoreceptor disk membranes. PLoS One 7:e30101?
18. Tam BM, Moritz OL, Hurd LB, Papermaster DS (2000) Identification of an outer segment targeting signal in the COOH terminus of rhodopsin using transgenic *Xenopus laevis*. J Cell Biol 151:1369–1380
19. Beck CW, Izpisua Belmonte JC, Christen B (2009) Beyond early development: *Xenopus* as an emerging model for the study of regenerative mechanisms. Dev Dyn 238:1226–1248
20. Casey ES, Tada M, Fairclough L, Wylie CC, Heasman J, Smith JC (1999) Bix4 is activated directly by VegT and mediates endoderm formation in *Xenopus* development. Development 126:4193–4200
21. Hyde CE, Old RW (2000) Regulation of the early expression of the *Xenopus* nodal-related 1 gene, Xnr1. Development 127:1221–1229
22. Karaulanov E, Knochel W, Niehrs C (2004) Transcriptional regulation of BMP4 synexpression in transgenic *Xenopus*. EMBO J 23:844–856
23. Mani SS, Besharse JC, Knox BE (1999) Immediate upstream sequence of arrestin directs rod-specific expression in *Xenopus*. J Biol Chem 274:15590–15597
24. Amaya E, Kroll KL (1999) A method for generating transgenic frog embryos. Methods Mol Biol 97:393–414
25. Kroll KL, Amaya E (1996) Transgenic *Xenopus* embryos from sperm nuclear transplantations reveal FGF signaling requirements during gastrulation. Development 122:3173–3183
26. Ogino H, McConnell WB, Grainger RM (2006) High-throughput transgenesis in *Xenopus* using I-SceI meganuclease. Nat Protoc 1:1703–1710
27. Ogino H, McConnell WB, Grainger RM (2006) Highly efficient transgenesis in *Xenopus tropicalis* using I-SceI meganuclease. Mech Dev 123:103–113
28. Pan FC, Chen Y, Loeber J, Henningfeld K, Pieler T (2006) I-SceI meganuclease-mediated transgenesis in *Xenopus*. Dev Dyn 235:247–252
29. Hamlet MR, Yergeau DA, Kuliyev E, Takeda M, Taira M, Kawakami K, Mead PE (2006) Tol2 transposon-mediated transgenesis in *Xenopus tropicalis*. Genesis 44:438–445
30. Ivics Z, Izsvak Z (2004) Transposable elements for transgenesis and insertional mutagenesis in vertebrates: a contemporary review of experimental strategies. Methods Mol Biol 260:255–276
31. Sinzelle L, Vallin J, Coen L, Chesneau A, Du Pasquier D, Pollet N, Demeneix B, Mazabraud A (2006) Generation of trangenic *Xenopus laevis* using the sleeping beauty transposon system. Transgenic Res 15:751–760
32. Allen BG, Weeks DL (2005) Transgenic *Xenopus laevis* embryos can be generated using phiC31 integrase. Nat Methods 2: 975–979
33. Allen BG, Weeks DL (2006) Using phiC31 integrase to make transgenic *Xenopus laevis* embryos. Nat Protoc 1:1248–1257
34. Groth AC, Olivares EC, Thyagarajan B, Calos MP (2000) A phage integrase directs efficient site-specific integration in human cells. Proc Natl Acad Sci USA 97:5995–6000
35. Chesneau A, Sachs LM, Chai N, Chen Y, Du Pasquier L, Loeber J, Pollet N, Reilly M, Weeks DL, Bronchain OJ (2008) Transgenesis procedures in *Xenopus*. Biol Cell 100:503–521
36. Amaya E, Kroll KL (1999) A method for generating transgenic frog embryos. Methods Mol Biol 97:393–414
37. Murray AW (1991) Cell cycle extracts. Methods Cell Biol 36:581–605
38. Batni S, Mani SS, Schlueter C, Ji M, Knox BE (2000) *Xenopus* rod photoreceptor: model for expression of retinal genes. Methods Enzymol 316:50–64
39. Thode S, Schafer A, Pfeiffer P, Vielmetter W (1990) A novel pathway of DNA end-to-end joining. Cell 60:921–928

Chapter 3

In Vivo Functional Analysis of Transcription Factor: Response Element Interaction Using Transgenic *Xenopus laevis*

Heithem M. El-Hodiri, Yi Pan, and Lisa E. Kelly

Abstract

Analysis of transcription factor–target interactions in vivo is important to the study of transcriptional regulation of gene expression. A key experiment involves analysis of the functional interaction between a *trans*-acting factor and its corresponding *cis*-acting element in the context of a target promoter in vivo. We describe a method for this analysis in transgenic *Xenopus* tadpoles in which expression of the *trans*-acting factor is knocked down using an shRNA-mediated approach.

Key words: Rx/rax, Transgenesis, *Xenopus laevis*, Transcription factor, Target gene, Cis-acting element, Trasngenesis

1. Introduction

The study of interactions between transcription factors and their corresponding response elements is key to understanding transcriptional regulation of gene expression. Identifying a target gene–transcription factor pair and a *cis*-acting response element is only the beginning. A next step is often to demonstrate binding of the transcription factor to its target in vivo and that the factor can regulate *cis*-element-dependent expression of an artificial reporter gene construct. However, a powerful result is one that demonstrates that the activity of the target promoter is dependent on the transcription factor of interest acting through its corresponding *cis*-acting element in vivo. Here we describe a method for demonstrating this result in transgenic *Xenopus laevis* tadpoles. To do so, we analyze the expression of a reporter transgene containing a wild-type or mutated version of the *cis*-element of interest in frog tadpoles in which we have knocked down expression of the *trans*-factor of

Shu-Zhen Wang (ed.), *Retinal Development: Methods and Protocols*, Methods in Molecular Biology, vol. 884,
DOI 10.1007/978-1-61779-848-1_3, © Springer Science+Business Media, LLC 2012

interest by a transgenic shRNA-mediated approach. We recently used this methodology to demonstrate that the retinal homeobox (Rx/RAX) gene product regulates the rhodopsin (RHO) promoter through the photoreceptor conserved element-1 (PCE-1), the Rx response element (1, 2).

2. Materials

2.1. General Supplies and Materials

2× PCR GoTaq® Green mix (Promega).

2× SYBR Green PCR master mix (Applied Biosystems).

pRNAT plasmid (Genscript).

qScript® cDNA SuperMix (Quanta BioSciences, Inc.).

Snowcoat X-tra™ microscope slides (Surgipath).

Tissue-Tek® O.C.T. compound (Sakura Finetek).

Vectashield® HardSet mounting medium (Vector Laboratories).

2.2. General Solutions

10× MMR: 1 M NaCl, 20 mM KCl, 10 mM $MgCl_2$, 20 mM $CaCl_2$, 50 mM HEPES, pH 7.5.

Anesthetic solution: 0.1% aminobenzoic acid ester (MS-222, Tricaine) dissolved in 0.1× MMR.

2.3. Solutions for Preparation of Sperm Nuclei

Stock Solutions

1.5 M sucrose (freeze in 10-ml aliquots so as not to freeze and thaw repeatedly).

10 mM spermidine trihydrochloride.

10 mM spermine tetrahydrochloride.

100 mM dithiotheritol.

500 mM EDTA.

1 M HEPES, titrated the pH with KOH to achieve pH 7.7 when diluted to 15 mM solution.

10% BSA (fraction V).

10 mg/ml Hoechst No. 33342.

Working Solutions

2× NPB (nuclear preparation buffer, see Note 1).

500 mM sucrose.

30 mM HEPES.

1 mM spermidine trihydrochloride.

0.4 mM spermine tetrahydrochloride.

2 mM dithiothreitol.

2 mM EDTA.

Prepare 25 ml of 2× NPB on day of sperm nuclei isolation and add protease inhibitors to 2×. For example, add one tablet Complete Protease Inhibitor Cocktail (Roche). Dilute this stock to 1× with water and BSA as required.

Sperm dilution buffer (SDB): 250 mM sucrose, 75 mM KCl, 0.5 mM spermidine trihydrochloride, 0.2 mM spermine tetrahydrochloride, adjust pH to 7.3–7.4 (add ~80 µl of 0.1 N NaOH per 20 ml SDB), store in 1-ml aliquots at −20°C.

2.4. Solutions for Transgenesis

General Supplies and Solutions

Tygon tubing (cut into 3 cm pieces)—ID: 1/32″ (0.8 mm); wall thickness: 1/32″ (0.8 mm); OD: 3/32″ (2.4 mm), such as Fisher Scientific Catalog # 14-169-1A.

Capillary tubes—6″ (152 mm) long, 1/0.75 OD/ID (mm), fire-polished, such as World Precision Instruments Catalog # TW100-6.

SigmaCote (Sigma Catalog #SL2).

Cysteine dejellying solution: 2.25% cysteine in 1× MMR, pH 7.9.

Injection Dishes

Injection dishes are 50-mm Petri dishes half-filled with 2.5% agarose, into which a depression has been formed. These dishes are designed to hold dejellied eggs for injection.

Melt agarose in 0.1× MMR. Pour into dishes. Overlay with a 1″ square of plastic cut from a pipette tip box or a 1″ square weigh boat. Remove after the agarose hardens. Seal with film and store at 4°C.

Siliconized Capillaries

Attach short pieces of Tygon tubing to a 200-µl pipette tip and two 19-G hypodermic needles. Attach each needle to a 10-ml syringe.

Draw up 100 µl of SigmaCote into the 200-µl tip.

Attach to a capillary and push the SigmaCote through capillary.

Detach capillary from pipette tip and tubing.

Use the syringe + needle + tubing to push 10 ml of water through the capillary.

Use the other syringe + needle + tubing to push air through the capillary.

Allow the capillaries to dry vertically in a beaker containing tissue paper in the bottom.

Injection Needles

Injection needles are pulled from siliconized capillaries using any needle puller programmed so that the taper is gradual. The tip is broken off using watchmaker forceps so as to leave a sharp,

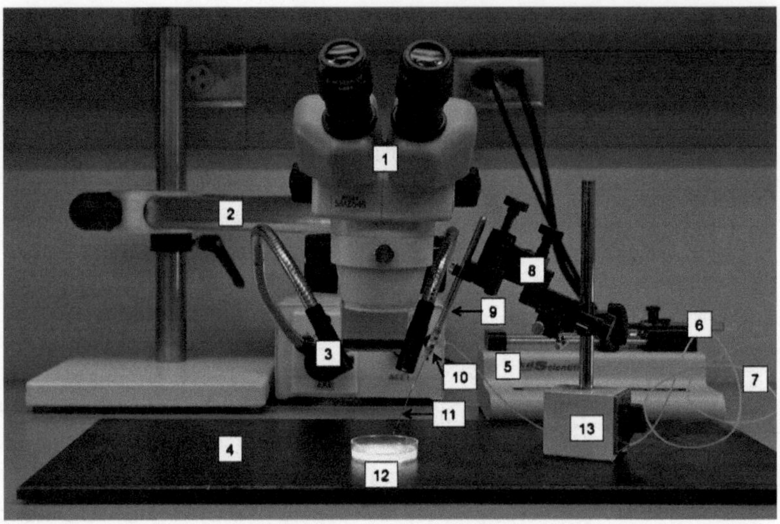

Fig. 1. Injection apparatus. Photograph of injection apparatus for transgenesis. Parts are numbered: (1) dissecting microscope, (2) boom stand, (3) light source, (4) steel base plate, (5) syringe pump, (6) syringe, (7) Luer-ended tubing, (8) micromanipulator, (9) rod, (10) electrode holder, (11) needle, (12) injection dish, (13) magnetic stand.

beveled edge and an internal diameter of at 80–100 μm. This internal diameter is necessary to avoid shearing of the sperm nuclei. To measure the diameter, use a stage or eyepiece micrometer. To achieve a beveled end, break the end of the drawn-out pipette at an angle using forceps with offset tips.

Pump, Syringe, Needle Holder, and Tubing

Syringe pump (such as World Precision Instruments Catalog # SP100i).

Hamilton gas-tight or plastic hypodermic syringe with male Luer end (volume depends on pump).

Double female Luer-ended tubing (Bio Rad Catalog # 732-8202).

Electrode holder (World Precision Instruments Catalog # MPH3).

Micromanipulator Nirishige M-152 (available from Tritech Research or other microscope suppliers).

Magnetic stand (World Precision Instruments Catalog # M-10).

Steel base (World Precision Instruments Catalog # 5479).

Dissecting microscope.

Mineral oil, embryo tested (Sigma M8410).

The injection apparatus is shown in Fig. 1. It is essentially a syringe pump that can be set at 10 nl/s fitted with a Hamilton syringe with a male Luer end. The volume of the syringe depends on the pump and the pump settings.

The syringe is attached to female Luer-ended tubing. The tubing is attached to an electrode holder that will hold the injection needle (described below). The syringe and tubing are filled with mineral oil.

The electrode holder is attached to a metal rod that is clamped into a micromanipulator. We use a Nirishige micromanipulator clamped onto a magnetic stand. We use a metal plate as a platform for attachment of the micromanipulator. Observation of the eggs and injections is facilitated by a dissecting microscope. To accommodate the metal plate, we mount the dissecting microscope on a boom.

Injection Solutions

0.4× MMR+6% ficoll.

0.1× MMR+6% ficoll.

These solutions should be filtered before use.

3. Methods

3.1. Gene Knockdown Late in Development by a Transgenic shRNA Approach

3.1.1. Generation of shRNA Transgene

1. To design the shRNA template, enter cDNA sequence (including untranslated regions) into the shRNAi Retriever Design Tool at http://cancan.cshl.edu/RNAi_central/RNAi.cgi?type=shRNA. This will yield the sequence of the shRNA template. The program yields the sequence of the sense portion of the template. Add the antisense portion as shown in Fig. 2 so that it will base-pair with the sense portion upon folding as a RNA (see Note 2).
2. Add linker sequences to ends of shRNA template sequence, including sites for restriction enzymes for subcloning (see Fig. 1). The HinDIII sites are used for subcloning and the BamHI site is used for determining orientation of the insert.
3. Obtain three primers: a single long primer to serve as one strand of the shRNA template and two amplification primers. In our scheme, the forward amplification primer is gene specific but the reverse amplification primer is common to all shRNA templates designed as described here. The shRNA template primer will be long (~94 nucleotides), so it should be ordered as a polyacrylamide gel electrophoresis (PAGE)-purified oligonucleotide.
4. Also design and order a similar oligonucleotide containing the reversed shRNA template sequence (NOT reverse complement!), to serve as a negative control. This will require a new forward amplification primer.

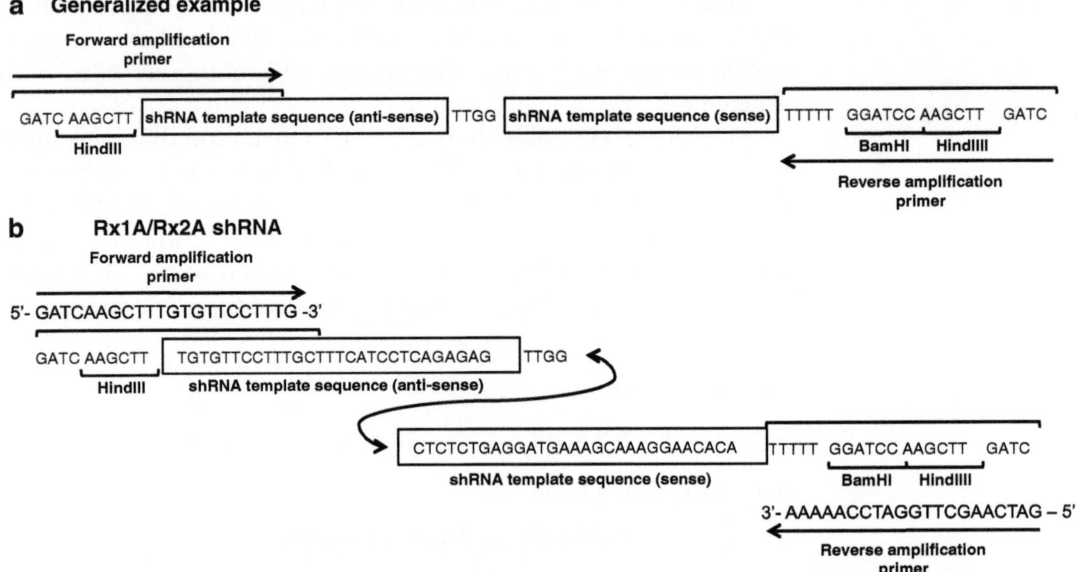

Fig. 2. Composition of shRNA template. (**a**) Schematic representation of shRNA template oligonucleotide. Invariant sequences are shown; gene-specific portions are depicted as *rectangles*. Output from the shRNA design tool is designated "shRNA template sequence (sense)". Also depicted are positions and sequences of amplification primers and restriction sites for subcloning of the double-stranded template. (**b**) Annotated sequence of the actual Rx shRNA template and amplification oligonucleotides we synthesized (1).

5. Synthesize the second strand of the test and control shRNA templates and amplify them. We used 5 nM template oligonucleotide, 50 nM each amplification primer, and 2× PCR mix from Promega.

Amplification	3 min at 95°C, followed by 20 cycles of
	30 s at 95°C
	30 s 55°C
	30 s 72°C, followed by 10 min at 72°C

6. Add A-overhangs (if necessary), clone into a T-vector (see Note 3), and verify the sequence of the templates.
7. Transfer the inserts from the T-vector to a shRNA expression plasmid such as pRNAT. It is important that the vector contain a selectable marker for identifying transgenic embryos, either a drug-resistance gene or a reporter gene expression cassette (such as a GFP expression cassette).
8. Digest plasmid to liberate transgene. Separate transgene from other digestion products by agarose gel electrophoresis. Purify transgene from agarose using GeneClean Kit (Bio101) or equivalent. Alternatively, linearize the transgenes and purify

transgene DNA (extraction with phenol:chloroform:isoamyl alcohol and ethanol precipitation is sufficient).

9. Carefully determine transgene concentration by nanodrop or by comparison with DNA of known mass on an agarose gel.

3.2. Sperm Nuclei Preparation

Sperm nuclei are prepared essentially as described previously (3) but using digitonin for demembranation (4).

1. Remove testes from male frog. First, anesthetize frog in 0.1% aminobenzoic acid ethyl ester (Tricaine, MS-222) in 0.1× MMR. Cut through ventral body wall and locate fat bodies—testes are attached to fat bodies, one on each side of midline. Remove testes and wash in cold 1× MMR. We usually perform sperm preparation from one to two testes.
2. Under the microscope, remove any adherent tissue, such as fat, connective tissue, or coagulated blood. Be careful not to tear or puncture testes.
3. Remove blood. To do this, puncture large blood vessels with sharp forceps and push blood out of vessels by mild pressure using flat side of forceps. Again, be careful to avoid puncturing testes.
4. Move cleaned testes to a clear 35-mm tissue culture dish with cold 1× NPB. Allow to cool on ice for 5 min.
5. Transfer testis to a dry 35-mm dish and mascerate by crushing and pulling apart with clean forceps until there are no more pieces visible to the naked eye. Do not allow this process to go on so long that the edges of the mascerated material begin to dry.
6. Add 2 ml of cold 1× NPB and resuspend macerated testes using a 5-ml pipette. Do not titurate.
7. Filter the suspension through four layers of cheesecloth placed in a small funnel into a 14-ml polyethylene culture tube. Rinse the dish with 3 ml of 1× NPB and add to the cheesecloth. Add an additional 5 ml of 1× NPB through the cheesecloth. Use a clean gloved hand to fold and squeeze the cheesecloth, allowing the liquid to go into the tube.
8. Centrifuge in a swinging bucket centrifuge 10 min at 1,500 × *g* at 4°C. The sperm should form a white pellet with a red center (due to red blood cells). During this spin, warm 1 ml of 1× NPB to room temperature.
9. Remove (decant) supernatant and gently resuspend the pellet in 9 ml of cold 1× NPB. Repeat centrifugation. During this spin, dissolve approximately 10 mg of digitonin in 100 μl of DMSO.
10. Decant supernatant and resuspend pellet in 1 ml of 1× NPB at room temperature. Add 50 μl of digitonin solution dropwise while gently swirling the suspension. Incubate at room temperature for 5 min.

11. Add 10 ml of cold 1× NPB + 3% BSA. Cover tube with Parafilm and mix by inverting. Centrifuge as before. Pellet should be wider and looser and somewhat more translucent than before. The red blood cell pellet should no longer be visible. The pellet now consists of sperm nuclei.
12. Gently resuspend pellet in 5 ml of 1× NPB + 0.3% BSA and repeat centrifugation.
13. Resuspend pellet in 250 μl of 1× NPB + 30% glycerol + 0.3% BSA.
14. Determine concentration of nuclei. Use a cut pipette tip to dilute 2 μl of sperm nuclei with 200 μl of SDB and 2 μl of 1:100 Hoechst stock. Count nuclei using a hemacytometer and a fluorescent microscope. Usually the 1:100 diluted sperm nuclei suspension contains $125–250 \times 10^4$ nuclei/ml, corresponding to $125–250 \times 10^6$ nuclei/ml or 125–250 nuclei/nl, for the undiluted stock. If the concentration is substantially lower than this, repellet the nuclei (or allow to settle over a few hours) and resuspend in a smaller volume.
15. Incubate at 4°C overnight. Resuspend and aliquot into 10-μl portions. Freeze in liquid nitrogen and store at −80°C.

3.3. Transgenesis

Transgenesis is performed by intra-cytosolic sperm injection (ICSI), as described previously (5). ICSI involves mixing the transgene(s) with sperm nuclei, prepared as described above, and injecting the mixture into *Xenopus* eggs. A description of the method follows:

1. The night before transgenesis, induce egg-laying by injecting adult female *X. laevis* frogs with 500–800 U human chorionic gonadotropin into the dorsal lymph sac.
2. Thaw an aliquot of sperm nuclei on ice. Keep on ice and do not refreeze. The nuclei should be usable all day. Prepare cysteine solution for dejellying eggs. Thaw SDB.
3. Prepare the following mixture:

   400,000 sperm nuclei.

   SDB to 4 μl.

   2.5 μl transgene DNA (100–250 ng DNA).

   Remember to use cut pipet tips for transferring or mixing sperm nuclei.
4. Incubate transgenesis mixture for 15 min at room temperature. During this time, strip eggs from hCG-injected females. Grasp the female with hands encircling abdomen and squeeze gently to strip eggs. Position the frog so that eggs are collected in a 50-mm Petri dish.

5. Dejelly eggs: cover with cysteine solution. Use a plastic pipette to detach eggs from the sides of the dish.

   Swirl gently until eggs are dejellied. Try not to let too many eggs come to the surface of the solution. (When the eggs are dejellied they will pack close together and no longer be connected by jelly.)

   Wash dejellied eggs 4–5 times in 1× MMR. Pick out broken or activated eggs. Activated eggs will display shrunken animal hemisphere—the animal hemisphere pigmentation will be concentrated in a noticeably smaller area.

6. Transfer eggs to injection dish containing 0.4× MMR + 6% ficoll (use a cutoff plastic transfer pipette). Tilt dish to one side and tap to pack the eggs. They will also largely reorient so that the animal hemispheres face upwards. Do not overpack!
7. Add 22.5 μl of SDB to transgenesis reaction. Transfer 2.5 μl of this dilution to 230 μl SDB.
8. Affix a piece of Tygon tubing to a cut 200-μl pipette tip. Gently pipette diluted transgenesis mixture ten times taking care not to introduce bubbles. Draw approximately 100 μl of transgenesis mixture into tip + tubing. Lay horizontally on lab bench and attach needle to tubing. Detach tip + tubing + needle from pipettor and turn to vertical (needle pointing down) to fill needle (by gravity). Return to horizontal and detach needle from tubing. Return pipette tip + tubing to tube containing remainder of transgenesis mixture. Attach needle to holder.
9. Start pump. Adjust micromanipulator so that tip of needle is in the dish containing the eggs. Watch for evidence of transgenesis mixture flow (visible as Schlieren, due to the difference in density of the transgenesis solution and the medium in the dish).
10. Position the tip of the needle close to an egg. Inject with a short sharp motion to puncture the egg with minimal stretching and damage. Make sure the motion is short so as not to run through the egg entirely. Gently retract needle approximately 1 s after insertion. Repeat with the next egg in a 3- to 4-s cycle (move dish to position needle near an egg—inject—retract needle—reposition dish to inject next egg). Inject in a pattern to more easily keep track of injected eggs. Eggs will show signs of activation (contracted pigmented animal hemisphere) within a minute of injection.
11. Incubate injected eggs for approximately 2 h at 16°C. Successfully injected eggs should be approximately at 4-cell stage. Carefully transfer normal embryos to a 35-mm dish or multiwell plate containing 0.1× MMR + 6% ficoll. Put no more than 15 embryos per dish or well. Incubate at 16°C.

12. Next day, remove dead or dying embryos and change solution to 0.1× MMR (without ficoll). Incubate at 16°C for an additional day and then at desired temperature (16°C—room temperature) with daily changes of buffer. Score for marker expression or drug resistance (see Note 4).

3.4. Validation of Target Gene Knockdown by qRT-PCR

1. Tadpoles are anesthetized using tricaine (MS222) and homogenized in a minimal volume of Trizol. In our case, we purified RNA from isolated eyes or heads.
2. RNA was purified according to Trizol manufacturer's protocol.
3. Expression levels of the gene product of interest (in our case, Rx) are determined by quantitative RT-PCR. Reverse transcription is performed using RNA from one head (usually about 1 μg) RNA and the qScript cDNA SuperMix Kit. Real-time PCR performed in a 25-μl amplification mixture containing 1 μl of cDNA product, 12.5 μl of 2× SYBR Green PCR master mix, and 100 nM forward and reverse primers (specific to each gene). The PCR conditions include a polymerase activation step at 95°C for 10 min followed by 40 cycles of 95°C for 15 s and 60°C for 60 s. A housekeeping gene, L8, is used as an internal control. The statistical significance of relative differences in expression levels is determined by Student's group *t*-test. We experience knockdown levels from 40 to 80%.

3.5. Regulation of Gene Expression by a trans-Acting Factor of Interest Through a Specific cis-Acting Element

1. Prepare a transgene containing the putative target promoter driving a reporter gene, such as a fluorescent protein. In our case we used dsRed Express as the reporter since our shRNA transgene expresses cGFP. Prepare a wild-type version and one harboring mutations in the *cis*-acting element of interest.
2. Generate transgenic embryos using the control or knockdown shRNA and wild-type (wt) or mutated reporter transgenes in all four combinations: control shRNA with wt or mutated reporter, knockdown shRNA with wt or mutated reporter.
3. Select embryos transgenic for the shRNA and reporter transgenes injected using the selectable markers included on the transgenes.
4. Culture tadpoles to the appropriate developmental stage for the gene of interest.
5. Fix embryos in 4% paraformaldehyde for 1 h at room temperature.
6. Wash with PBS and transfer to 30% sucrose at 4°C overnight or until the embryos have settled to the bottom of the vial.
7. Transfer the embryos to an embedding mold and fill with fresh OCT. Rapidly submerge the mold into 95% ethanol cooled with dry ice. After the material is frozen wrap the block in aluminum foil and store at −80°C.

8. Section onto charged glass slides at 10 μm using a cryostat. Mount with Vectashield HardSet with DAPI and a coverslip.
9. Capture fluorescent images as TIFF files.
10. Quantify immunofluorescence using IMAGEPRO 6.2 (Media Cybernetics). For each image, select the whole organ of interest. Green (shRNA transgene) or red (reporter transgene) cells or cell clusters will be picked automatically by the program after adjusting the intensity value for each color, threshold to 255 (0 = black and 255 = saturated green or red). Threshold is set to cover all the labeled cells in the organ. The average pixel intensity is calculated for all pixels within threshold regions. These calculations are determined for each region sampled from two to three different embryos for each experiment conditions.

4. Notes

1. Prepare NPB buffers on day of sperm nuclei preparation from stock solutions.
2. Many genes in *X. laevis* are duplicated. We try to discover duplicates by searching EST databases. We then select candidate shRNAs that have good matches to both forms with minimal number of mismatches. For example, in case of Rx, the shRNA we selected was 29 nucleotides long, matched Rx1A perfectly, and had only two mismatches with Rx2A (1).
3. We use a version of pRNAT containing the *X. tropicalis* U6 promoter (6). The original version should work, however.
4. We use fluorescent markers for selection, as they are useful for the quantification steps to follow. In the case of the shRNA transgene, we use a coral GFP expression cassette contained in pRNAT. However, G418 resistance can also be used (7).

References

1. Pan Y, Martinez-De Luna RI, Lou CH, Nekkalapudi S, Kelly LE, Sater AK, El-Hodiri HM (2010) Regulation of photoreceptor gene expression by the retinal homeobox (Rx) gene product. Dev Biol 339:494–506
2. Kimura A, Singh D, Wawrousek EF, Kikuchi M, Nakamura M, Shinohara T (2000) Both PCE-1/RX and OTX/CRX interactions are necessary for photoreceptor-specific gene expression. J Biol Chem 275:1152–1160
3. Sive HL, Grainger RM, Harland RM (2000) Earlydevelopment of *Xenopus laevis*: a laboratory manual. Cold Spring Harbor Laboratory Press, Cold Spring Harbor, NY
4. Huang H, Marsh-Armstrong N, Brown DD (1999) Metamorphosis is inhibited in transgenic *Xenopus laevis* tadpoles that overexpress type III deiodinase. Proc Natl Acad Sci USA 96:962–967
5. Sparrow DB, Latinkic B, Mohun TJ (2000) A simplified method of generating transgenic *Xenopus*. Nucleic Acids Res 28:E12
6. Li M, Rohrer B (2006) Gene silencing in *Xenopus laevis* by DNA vector-based RNA interference and transgenesis. Cell Res 16:99–105
7. Moritz OL, Biddle KE, Tam BM (2002) Selection of transgenic *Xenopus laevis* using antibiotic resistance. Transgenic Res 11:315–319

Chapter 4

Subretinal Delivery and Electroporation in Pigmented and Nonpigmented Adult Mouse Eyes

John M. Nickerson, Penny Goodman, Micah A. Chrenek, Christiana J. Bernal, Lennart Berglin, T. Michael Redmond, and Jeffrey H. Boatright

Abstract

Subretinal injection offers one of the best ways to deliver many classes of drugs, reagents, cells and treatments to the photoreceptor, Müller, and retinal pigment epithelium (RPE) cells of the retina. Agents delivered to this space are placed within microns of the intended target cell, accumulating to high concentrations because there is no dilution due to transport processes or diffusion. Dilution in the interphotoreceptor space (IPS) is minimal because the IPS volume is only 10–20 μl in the human eye and less than 1 μl in the mouse eye. For gene delivery purposes, we wished to transfect the cells adjacent to the IPS in adult mouse eyes. Others transfect these cells in neonatal rats to study the development of the retina. In both neonates and adults, electroporation is found to be effective. Here we describe the optimization of electroporation conditions for RPE cells in the adult mouse eye with naked plasmids. However, both techniques, subretinal injection and electroporation, present some technical challenges that require skill on the part of the surgeon to prevent untoward damage to the eye. Here we describe methods that we have used for the past 10 years (Johnson et al. Mol Vis 14: 2211–2226, 2008).

Key words: Subretinal, Injection, Electroporation, Interphotoreceptor Space, Transfection, Dilation, Sclera, Cornea, Reporter Gene Expression, Subretinal bleb

1. Introduction

The subretinal space is a useful target for drug delivery (1–5) and gene therapy purposes (6–20) because subretinal delivery places injected material within microns of the plasma membranes of the photoreceptor (PhR), Müller, and the retinal pigment epithelium (RPE) cells. It is important that in many cases, detached retina rejoins the RPE sheet quickly, a process called bleb regression.

Shu-Zhen Wang (ed.), *Retinal Development: Methods and Protocols*, Methods in Molecular Biology, vol. 884, DOI 10.1007/978-1-61779-848-1_4, © Springer Science+Business Media, LLC 2012

Once the bleb has regressed, the reattached retina functions again. Subretinal injection surgery is used clinically (cf., tPA injection for submacular or subretinal hemorrhage) (21, 22) and has been demonstrated in many animal models (3–5).

Genetics and genomic modifications in the mouse are facile and highly informative, making the mouse the "go-to" animal in much of biomedical research and in particular in vision research. It is the smallest mammal that has an eye resembling the human counterpart. However, the small size of the eye and the relatively large size of the lens make subretinal surgery difficult in mice. Several surgical approaches have evolved for the mouse.

Many research groups have reported a trans-scleral route for subretinal injections. In this route, a needle is advanced through the sclera, entering at the limbus or pars plana, crossing through the vitreous, penetrating through the diametrically opposite retina into the subretinal space. Another route is a transscleral–transchoroidal–Bruch's membrane approach without penetrating the retina (23–26). Both routes are effective for injecting many materials in fluid form, and in collecting the contents of the interphotoreceptor (subretinal) space. However, the small size of the mouse eye and the comparative toughness of the sclera increase the risk of accidentally induced hemorrhages at the ciliary body or choroid. These hemorrhages cause autofluorescence and retinal damage, rendering further treatment or experimentation futile. To solve the problem of hemorrhages, Timmers et al. (27) developed a subretinal injection approach in rats via a transcorneal route. We adapted this route to the mouse as described here.

There are many ways to transfect DNA into a target cell, including viruses (28, 29), physical (electroporation, ballistic, and sonication) (30–34), chemical (liposomes (35, 36), DNA compaction (37), dendrimers (38), and precipitates—e.g., calcium phosphate) (39). Of these, viruses achieve 100% transduction efficiency in cultured PhR cells (16) and electroporation achieves up to 90% transfection efficiency (40). Other chemical-based agents may be highly successful but often require serum-free conditions (41), a state that is impossible in vivo.

Electroporation is inexpensive, safe, and easy to replicate under well-controlled conditions (31, 32, 40). Initially we found electroporation to work erratically in mice, but once we standardized our protocols (27), we found it to work well. Here we detail steps that we found essential for consistent results. A key step was to evaluate the fundus after the subretinal injection. Here, we present videos of the surgical technique so that others can more readily learn subretinal injection, as initially described in rats by Timmers et al. (27). Given more reliable surgery, it was possible to optimize electroporation to deliver plasmids to RPE cells. These conditions show high-level reporter gene expression from plasmids in the RPE of the living adult mouse (42).

2. Materials

2.1. Mouse Strains

We have used pigmented and nonpigmented mice aged between birth and 9 months old at time of surgery. We do not think that there would be a need to modify the protocol for older mice, but we have not tested mice older than 10 months. For neonatal mice, the procedure of ref. (23) is fast and easy

2.2. A HOBO Datalogger

To be used to record data every 5 min for a week or a month (Lab Safety Supply; U12-012) and to measure light level, temperature, and humidity in animal care rooms and elsewhere.

2.3. Reporter Gene Plasmid

The reporter expression plasmid, called pVAX-tdTomato (43), contained the CMV immediate early promoter driving expression of tdTomato. This plasmid contains a bovine growth hormone poly(A) signal on the 3′ flanking side of the tdTomato cDNA. The plasmid contains the Kanamycin-resistance gene for selection and growth. This plasmid was a kind gift from Dr. Ton N.M. Schumacher of the Department of Immunology, The Netherlands Cancer Institute, Amsterdam, The Netherlands. Plasmid is isolated from transformed DAM⁻/DCM⁻ cells (catalog no. C2925H) from New England Biolabs. *Escherichia coli* are grown overnight in Luria broth using a Qiagen Endotoxin Free GIGAprep kit following the manufacturer's protocol. Plasmid pellet is dissolved using molecular grade water. Fifty microgram of plasmid is dried using a speedvac system. Pellets are stored in the –20°C until needed.

2.4. Dissecting Microscope System

We use an Olympus SZX2-ZB16 stereo microscope (Hunt Optics; Pittsburgh, PA), which is equipped with a halogen lamp mounted to an epifluorescence adapter with no fluorescence filters (we have also tried a ring light and a co-axial illuminator). This light source gives true coaxial illumination through the objective lens which is far better than a ring light for illuminating the back of the mouse eye. The new light is about eight times brighter than the previously described coaxial light source (42). The improved light source is important for the subretinal injection of pigmented mice. It is bright enough to make the injection of pigmented mice easy, fast, and practical.

2.5. Video Recording

These are conducted with a Panasonic GPUS932HT HD Video camera (Hunt Optics and Imaging). The camera was interfaced to a KONA LHe HD-video capture card (AJA Video Systems) installed in a Mac Pro (Apple Computer) running OSX Snow Leopard 10.6.8. Videos were edited with Final Cut Pro (version 6; Apple Computer). Video recordings are optional, but helpful for training and demonstration purposes.

2.6. Injection Equipment

We use a NanoFil™ Sub-microliter injection system with a UMP-II microsyringe pump and Micro4 controller with a foot-switch [World Precision Instruments (WPI)].

2.7. Warm System

It is important to maintain the mouse at body temperature during and after anesthesia. We use two systems to maintain the mouse at 37°C. One is an in-house developed aluminum block having channels cut in it for warm water circulation, with the block temperature controlled with a Lauda Circulating water bath. The second system is a commercial product, a T/Pump TP500 (Gaymar).

2.8. Electroporator

We use BTX model ECM830 (Harvard Apparatus), a commercial square wave generator. Others are acceptable, too.

2.9. Fine Tools

1. Microknife. Blade size 15° stab knife straight (Sharpoint, catalog no. REF72-1501).
2. Beveled 34-G needles (catalog no. NF34BV-2), blunt 35-G needles (catalog no. NF35BL-2), and curved forceps (catalog no. 15915) are obtained from WPI.
3. Most other incidental equipment and tools are from Fisher Scientific or VWR.
4. Platinum–iridium 20-G wire (catalog no. 50822164; Fisher).
5. Small test-jumper leads (catalog no. 278-001; Radio Shack Corporation).

2.10. Regents

1. Transparent clear ocular hydro-gel. We use Vidisic® Augengel (catalog no. 1-19006, distributed by Dr. Mann Pharma), which contains high molecular weight polyacrylic acid and was a kind gift of Dr. Philipp Lirk, Department of Anesthesiology and Critical Care Medicine, Medical University Innsbruck, Austria. Other clear ophthalmic grade hydrogels, including viscous methylcelluloses are acceptable as well.
2. Quantum dots with a 600-nm fluorescence emission maximum (EviTags, E2-C11-NF2-0600; Evident Technologies). Quantum dots are injected as the stock preparation (see Note 1). This source is no longer available, but similar quantum dots and other fluorescent spheres can be obtained from Invitrogen to mark the bounds of a subretinal injection.
3. Apoptotic cells are detected with a DeadEnd TUNEL kit (product number G3250; Promega).
4. Pre-mixed 80 and 12 mg/ml Ketamine and Xylazine solution (K-113; Sigma-Aldrich).
5. Phenylephrine (2.5% w/v; ophthalmic grade; Bausch & Lomb).
6. Betadine (5%, ophthalmic grade, Alcon, NDC 0065 0411 30).
7. Fast Green (catalog no. BP123-10, Fisher).

8. Triple antibiotic ointment (Taro Pharmaceuticals, Inc.), which contains bacitracin, neomycin sulfate, and polymyxin B.
9. Proparacaine hydrochloride ophthalmic solution USP, 0.5% [National Drug Code (NDC) 17478-263-12; Akorn Inc.].
10. Optispears (Ocusoft, Inc.).
11. "Refresh" eye drops (Allergan).

3. Methods

3.1. Mouse Husbandry

Mice should be used according to regulatory agents' guidelines and must be approved by a local Institutional Animal Care and Use Committee. Mice for our experiments are housed at 22°C in facilities managed by the Emory University Division of Animal Resources and given standard mouse chow (Lab Diet 5001; PMI Nutrition Inc., LLC) and water ad libitum. They are maintained on a 12 h:12 h light-dark cycle, with daytime lighting ranging 200–750 lux outside the cage depending on lower, middle, or top shelf position of the cage rack. We have found it useful to monitor light levels, temperature, and humidity with a datalogger. This can identify unexpected changes in lighting, humidity, or temperature that can alter the outcomes of any in vivo study (a sample record showing a stuck light switch, a light bulb burning out, and a humidity spike are illustrated in Fig. 1).

3.2. Injection Material Preparation

1. Resuspend plasmid DNA (pVAX-TdTomato) in sterile water at 2 mg/ml.
2. Add marker dye Fast Green in all injected solutions at 0.1% (w/v).
3. Centrifuge the plasmid solution at 10,000 × *g* for 5 min to sediment any particulates from the solution that might clog a 35-G needle. Do this immediately before loading the needle and injection syringe (Fig. 2).
4. Surgical sterility: use sterile surgical technique throughout. Sterilize fluid lines, surgical instruments, and needles by repeated rinsing with 70% ethanol and sterile water.
5. Prepare a mixture of Ketamine and Xylazine with working concentration as 16 μg/μl Ketamine and 2.4 μg/μl Xylazine by diluting the stock 1:5 with sterile dPBS. Keep the working solution ice cold at all times.
6. Calculate the injection volume. We use 80 mg ketamine and 12 mg xylazine per kilogram body weight to achieve adequate anesthesia in 3–5 min. Since the working concentration is 16 μg/μl Ketamine and 2.4 μg/μl Xylazine, use 100 μl of the mixture for a for a 20 g mouse.

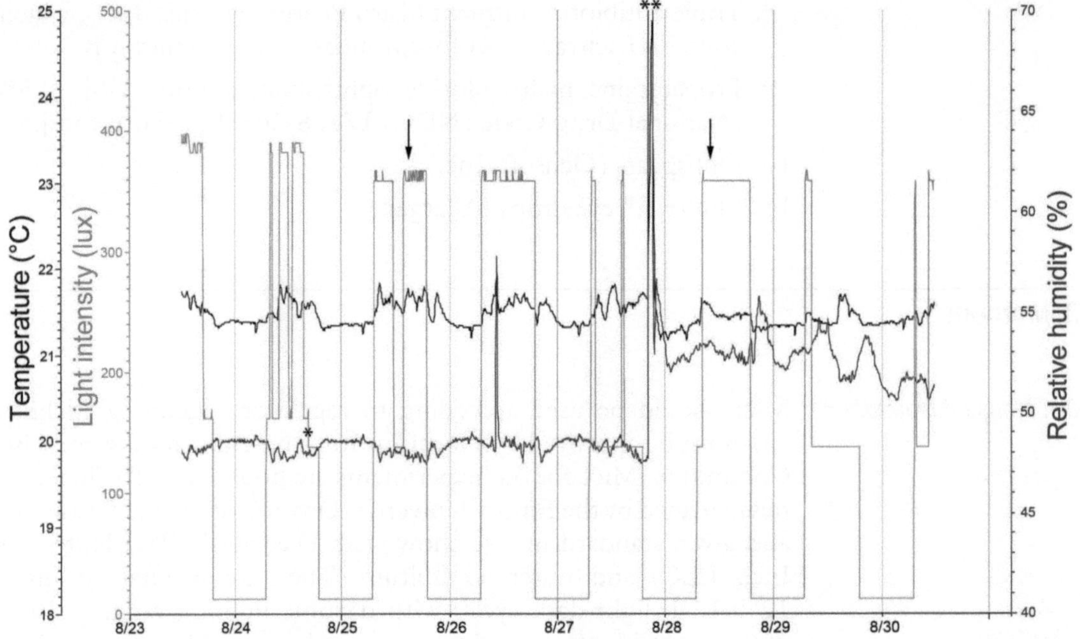

Fig. 1. HOBO traces. This set of traces represents a time period of 1 week in an animal room. A datalogger was positioned at the level of a middle row of cages. The *light gray* tracing represents light level. The *dark gray* tracing represents relative humidity, and the *black* tracing represents temperature. As an example of quirks that can go wrong in animal rooms that are difficult to detect, we found that a light switch was intermittently stuck in the high position, used when technicians work in this room. The switch should return the room to a light level of about 160 lux when the workers exit the room. With the switch in the high position, the middle row of cages receives about 380 lux. In this circumstance, cages at the top level, the light level is about 800 lux. This level on a 12 h on:12 h off cycle is sufficient for mice to become preconditioned to high light levels. The light switch stuck in the "high" position is illustrated by *arrows*. A single *asterisk* represents a fluorescent light bulb burning out (shifting the normal lights on level from 160 lux to about 140 lux). A humidity spike (*double asterisks*) illustrates a substantial but short rain downpour outside the building. The temperature at about 21.5°C is quite consistent but the daytime temperature is generally higher than night by a fraction of a degree due to warming from fluorescent lights and normal building operations.

3.3. Anesthesia and Pupil Dilation (See Note 2)

1. Inject the mixture of Ketamine and Xylazine (into the hind leg muscle with a 29-G needle. Be sure to record these drugs accurately in a suitable logbook following DEA regulations.
2. Anesthetize the cornea topically with one drop of proparicaine for 2 min. Remove excess with an optispear. Add a drop of Refresh artificial tears on the eye to prevent drying, and remove excess.
3. Add one drop of Phenylephrine, and wait 2–3 min. Place the animal on the heating pad and cover to shield from light to help dilate the pupils. A cardboard boxtop covering the entire mouse that does not impede airflow is adequate. The pupil should be fully dilated within 90 s. If not dilated, apply another drop of phenylephrine.
4. When moving to the injection scope, remove excess phenylephrine with an Optispear. Place the mouse on the heated stage to

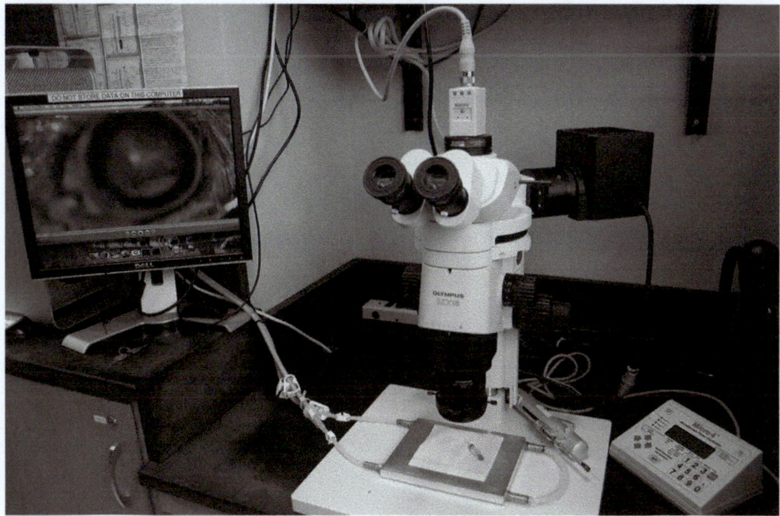

Fig. 2. Injection and microscope setup. A conventional dissecting microscope is used with an epifluorescence halogen light source. A video camera is mounted to the microscope. A nanoliter injection system from WPI is employed. An air filter to the *right* suppresses dust and air currents at the injection station. A computer for control of the video camera and for video editing is partially pictured to the *left*. A homemade aluminum stage warmer is shown on the stage. Temperature is controlled with a Lauda circulating water bath. During surgery, the aluminum block is covered with a *small rectangle* of fresh spill paper.

maintain body temperature during surgery. Mice should be monitored carefully for signs of pain and distress during and after surgery.

3.4. Subretinal Injection (See Notes 3–7)

Static images of the subretinal injection technique are provided in Fig. 3. A video of the surgical procedure is given on the Web site (video 1, Fig. 4), and it highlights the correct techniques from Johnson et al. (42).

1. Position the mouse with its nose pointing away from the surgeon and its left eye facing up toward the light and the microscope.
2. Place a drop of hydrogel (Vidisic®) on the mouse cornea.
3. Grasp the corner of a 22 × 40 microscope coverslip by hand and adjust the coverslip on the Vidisic® eye gel in such a way that the fundus, its blood vessels, and the optic nerve head can be seen.
4. Verify that the pupil is fully dilated. This fundus exam serves to assess the condition of the eye before injection and as a comparison for the postoperative condition of the retina. Note the structure and appearance of blood vessels and optic disc. [For pigmented animals, the lighting should be about 54,000 lux, and for nonpigmented the light source should be about 4,800 lux (adjust with neutral density filters to reduce the light level, or use another source)].

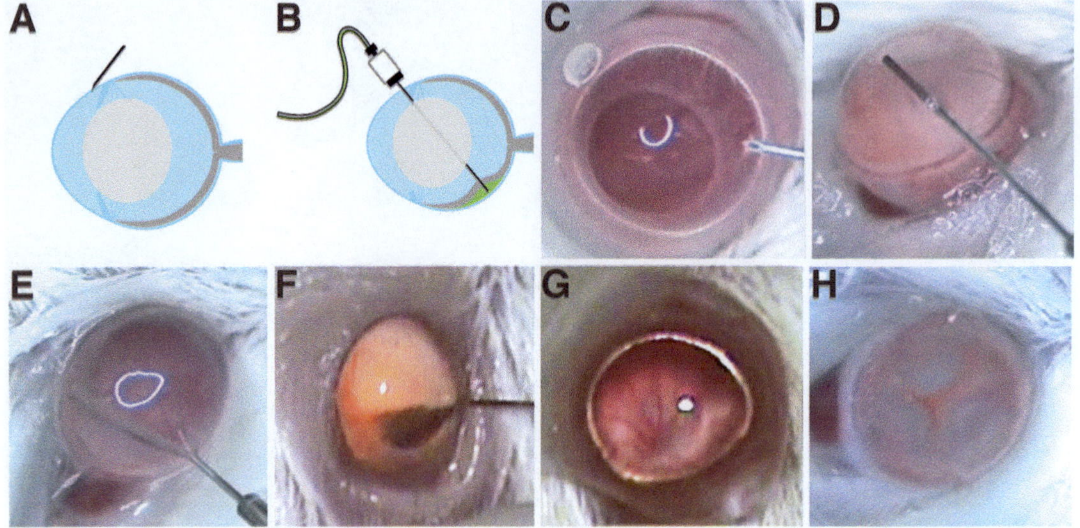

Fig. 3. The subretinal injection technique. (**a**) Position of the 34-G beveled needle is shown nearly tangential just before lancing the cornea. (**b**) Position of the 35-G blunt needle after puncturing the neural retina and partially inflating the interphotoreceptor space (the subretinal space) to produce subretinal blebs. Note that the needle is not running through the lens, but rather is running behind it. (**c**) Presented is a still image from a video illustrating penetration of the cornea. (**d**) This *panel* shows the positioning of the 35-G blunt needle in the center of the anterior chamber. (**e**) The 35-G needle penetrates through the retina into the subretinal space. (**f**) The 35-G needle is removed from the vitreous after subretinal injection of quantum dots. A small number of quantum dots are evident in the vitreous that generate a *reddish-orange* color. (**g**) Illustrated is a fundus before subretinal injection. The retinal vessels can be readily detected in the fundus image. A *ruddy red* background color can be observed before injection. (**h**) Shown is the fundus immediately after subretinal injection. The positions of three blebs surrounding the optic nerve head are located at clock face positions 4, 8, and 11. Each bleb appears *puffy* and *gray* in color with *red vessels* between the blebs. The optic nerve head is nearly centered in the image of the fundus. The imaged mouse eyes are about 3 mm in diameter. The caption and figure image are from Johnson et al. (42). Reprinted with permission.

5. Adjust the position of the mouse on a heating pad as necessary for the surgical procedure. It is not necessary or desirable to restrain the mouse.
6. Remove excess Vidisic with an Optispear.
7. Apply Betadine (5%, ophthalmic grade) to eye, lids, and fur surrounding eye. Remove excess with an Optispear. At this point, it should take about 90 s to complete the subretinal injection.
8. Grasp the left eye with curved forceps held in the surgeon's left hand so that the eye is slightly proptosed by partially and gently closing the forceps.
9. With a 34-G beveled needle (held in the surgeon's right hand) lance the cornea near the limbus penetrating into the anterior chamber at an oblique (nearly tangential) angle (Fig. 3c). Some stretching of cornea is advantageous, allowing the wound to effectively reseal itself at the end of surgery.
10. Remove the beveled needle. Replace it with a blunt 35-G needle connected to the injection system. Advance the needle into the

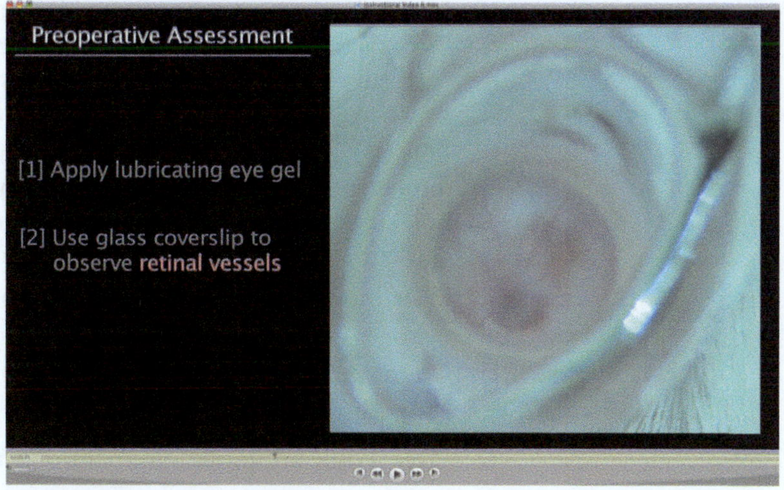

Fig. 4. Video of subretinal injection by transcorneal route. This video was created on an Olympus dissecting microscope equipped with a ring light and an HD video camera. Double-click on the image to play the video. The *orange color* upon subretinal injection comes from the fluorescence of *quantum dots*, which demarcate the extent of the subretinal bleb. The *slide bar* at the bottom of the Quicktime movie can be used to manually control the flow of the movie. The caption and figure image are from Johnson et al. (42), reprinted with permission.

anterior chamber until the tip is centered on the optical axis (Fig. 3d). With a sweeping motion, move the tip of the needle through the pupil, around the lens, and into the vitreous (Fig. 3e). While rare, if the lens is nicked, the surgery should be abandoned.

11. Advance the needle tip to puncture the retina. The lens magnifies the view of the needle (Fig. 3f). Once you encounter a slight resistance to the needle, STOP! You have reached the RPE layer. Apply a gentle amount of pressure (the touch of which must be learned by experience) to penetrate the neural retina into the subretinal space, but not so much that the tip penetrates or damages the RPE sheet (see Note 3).
12. Use the foot pedal to inject fluid into the subretinal space, being careful of the slightest of movement. The nanojector system should be set to deliver 1,000 nl at a rate of 170 nl/s. It can be useful for an assistant to press the injection button on the face of the nanojector, but the footpedal is just as easy. Faster or slower injection rates have not been systematically investigated, but this rate provides acceptable filling of the subretinal space in our hands. Blood in the fundus means that the tip pressure is too great and the choroid is damaged. Too little pressure and the retina is not penetrated. Marker dye in the vitreous indicates the pressure is insufficient and the needle does not enter the subretinal space (see Note 4).

13. Leave the injection needle in the subretinal space for a few seconds to allow pressure in the injection system to equilibrate with subretinal bleb pressure, otherwise large fractions of the subretinal injection material will leak into the vitreous.
14. Pull the needle out slightly, wait another second, and then remove needle entirely. Remove the needle slowly to allow the hole in the retina to reseal and to avoid damaging the lens, iris, and corneal endothelium during removal.
15. Place more hydrogel on the cornea to examine the fundus (see Note 5).
16. Count the number of large blebs (Fig. 3h) on the fundus.
17. Complete the post-op exam to confirm the presence of the blebs.
18. Place Triple antibiotic ointment on the eye. Verify the ear tag.
19. Recovery: cover the animal on a 37°C heating pad until it is awake and actively moving. Transfer it to a clean cage by itself for 1–2 h. Apply more Refresh every 10 min to keep corneas from drying out. Afterwards, return the mouse to its home cage until analysis, usually several days after treatment.

3.5. Worksheet

A record of the subretinal injection procedure should be kept (see Note 5). Documentation should include ear tag number, date of birth, sex, amount of anesthesia, bleb size (small, medium, or large), and, especially important, the number of blebs. Also record any complications including backflow through the retinotomy, hemorrhage, lens damage, corneal clouding, or the presence of air bubbles. A qualitative comment on the outcome of each injection is encouraged.

3.6. Electroporation

Immediately following subretinal injection, any plasmid-treated mouse eyes or control (vehicle only) eyes are electroporated. Typical negative controls include omitting plasmid (vehicle-only subretinal injection) or omitting electroporation in different mice. The contralateral eye served as an uninjected control in all mice.

1. Make electrodes by wrapping 20-G platinum–iridium wire around a sharpened pencil tip, creating a 1.5–2-mm loop. Clip the loops to jumper leads and then to the BTX electroporator.
2. Position one platinum loop (anode) directly underneath the retina bleb site on the scleral surface of the mouse globe, and the other loop (cathode) should be positioned diametrically opposite from the retinotomy. Space the electrodes 1.5–2 mm apart. Try to be as consistent as possible in this spacing, as it determines the potential difference per cm, one of the critical variables in consistent transfection efficiency.
3. Optimal conditions and minimum requirements ought to be investigated by varying the voltage, pulse length, number of pulses, and number of pulse trains. With our apparatus, an

optimum was found with 80 V, five pulses, 5 ms pulse duration, 1 s interval between pulses, and two pulse trains. The range of conditions that we tested were: 0.1–100 ms for pulse length, 0–200 V for potential difference, 5–20 pulses, 0.125 and 1 s interval between pulses, and one or two pulse trains.

3.7. Euthanization

After experimentation, mice are euthanized by CO_2 asphyxiation.

3.8. Outcome Measures of Safety

For RPE cell sheets, TUNEL staining as a marker of apoptosis is an effective tool to assess damage (see Note 3). Flatmounts are created (44). Essentially, the eyecup flatmount includes all the cornea and sclera, but the neural retina, iris, ciliary body, and lens are removed (10, 45). A puncture is made in the cornea with a microknife, and iridectomy scissors (WPI) are used to make four radial cuts, starting at the center of the cornea, and extending toward the optic nerve. The flattened eyecups are placed on microscope slides in 100 μl of dPBS. Primary and secondary antibody staining solutions are pipetted on and off with handheld pipettors, but otherwise do not differ from standard immunostaining procedures. They are mounted in Vectashield hardset, and ought to be examined as soon as the hardset has fully solidified. We normally allow the mountant to set overnight and image the slides the next day.

3.9. Outcome Measures of Efficacy

Fluorescence detection of reporter gene expression. The tdTomato reporter gene has an excitation optimum at 554 nm and an emission maximum at 581 nm. Compared to other naturally fluorescent proteins, tdTomato has reduced photobleaching and provides excellent fluorescence (46). TdTomato is excited using 561-nm laser line and emissions are filtered using a 605/75 bandpass filter (see Notes 6–8).

4. Notes

1. Quantum dots have a tendency to aggregate, clogging a 35-G needle and tubing in the injector system. To prevent clogging it has been suggested that they be mixed with serum albumin at 0.5 mg/ml (47).
2. Perform injections between 08:00 and 18:00 h (i.e., during lights-on of the photoperiod). Keep each mouse on a 37°C pad during and after surgery until it regains consciousness and mobility.
3. Tissue death under optimized conditions should be <1% of RPE cells. An internal positive control for cell death is apoptosis in the corneal endothelium.
4. Potential pitfalls: the costs (in time and effort) associated with optimizing this procedure can be significant. Table 1 presents

Table 1
Troubleshooting guide for subretinal injections

Problem	Probable cause	Solutions
Lens cloudy	Lens capsule nicked during surgery	Avoid the lens
No blebs	Not penetrating the retina	Press a little harder
No blebs	Torn retina or retina hole; fluid leaks out quickly into vitreous	Penetrate retina in a single motion
No blebs	Fluid leaks out quickly into vitreous or a poorly inflated bleb	Pause for 5–10 s before removing the blunt needle from the subretinal bleb. This allows pressure equilibration
Blood	Penetrating into the choroid	Press gently
Blood	Nicking the ciliary body	Sweep closer to lens
Cloudy cornea	Eye was not kept moist before surgery	Apply lubricating eye drops between proparicaine and phenylephrine application
Air bubbles	Air in lines or solutions not degassed	Degas solutions. Flush the lines and prime them with water before filling with delivery solution

some of the problems we encountered with the injection technique and solutions we found.

5. Signs of a successful procedure: careful examination of the adult mouse fundus, after surgery, is a good way to evaluate the success or failure of the injection. A thorough look at the fundus is a critical step.

   For nonpigmented mice, three blebs underlying the neural retina is a sign of proper inflation of the subretinal space in the adult mouse eye. These blebs are clearly visible, and they demonstrate that the injected material was well confined within the subretinal space. We have not observed any cases of more than three blebs in nonpigmented mice. Eyes with only one or two blebs were usually accompanied by evidence of a torn or damaged retina, as seen on funduscopic examination. Nonpigmented mice without three blebs were excluded from experimental test groups.

   For pigmented mice, we found that there were four blebs (Fig. 5). The difference may lie in the more intense light source used for pigmented mice, or it may be a fundamental characteristic of the pigmented strains usually C57BL/6J that we routinely use.

   There are several causes of incorrect inflation of the subretinal space: (a) material is injected into the vitreous; (b) fluid rapidly leaks out of the subretinal space into the vitreous through a hole in the retina; (c) retina is hopelessly torn; (d) material is injected elsewhere (cf., suprachoroidally or subchoroidally); (e) the pump does not operate correctly; or (f) the

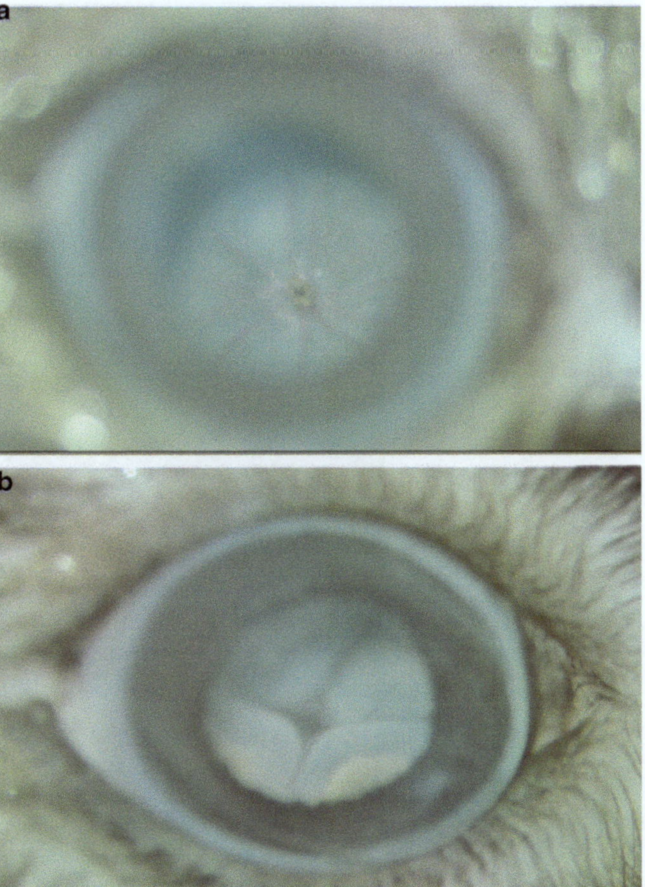

Fig. 5. Four blebs in the fundus from a pigmented mouse eye after subretinal injection. *Panel* (**a**): Fundus appearance before subretinal injection, and panel (**b**): after injection. Four blebs are readily apparent.

needle is clogged. Evidence of blood in the vitreous or aqueous and nicking of the lens are major indicators of problems, and mice with these complications should be immediately excluded from subsequent experiments and analysis. All signs can be readily observed during fundus examination.

6. Expected results for expression of tdTomato: about 30–40% of RPE cells should express the reporter gene, in this case the red fluorescence from TdTomato (Fig. 6). The polygonal pattern of RPE cells is observed in both transfected and nontransfected cells, and there is no characteristic bias in the transfection of cells of lesser or greater polygonality.
7. Expected injection success rate: this technique is successful in our hands about 80% of the time. It requires a significant amount of time to learn. Depending on the amount of practice and prior surgical skills, we find it takes 50–100 surgeries to become proficient in the subretinal injection of the adult mouse eye.

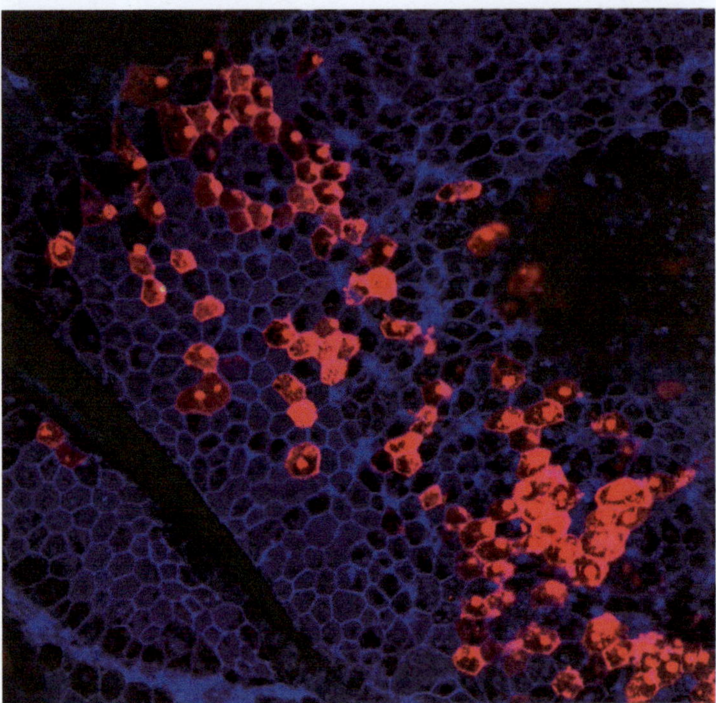

Fig. 6. A typical result of subretinal injection and electroporation of the mouse eye. Transfection and expression of TdTomato were detected 2 days after injection. *Red* represents TdTomato protein fluorescence and *blue* represents actin rings stained with AlexaFluor635-phalloidin that mark the outline of the RPE cells. Both *red hexagons* and *blue hexagons* are readily detected in abundance near the injection site.

This is in contrast to the relative ease in the subretinal injection of neonatal rat pups (27). We found the outcome of the surgery could be rapidly determined postoperatively by examining the mouse fundus. The subretinal injection raised blebs, indicating that the retina was elevated from the RPE sheet, and the persistence of these blebs demonstrated the retinal hole had sealed. No bleb meant that the surgery had failed, and the mouse should not be used for further experimentation. Initially, we expected only one large bleb, because that is what has traditionally been observed in subretinal injections of neonatal rat pups. However, we found the best outcomes were when there were three blebs in nonpigmented mice and four blebs in pigmented mice, and when the perimeters of these blebs, as viewed by fundoscopy, were demarcated by major blood vessels. It should be noted that bleb formation is also used as a sign for successful subretinal injection in human patients (7, 9, 13, 47).

8. Expected results from electroporation: after electroporation, tdTomato fluorescence should be found in a tight patch of RPE cells nearest to the positive electrode (anode) (Fig. 6).

No fluorescence was detected at the negative electrode (cathode), which is nearer to the cornea. Accurate placement of the anode circumscribing the bleb and highly concentrated plasmid are necessary to achieve delivery of the plasmid into RPE cells and expression of tdTomato. Under suboptimal conditions, tdTomato fluorescence was detected in the cornea, ciliary body, and iris. This result occurred with much longer pulse durations (25–50 ms). Voltages beyond about 100 V resulted in immediate evidence of burn damage and were not tested further. Pulse lengths were varied from 1 to 50 ms at different voltages. At 10 ms or longer, the area of the flatmount that showed evidence of transfection extended beyond the immediate location of the electrodes as far as the cornea.

Acknowledgment

This work was supported by the National Eye Institute (R01EY016470, R01EY014026, P30EY006360, R24EY017045, T32EY007092), an unrestricted grant to the Department of Ophthalmology at Emory University from Research to Prevent Blindness, Inc., the Foundation Fighting Blindness, Fight for Sight, The Katz Foundation, and the Intramural Research Program of the National Eye Institute.

References

1. Yasukawa T, Ogura Y, Sakurai E, Tabata Y, Kimura H (2005) Intraocular sustained drug delivery using implantable polymeric devices. Adv Drug Deliv Rev 57:2033–2046
2. Schorderet DF, Manzi V, Canola K, Bonny C, Arsenijevic Y, Munier FL, Maurer F (2005) D-TAT transporter as an ocular peptide delivery system. Clin Exp Ophthalmol 33:628–635
3. Maia M, Kellner L, de Juan E Jr, Smith R, Farah ME, Margalit E, Lakhanpal RR, Grebe L, Au Eong KG, Humayun MS (2004) Effects of indocyanine green injection on the retinal surface and into the subretinal space in rabbits. Retina 24:80–91
4. Shen WY, Rakoczy PE (2001) Uptake dynamics and retinal tolerance of phosphorothioate oligonucleotide and its direct delivery into the site of choroidal neovascularization through subretinal administration in the rat. Antisense Nucleic Acid Drug Dev 11:257–264
5. Kimura H, Spee C, Sakamoto T, Hinton DR, Ogura Y, Tabata Y, Ikada Y, Ryan SJ (1999) Cellular response in subretinal neovascularization induced by bFGF-impregnated microspheres. Invest Ophthalmol Vis Sci 40: 524–528
6. Maguire AM, Simonelli F, Pierce EA, Pugh EN, Mingozzi F, Bennicelli J, Banfi S, Marshall KA, Testa F, Surace EM, Rossi S, Lyubarsky A, Arruda VR, Konkle B, Stone E, Sun J, Jacobs J, Dell'Osso L, Hertle R, Ma J-X, Redmond TM, Zhu X, Hauck B, Zelenaia O, Shindler KS, Maguire MG, Wright JF, Volpe NJ, McDonnell JW, Auricchio A, High KA, Bennett J (2008) Safety and efficacy of gene transfer for Leber's congenital amaurosis. N Engl J Med 358:2240–2248
7. Bainbridge JW, Smith AJ, Barker SS, Robbie S, Henderson R, Balaggan K, Viswanathan A, Holder GE, Stockman A, Tyler N, Petersen-Jones S, Bhattacharya SS, Thrasher AJ, Fitzke FW, Carter BJ, Rubin GS, Moore AT, Ali RR (2008) Effect of gene therapy on visual function in Leber's congenital amaurosis. N Engl J Med 358:2231–2239
8. Hauswirth WW (2005) The consortium project to treat RPE65 deficiency in humans. Retina 25:S60

9. Cideciyan AV, Aleman TS, Boye SL, Schwartz SB, Kaushal S, Roman AJ, Pang J-J, Sumaroka A, Windsor EAM, Wilson JM, Flotte TR, Fishman GA, Heon E, Stone EM, Byrne BJ, Jacobson SG, Hauswirth WW (2008) Human gene therapy for RPE65 isomerase deficiency activates the retinoid cycle of vision but with slow rod kinetics. Proc Natl Acad Sci USA 105:15112–15117
10. Berglin LC, Schmack I, Holley G, Nie X, Yang H, Grossniklaus HE, Edelhauser HF (2005) Human RPE ex vivo 'flatmount technique' for comparative morphometric and tissue culture survival analysis (mouse) using alizarin red staining, live/dead cell analysis and epifluorescent microscopy. Invest Ophthalmol Vis Sci 46:3064
11. Lai CM, Yu MJ, BrankovM BNL, Zhou X, Redmond TM, Narfstrom K, Rakoczy PE (2004) Recombinant adeno-associated virus type 2-mediated gene delivery into the Rpe65-/- knockout mouse eye results in limited rescue. Genet Vaccines Ther 2:3
12. Pang J-J, Chang B, Hawes NL, Hurd RE, Davisson MT, Li J, Noorwez SM, Malhotra R, McDowell JH, Kaushal S, Hauswirth WW, Nusinowitz S, Thompson DA, Heckenlively JR (2005) Retinal degeneration 12 (rd12): a new, spontaneously arising mouse model for human Leber congenital amaurosis (LCA). Mol Vis 11:152–162
13. Nusinowitz S, Ridder WH III, Pang JJ, Chang B, Noorwez SM, Kaushal S, Hauswirth WW, Heckenlively JR (2006) Cortical visual function in the rd12 mouse model of Leber congenital amarousis (LCA) after gene replacement therapy to restore retinal function. Vision Res 46:3926–3934
14. Batten ML, Imanishi Y, Tu DC, Doan T, Zhu L, Pang J, Glushakova L, Moise AR, Baehr W, Van Gelder RN, Hauswirth WW, Rieke F, Palczewski K (2005) Pharmacological and rAAV gene therapy rescue of visual functions in a blind mouse model of Leber congenital amaurosis. PLoS Med 2:e333
15. Bainbridge JW, Mistry A, Schlichtenbrede FC, Smith A, Broderick C, De Alwis M, Georgiadis A, Taylor PM, Squires M, Sethi C, Charteris D, Thrasher AJ, Sargan D, Ali RR (2003) Stable rAAV-mediated transduction of rod and cone photoreceptors in the canine retina. Gene Ther 10:1336–1344
16. Dinculescu A, Glushakova L, Min S-H, Hauswirth WW (2005) Adeno-associated virus-vectored gene therapy for retinal disease. Hum Gene Ther 16:649–663
17. Balaggan KS, Binley K, Esapa M, MacLaren RE, Iqball S, Duran Y, Pearson RA, Kan O, Barker SE, Smith AJ, Bainbridge JW, Naylor S, Ali RR (2006) EIAV vector-mediated delivery of endostatin or angiostatin inhibits angiogenesis and vascular hyperpermeability in experimental CNV. Gene Ther 13:1153–1165
18. Bemelmans A-P, Bonnel S, Houhou L, Dufour N, Nandrot E, Helmlinger D, Sarkis C, Abitbol M, Mallet J (2005) Retinal cell type expression specificity of HIV-1-derived gene transfer vectors upon subretinal injection in the adult rat: influence of pseudotyping and promoter. J Gene Med 7:1367–1374
19. Doi K, Hargitai J, Kong J, Tsang SH, Wheatley M, Chang S, Goff S, Gouras P (2002) Lentiviral transduction of green fluorescent protein in retinal epithelium: evidence of rejection. Vision Res 42:551–558
20. Bloquel C, Bourges JL, Touchard E, Berdugo M, BenEzra D, Behar-Cohen F (2006) Non-viral ocular gene therapy: potential ocular therapeutic avenues. Adv Drug Deliv Rev 58:1224–1242
21. Kamei M, Tano Y (2009) Tissue plasminogen activator-assisted vitrectomy: surgical drainage of submacular hemorrhage. Dev Ophthalmol 44:82–88
22. Sandhu SS, Manvikar S, Steel DHW (2010) Displacement of submacular hemorrhage associated with age-related macular degeneration using vitrectomy and submacular tPA injection followed by intravitreal ranibizumab. Clin Ophthalmol 4:637–642
23. Price J, Turner D, Cepko C (1987) Lineage analysis in the vertebrate nervous system by retrovirus-mediated gene transfer. Proc Natl Acad Sci USA 84:156–160
24. Gekeler F, Kobuch K, Schwahn HN, Stett A, Shinoda K, Zrenner E (2004) Subretinal electrical stimulation of the rabbit retina with acutely implanted electrode arrays. Graefes Arch Clin Exp Ophthalmol 242:587–596
25. Pfeffer B, Wiggert B, Lee L, Zonnenberg B, Newsome D, Chader G (1983) The presence of a soluble interphotoreceptor retinol-binding protein (IRBP) in the retinal interphotoreceptor space. J Cell Physiol 117:333–341
26. Gerding H (2007) A new approach towards a minimal invasive retina implant. J Neural Eng 4:S30–S37
27. Timmers AM, Zhang H, Squitieri A, Gonzalez-Pola C (2001) Subretinal injections in rodent eyes: effects on electrophysiology and histology of rat retina. Mol Vis 7:131–137
28. Pachnis V, Pevny L, Rothstein R, Costantini F (1990) Transfer of a yeast artificial chromosome carrying human DNA from *Saccharomyces cerevisiae* into mammalian cells. Proc Natl Acad Sci USA 87:5109–5113

29. Reeves RH, Cabin DE, Lamb B (2001) Introduction of large insert DNA into mammalian cells and embryos. Curr Protoc Hum Genet/editorial board, Jonathan LH et al. Chapter 5: Unit 5.12
30. Tsong TY (1991) Electroporation of cell membranes. Biophys J 60:297–306
31. Neumann E, Schaefer-Ridder M, Wang Y, Hofschneider PH (1982) Gene transfer into mouse lyoma cells by electroporation in high electric fields. EMBO J 1:841–845
32. Wong TK, Neumann E (1982) Electric field mediated gene transfer. Biochem Biophys Res Commun 107:584–587
33. Taniyama Y, Tachibana K, Hiraoka K, Namba T, Yamasaki K, Hashiya N, Aoki M, Ogihara T, Yasufumi K, Morishita R (2002) Local delivery of plasmid DNA into rat carotid artery using ultrasound. Circulation 105:1233–1239
34. O'Brien JA, Holt M, Whiteside G, Lummis SC, Hastings MH (2001) Modifications to the hand-held gene gun: improvements for in vitro biolistic transfection of organotypic neuronal tissue. J Neurosci Methods 112:57–64
35. Felgner PL, Gadek TR, Holm M, Roman R, Chan HW, Wenz M, Northrop JP, Ringold GM, Danielsen M (1987) Lipofection: a highly efficient, lipid-mediated DNA-transfection procedure. Proc Natl Acad Sci USA 84:7413–7417
36. Langer R (1990) New methods of drug delivery. Science 249:1527–1533
37. Boussif O, Lezoualc'h F, Zanta MA, Mergny MD, Scherman D, Demeneix B, Behr JP (1995) A versatile vector for gene and oligonucleotide transfer into cells in culture and in vivo: polyethylenimine. Proc Natl Acad Sci USA 92:7297–7301
38. Luo D, Saltzman WM (2000) Synthetic DNA delivery systems. Nat Biotechnol 18:33–37
39. Gorman CM, Moffat LF, Howard BH (1982) Recombinant genomes which express chloramphenicol acetyltransferase in mammalian cells. Mol Cell Biol 2:1044–1051
40. Zheng QA, Chang DC (1991) High-efficiency gene transfection by in situ electroporation of cultured cells. Biochim Biophys Acta 1088:104–110
41. Li S, Tseng WC, Stolz DB, Wu SP, Watkins SC, Huang L (1999) Dynamic changes in the characteristics of cationic lipidic vectors after exposure to mouse serum: implications for intravenous lipofection. Gene Ther 6:585–594
42. Johnson CJ, Berglin L, Chrenek MA, Redmond TM, Boatright JH, Nickerson JM (2008) Technical brief: subretinal injection and electroporation into adult mouse eyes. Mol Vis 14:2211–2226
43. Bins AD, van Rheenen J, Jalink K, Halstead JR, Divecha N, Spencer DM, Haanen JB, Schumacher TN (2007) Intravital imaging of fluorescent markers and FRET probes by DNA tattooing. BMC Biotechnol 7:2
44. Bodenstein L, Sidman RL (1987) Growth and development of the mouse retinal pigment epithelium. I. Cell and tissue morphometrics and topography of mitotic activity. Dev Biol 121:192–204
45. Berglin L, Mandell K, Schmack I, Holley G, Grossniklaus H, Parkos C, Edelhauser H (2006) Reduction of retinal pigment epithelium (RPE) background autofluorescence with sudan black enhances visualization of fluorescently-labeled proteins in ex vivo RPE flatmounts. Invest Ophthalmol Vis Sci 2006, 46:2880
46. Shaner NC, Campbell RE, Steinbach PA, Giepmans BN, Palmer AE, Tsien RY (2004) Improved monomeric red, orange and yellow fluorescent proteins derived from *Discosoma* sp. red fluorescent protein. Nat Biotechnol 22:1567–1572
47. Hanaki K, Momo A, Oku T, Komoto A, Maenosono S, Yamaguchi Y, Yamamoto K (2003) Semiconductor quantum dot/albumin complex is a long-life and highly photostable endosome marker. Biochem Biophys Res Commun 302:496–501

Chapter 5

In Ovo Eye Electroporation

Teri L. Belecky-Adams, Scott R. Hudson, and Sarika Tiwari

Abstract

Electroporation has been used successfully to introduce macromolecules such as DNA into the chick embryo for at least 15 years. Purified plasmid DNA is microinjected into embryo and then a series of low voltage electrical pulses are applied to the embryo which allows naked DNA to enter cells. Following entrance into the cytoplasm, the DNA is transported to the nucleus where it is transiently expressed. This powerful technique is useful for studies involving overexpression, misexpression, and knockdown of genes of interest at a variety of developmental timepoints.

Key words: Electroporation, Microinjections, Retina, Chick embryos, Optic cup

1. Introduction

The chicken embryo is a time-honored model used for studying development of various organs (1). One of the most exciting advances in the study of molecular mechanisms of development is the advent of in ovo electroporation. This is a method whereby DNA is introduced into cells through the delivery of a series of electrical pulses. The pulses appear to be critical in two aspects of electroporation; they (1) disrupt the cell membrane, possibly opening pores through which the DNA can travel and (2) provide an electrophoretic field which drives the negatively charged DNA into the cells (2).

Up until the early 1990s, electroporation was used to introduce DNA and other macromolecules into cultured cells. The first article using electroporation in vivo was Titomirov et al. (3), to introduce plasmid DNA into the skin cells of newborn mice. Muramatsu et al. (4) was the first to use electroporation in the chick embryo. Subsequently, the number of studies using electroporation has skyrocketed and it is now a commonly used technique in analyses of gain- and loss-of-function. Recently, a genome-wide analysis was

Shu-Zhen Wang (ed.), *Retinal Development: Methods and Protocols*, Methods in Molecular Biology, vol. 884,
DOI 10.1007/978-1-61779-848-1_5, © Springer Science+Business Media, LLC 2012

completed comparing control tissues with those electroporated with a transcription factor (5). This analysis showed that electroporation itself had minimal impact on gene expression patterns, lending validity to the use of the technique in investigating developmental phenomenon.

Electroporation is advantageous in many respects in comparison to other methods used to transfer DNA into cells in ovo and in vivo in species other than chick. Viruses and retroviruses have commonly been used to obtain expression of various genes both in embryonic and adult tissues (6–8). Common problems associated with using viral particles to introduce DNA are: (1) the cost and time associated with making viral constructs, making and concentrating viral stocks, and testing infective units, (2) size limitations in the length of foreign DNA that can be inserted into the viral particle before it becomes unstable, (3) unintended consequences of the remaining viral genome (reviewed in Nayak and Herzog (9)), and (4) a decrease in the efficiency of gene transfer due to potential prior host immunity to the virus (10–12). Further, those investigators using the chicken-specific retrovirus, replication-competent ALV LTR with splice acceptor virus (RCAS), should keep in mind that endogenous viral proteins have been detected in lines thought to be pathogen-free, which could potentially make for uninterpretable results should embryos not be consistently screened for the presence of viral proteins (13). Lipofection is another commonly used technique for establishing expression of genes in vivo (14, 15). In a direct comparison of electroporation with lipofection and particle bombardment, Muramatsu et al. (4) showed electroporation was more efficient at gene delivery than the other methods, at least in ovo. There are also new methods for introducing macromolecules into cells, such as sonication and nanoparticle delivery; however, to our knowledge there have not been any direct comparisons of these methods with electroporation in ovo or in vivo (16, 17). Finally, the means to create transgenic chickens to study gene regulation and/or protein function are being developed; however, it is not a widely used technique at this point (18–26).

There have been many new variations on the "standard" in ovo electroporation technique in the last 5–10 years. For instance, several groups have been working on obtaining stable integration of constructs through the use of transposons or conditional expression of genes using tet-indicuble vector or the Cre-loxP systems (27–30). Other investigators have focused on combining electroporation with other known techniques to obtain specific advances, such as (1) electroporation of RNA interfering molecules (28, 31–34), (2) ex ovo electroporation on older chick embryos (35), (3) electroporation of DNA encoding retroviral vectors (36), (4) electroporation of photoactivatable fluorescent proteins and activation of those proteins in small groups or single cells, and (5) electroporation of DNA introduced into tissues using agarose beads in order to confine subsequent expression to a smaller field (37). In addition, electroporation is now being used as a means to deliver therapeutic intervention in humans, including DNA vaccines, introduction of DNA with vaccines to enhance immunogenicity, and introduction of antitumor agents to cancers (38–42).

What follows in this chapter is a typical electroporation that one might perform in a Hamburger and Hamilton stage 10 embryos (43). This stage is marked by the emergence of the two regions that will give rise to the eyes, known as the optic vesicles, hence stage 10 is also frequently referred to as the optic vesicle stage. While electroporation to obtain expression in the eyes is commonly done at this stage because it yields a larger region expressing the gene of interest, the technique can be easily adapted to earlier or later stages.

2. Materials

2.1. Equipment

1. 16°C Biochemical Oxygen Demand (BOD) incubator (Jeio Tech: IL-11A).
2. 37°C Egg incubator (Kuhl: B-LAB-600-110).
3. Thermal Air Hova-Bator incubator (G.Q.F. Manufacturing Co.: 1602N).
4. Pico-injector (Harvard Apparatus: PLI-100).
5. Zeiss stemi SV11 Microscope.
6. Flaming/Brown micropipette puller (Sutter instrument Co: P97).
7. Fiber light high-intensity illuminator series 180 (Doaln–Jenner Industries).
8. Electro Square Porator ECM 830 (BTX Harvard Apparatus).
9. BTX Genepaddles (BTX Harvard Apparatus: 45-0169).
10. Nikon SMZ 1500 Fluorescent Microscope w/Nikon Digital Camera DXM 1200.
11. Hot bead sterilizer (FST 250 Bench top Hot Bead Sterilizer-18000-45).
12. K T Brown Type micropipette beveler (Sutter instrument Co Model BV-10).

2.2. Supplies and Reagents

All reagents should be made in Milli-Q filitered water (0.2 μm filter).

1. Ink solution: mix ¼ part trypan blue solution (0.4%) (Sigma) and ¾ part Milli Q filtered water.
2. DNA injecting solution: mix expression plasmid of interest purified using a Qiagen plasmid prep kit with either a crystal of

fast green dye or a solution of 0.25% fast green dye. The final concentration of the DNA after mixing with the fast green dye should be 2 μg/μl (44).

3. Leibovitz's L-15 medium.
4. Syringe needles (Becton Dickinson) precision Glide 20G 1½ and 18G 1.
5. 3 M Transpore surgical tape.
6. Para film (American National Can™).
7. Clear packaging tape.
8. 10-ml Syringe (Becton and Dickinson).
9. Alcohol wipes (Kendall Webcol).
10. 70% Ethanol.
11. Forceps and Scissors (Fine science tools 14569-12, 11295-10).
12. Borosilicate glass pipettes (Sutter Instrument: OD: 1.0 mm, ID:0.75 mm, 10 cm length).

3. Methods

1. Purify DNA using Qiagen Mega Kit (Qiagen, Valencia, CA, USA). OD 260/280 ratio should be 1.8 or better. If the 260/280 ratio is not 1.8, then purification should be done again. DNA should be eluted so that the concentration is between 2 and 8 μg/μl.
2. Prepare pipettes, with an opening of approximately 10–15 μm. The glass pipettes are first pulled using a Flaming-Brown micropipette puller (Fig. 1). This will give you a pipette with a taper at one end that will be used for injection. You will need to work with the settings on the puller to obtain the exact shape you want as the taper you obtain is highly dependent on the type of borosilicate pipettes used, the heating element used, and the age of the heating element (Fig. 2). Once we have pulled the pipettes, we routinely break pipette tips with forceps to obtain a 10–15 μm opening. However, a beginner may need to measure the size of their pipette tips as they break them or use a pipette beveler to obtain the correct size opening (see Note 1) (Fig. 3).
3. Store fertilized chick eggs in a 16°C BOD incubator until they are to be placed in the 37.5°C incubator. The BOD incubator will keep the embryos alive but will not allow them to develop rapidly. We generally do not keep eggs longer than 7–10 days prior to incubating them at 37.5°C. The longer the eggs are stored in the BOD incubator, the fewer the number of embryos that will develop.

Fig. 1. Flaming-Brown micropipette puller. The micropipette puller heats the borosilicate glass tubing and applies pressure to pull the tubing into two pieces, each with a tapered end. The tapered ends are then broken to create needles to be used for microinjections that have an opening of a specific size. The taper of the needle can be controlled by the type of glass tubing ordered, the type of heating filament used, and the program used to heat the capillary.

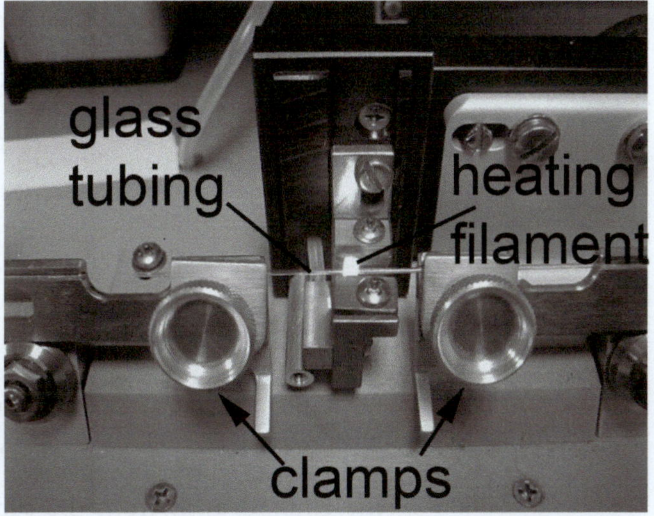

Fig. 2. Heating and pulling a borosilicate glass tubing to create needles for microinjection. A 10-cm piece of borosilicate glass tubing with an inner diameter of 0.75 mm and an outer diameter of 1.0 mm is being heated in a Flaming-Brown micropipette puller. The pipette is held on with side with a clamp that applies constant pressure so that when the filament in the center of the tubing is heated, as in the figure, the tubing is pulled into two pieces.

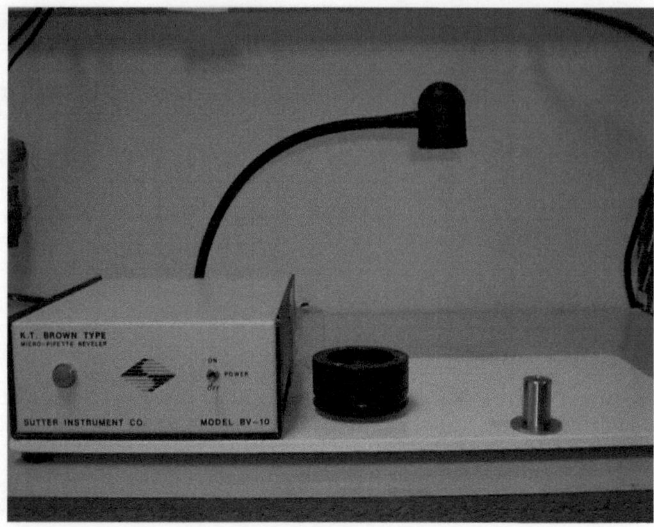

Fig. 3. Pipette beveler. The pipette beveler lowers the tapered tips of the pulled glass tubing to a rotating grinding stone to create a sharpened needle tip of a specific size. For our application, the tips should be between 10 and 15 μm. Alternatively, one can also break the tip of the pulled glass tubing with a pair of #5 sharpened forceps to create an opening. This takes a little bit of skill and practice. The size of the openings can be viewed with a microscope.

4. Place chick eggs in a 37.5°C incubator 46 h before experimental start time. This time will allow the embryonic development process to reach stage Hamburger and Hamilton stage 10 (when the embryo has developed ten somites) (43).
5. One to two hours prior to starting egg-opening, rotate the eggs to their side to allow the embryo to move to the side where a window will be opened (see Note 2) (Fig. 4). Place an X on the surface of the egg facing up with a pencil so that it can be identified later.
6. Following the 1–2 h incubation on their side, take the eggs out of the incubator and place them in a 37°C Hova-Bator benchtop incubator (Fig. 5). A small opening, called a window, must be made in the egg to facilitate microinjections and electroporation. In order to make the window without injuring the embryo, some of the egg's albumin must be removed to drop the embryo away from the surface where a window will be made. To drop the embryo away from the region where the window will be made, perform the following steps:
   (a) Sterilize the larger rounded end of the egg by wiping with a sterile ethanol wipe (Fig. 6).
   (b) Place a ¼″ piece of transpore surgical tape on the larger rounded side of the egg (see Note 3) (Fig. 7).
   (c) Using an 18G 1 needle, gently poke a hole through the tape being careful not to go too far and puncture the yolk (see Note 4) (Fig. 8).

Fig. 4. Eggs rotated to their sides. Eggs are typically incubated with the smaller tips of the eggs facing down into the egg carton. One to two hours prior to the start of injections, the eggs should be rotated to their sides, as shown in the figure. This allows the embryos to move to the region where an opening to allow access to the embryo is made.

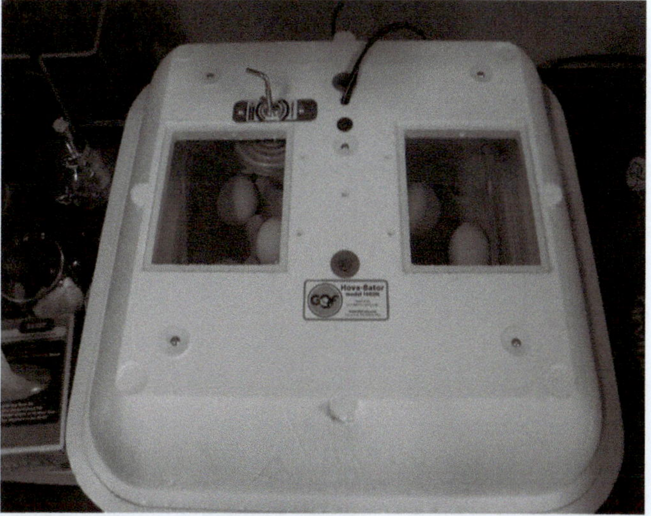

Fig. 5. A bench-top Styrofoam incubator for keeping eggs. For easy access to eggs during the opening and injection procedures, we keep the eggs in a small Styrofoam incubator on the bench top.

(d) Place a sterile 20G 1½ needle into a 10 ml syringe with a Luer-Lok tip. Carefully place the tip of the needle into the previously poked hole and insert needle at approximately 30° angle to avoid hitting the yolk (see Note 5). Remove approximately 1–3 ml of albumin from each egg (Fig. 9).

(e) Following albumin removal, place a ¼″ piece of clear packaging tape to cover the hole (Fig. 10).

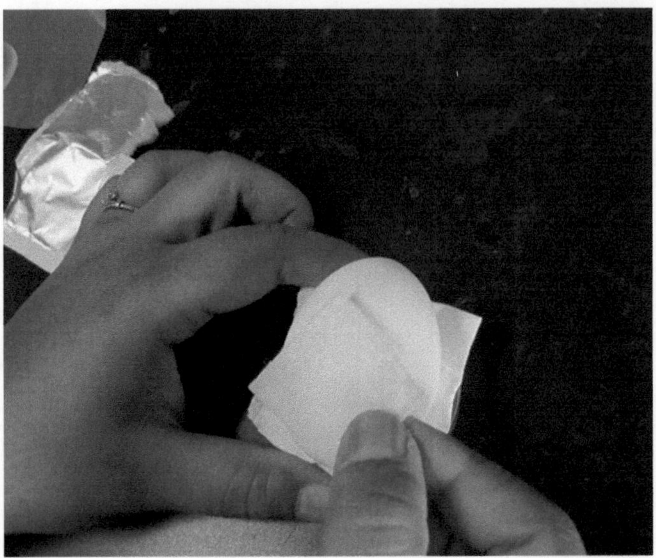

Fig. 6. Sterlizing the egg with an ethanol wipe. Each egg is sterilized in two places; (1) where a hole is punctured on the larger rounded side of the egg, and (2) on the flattened surface at the top of the egg where an opening, referred to as the window, will be placed. We sterilize each surface using a sterile 70% ethanol wipe which can be purchased from a number of suppliers.

Fig. 7. Placing transpore surgical tape over the larger rounded surface of the egg. To prepare the egg for puncturing with a syringe needle so that albumin can be removed, a piece of surgical transpore tape is placed on the larger rounded surface of the egg. This will help keep the egg from cracking during the puncturing process.

Fig. 8. Making a hole using an 18-gauge needle. Once the transpore tape has been placed over the end of the egg, an 18-gauge syringe needle is used to puncture a hole in the egg. Gentle pressure is applied to the needle along with small rotating movements until the needle punctures the eggshell. Be careful not to penetrate the egg too far or you will puncture the yolk. This can be avoided by placing a small black line near the tip of the needle for guidance.

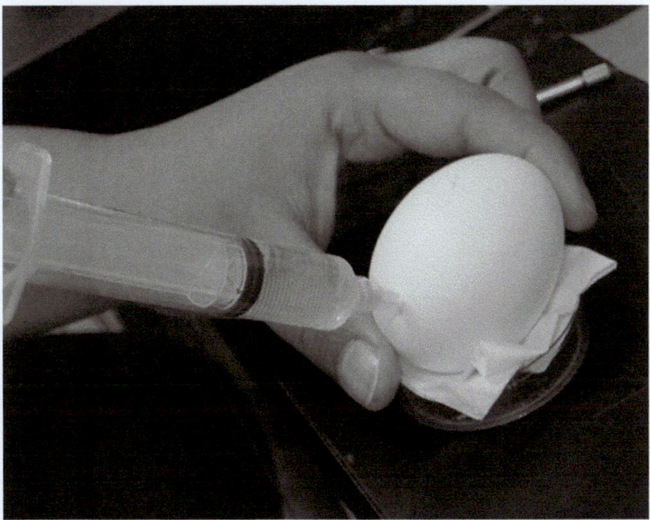

Fig. 9. Extracting albumin with a syringe. In order to open a window that allows access to the embryo for microinjection and electroporation, the embryo must be dropped away from the shell. This will keep the embryo from being injured during the process of making the window. The embryo is dropped away from the shell by removing some of the albumin. As seen in the figure, albumin is removed by placing a 1½″ 20-gauge syringe needle coupled to 10-ml syringe and drawing out 1–3 ml of albumin. Care must be taken to avoid penetrating the yolk. The albumin has a slightly yellow clear appearance, whereas, if the yolk has been penetrated, the liquid is bright yellow and opaque in appearance.

Fig. 10. Covering the hole with clear packaging tape. Once the albumin has been removed from the egg, the hole can be sealed using a small piece of clear packaging tape. There are a variety of packaging tapes and we have found that more pliable tapes are better for this process. The stiffer tapes have a tendency to pull away from the curved surfaces of the egg, leaving the egg exposed to the forced air present in the egg incubator. This can lead to the embryos getting dried out, which ultimately leads to death. In addition, it can also lead to mold spores and bacteria entering the egg to create an infection.

7. Now that the embryo is a safe distance from the wall of the egg, the window can now be cut into the surface.
   (a) First wipe the surface of the egg with an ethanol wipe where you placed the "X" earlier.
   (b) Place a piece of transpore surgical tape on the top side of the egg in the region where the egg had been cleaned with the ethanol wipe (Fig. 11).
   (c) Using a pair of curved dissecting scissors, cut a hole in the middle of the piece of transpore surgical tape (see Note 6) (Fig. 12).
   (d) Place a piece of packing tape over the hole and make sure the hole is well sealed.
   (e) Place the egg back in the hova-bator and complete opening windows in the rest of the eggs.
8. Once windows have been placed in all the eggs, you are ready to start the microinjection and electroporation procedure.
   (a) Take a previously prepared egg and remove the packaging tape (alternatively you can cut an opening in the packing tape).
   (b) Using a stereomicroscope, identify the area where the embryo is located (see Note 7) (Fig. 13).

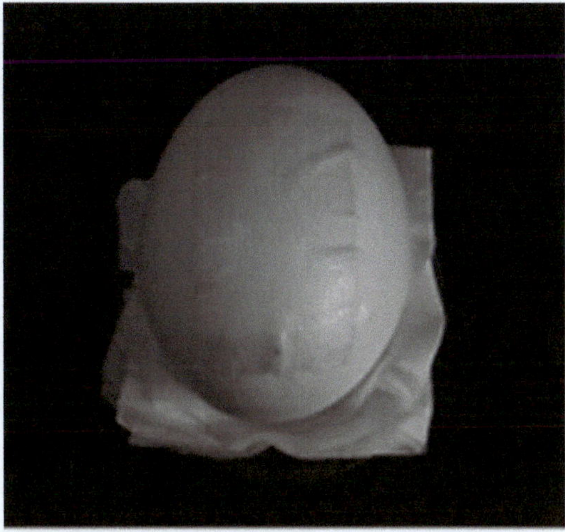

Fig. 11. Placing transpore surgical tape on the top side of the egg. A piece of transpore tape is placed on the top side of the egg, directly over where the embryo will be. The tape helps keep the pieces of eggshell from falling into the egg while an opening is made to access the embryo. We use transpore tape for opening the window, and then in the next step use clear packaging tape to cover the hole. In our hands, the packaging tape appears to adhere better to the transpore tape than other types of tape, creating a better seal.

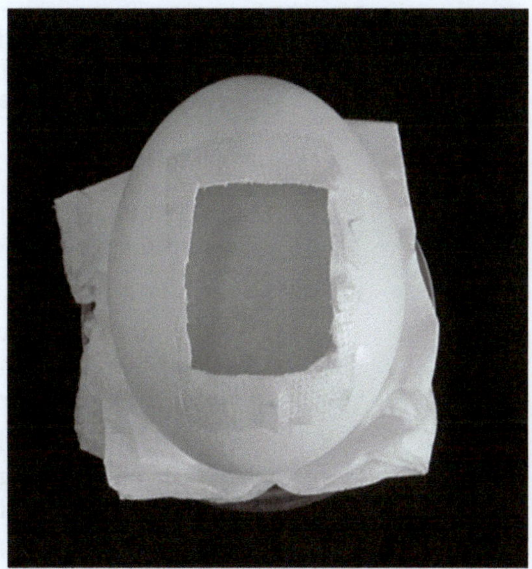

Fig. 12. Making a window in the egg. Using a pair of curved scissors, the egg is first punctured and then a window is cut directly above the embryo. This step takes a little bit of practice. You must be able to puncture the egg and cut a hole, without making any large cracks in the egg. Cracks will compromise the egg, making it more difficult to handle and more susceptible to drying out. Once you have made the hole, determine whether you will have easy access to the embryo. If the embryo is off to the side of the hole, you may have to enlarge it slightly to give you better access to the embryo. In general, the smaller the hole, the better, as long as it is large enough for you to complete the procedure.

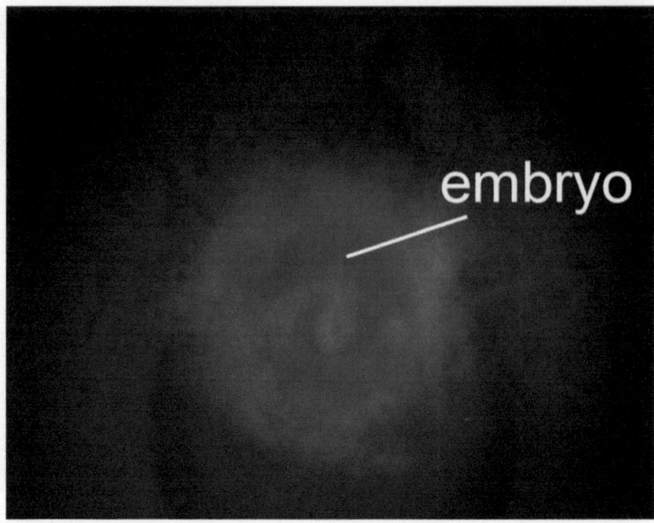

Fig. 13. The embryo as viewed under the stereomicroscope. Once you have made a window in the egg, place the egg under a stereomicroscope. Depending on the placement of the light, you will see something that looks like this. At this magnification, you cannot see the specifics of the embryo (such as the stage), but this magnification will allow you to first determine whether there is an embryo in the egg and the general region. Generally, look for the clear halo surrounding the embryo at this magnification. If you are certain there is no embryo, then you can discard the egg. Depending upon the supplier, the age of the eggs prior to incubation, and the season, you may have as many as 5—50% of eggs without an embryo. Once you have ascertained there is an embryo, then you can increase the magnification to determine the stage of the embryo. It is best to count the somites at this stage to get an accurate determination of the stage (see later figures). Eggs that are past stage 10 (for our procedure) are discarded.

(c) Place a pulled pipette into the Harvard Apparatus pico-injector pipette holder (Fig. 14).

(d) Draw up the trypan blue ink into the pipette.

(e) Place the needle underneath the area where the embryo is located and inject. This will allow the outline of the embryo to become more apparent (Fig. 15).

9. Change the pipette in the holder. Draw the DNA solution up into the fresh pipette (Fig. 16). Place the needle inside the lumen of the neural tube of the embryo for injection (see Note 8). Inject the plasmid/fast green dye solution. Successful injection will have green dye solution present in the head region and in the optic vesicles (Fig. 17).

10. Pipette approximately 100 μl of L-15 solution directly on top of the embryo. This solution will keep the electrodes from sticking to the membranes.

11. Place the electrodes on either side of the chick optic vesicles (Fig. 18), with the positive electrode (anode) on the side you

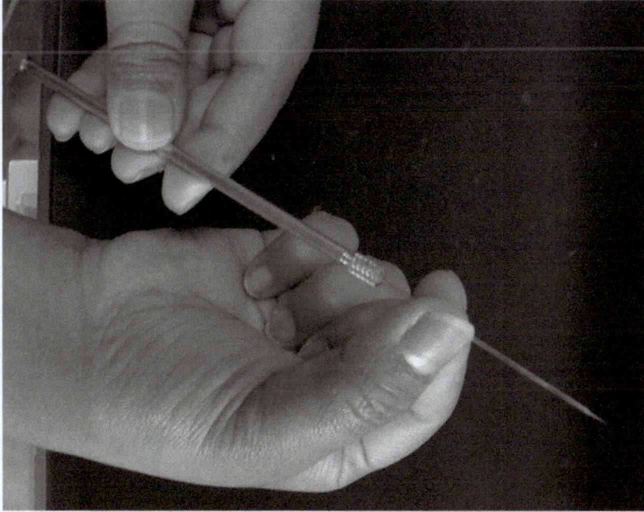

Fig. 14. Pipette inserted into the pico-injector pipette holder. Once windows have been made in the eggs and you have ascertained that the emrbyos are at the correct stage, you are ready to start the microinjection. The micropipette is inserted, as shown above, into the pipette holder, so that the tapered end is protruding out from the holder. This will enable you to draw up dye or the DNA solution into the pipette.

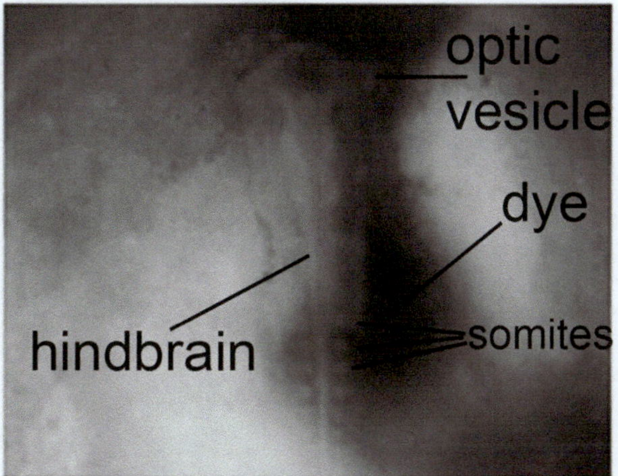

Fig. 15. Injecting dye under the embryo. Prior to injections into the neural tube, you may want to inject dye under the embryo to make it more visible. We use a trypan blue solution for this procedure. Other labs may use other dyes, such as India ink. The ink solution is drawn up into the micropipetter, then the pipette is gently inserted under the embryo and the several pulses of dye injected. The figure above shows an embryo in which trypan blue has been injected. This makes the outline of the neural tube and somites easier to see. Once you have become experienced, you will be able to do DNA injections without having the dye present.

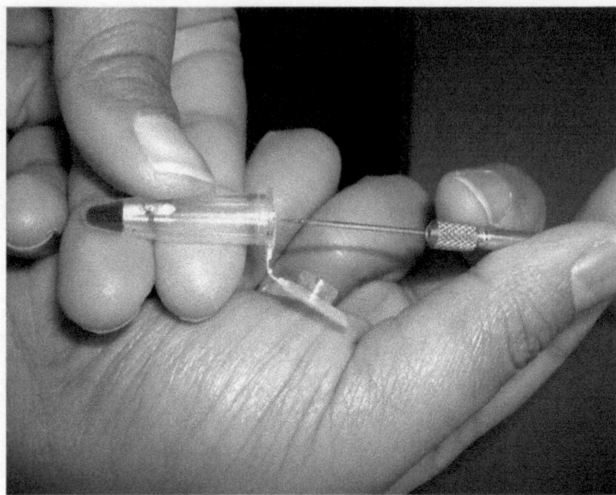

Fig. 16. Loading DNA solution into the micropipette. DNA is back-loaded into the micropipette by placing the tip into the DNA solution and drawing it up into the micropipette. If you have trouble with accidentally breaking the micropipette tip while trying to load it, you can place a small drop of the solution onto a piece of parafilm and draw it up into the needle from the drop.

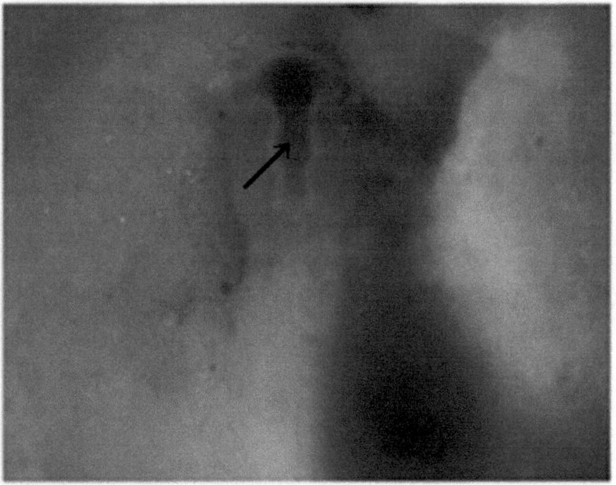

Fig. 17. Embryo injected with DNA solution. Place the micropipette tip into the region of the midbrain or hindbrain (see Fig. 15), with the tip pointing towards the anterior portion of the neural tube. The *arrow* indicates the region where the pipette should penetrate. Inject the DNA solution into the anterior part of the neural tube. Be careful when penetrating the neural tube not to accidentally penetrate all the way through the embryo. If this has happened, then you will see the DNA solution under the embryo, rather than filling the neural tube. Further, be careful not to overfill the neural tube, which will cause it to burst and the DNA solution to leak out of the neural tube. Once the DNA has been injected, remove the micropipette and electroporate.

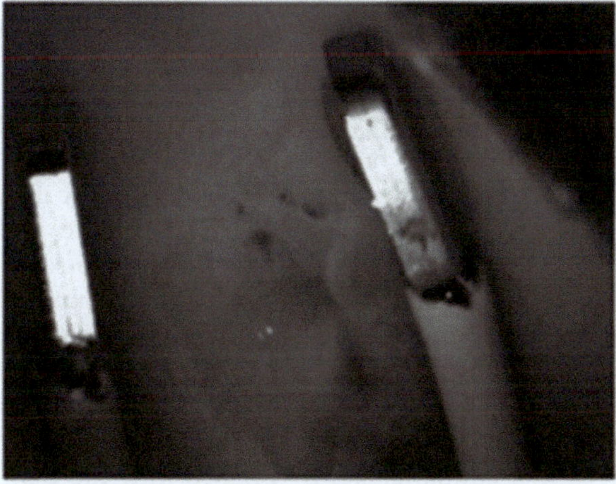

Fig. 18. Placing the electroporator electrodes. Following injection of the DNA solution into the neural tube, the anode and cathode of the embryo are placed on either side of the embryo with the positive electrode (anode) placed near the optic vesicle that you want to be electroporated. We generally electroporate the same side every time, so that there is no need to keep a record of which side was injected. We wet the electrodes prior to placement so that they do not stick to the membrane surrounding the embryo. This allows for easier placement of the electrodes. We also drop some L-15 medium on the top of the embryo to combat any dried membranes. When the electrodes are properly placed, a series of electrical pulses are delivered to the embryo which allows the DNA to enter the cells closest to the anode.

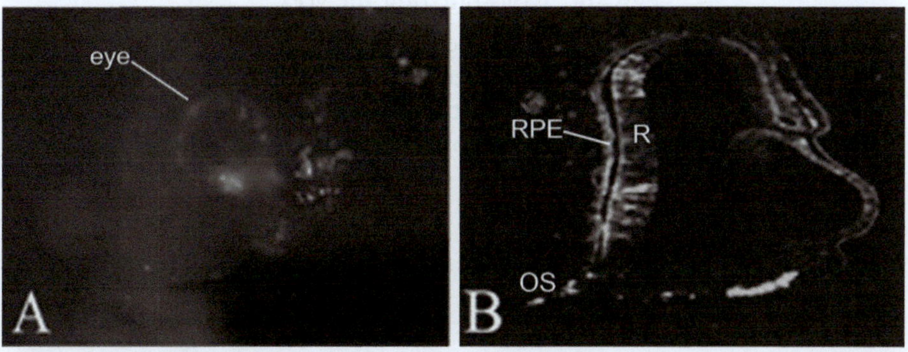

Fig. 19. Green fluorescent protein (GFP) expression in the embryonic eye. In embryos that have been electroporated with GFP, the electroporated region can be visualized in ovo the next day using a stereomicroscope with fluorescence capability (**a**). You can also visualize the electroporated GFP in sections through the optic cup (**b**). *R* retina, *RPE* retinal pigmented epithelium, *OS* optic stalk.

wish to electroporate. Using an Electro Square Porator, electroporate the embryo with 5–15 V for 50 ms pulses (see Note 9). Cover the opening on top of the chick egg with a piece of clear packaging tape and place back in the 37°C incubator (see Note 10).

12. If you electroporated with a construct that contains one of the fluorescent proteins, you will be able to identify electroporated cells the next day (see Note 11), using an epiflourescent stereomicroscope (Fig. 19).

4. Notes

1. Pulled pipette tips need to be broken to create a sharp tip through which the DNA can be delivered to the lumen of the embryonic neural tube. We break the tips using a #5 sharpened Dumont forcep. As stated above, the opening of the tip should be between 10 and 15 μm. An inexperienced pipette maker can determine the size by measuring the tip with an eyepiece micrometer under the microscope. You may also wish to bevel the tips using a beveler that will slowly grind the pipette tip to the recommended size by pressing it to a rotating grinding stone. The freshly broken or beveled pipettes can be stored in a container lightly pressed into the dental wax to keep them from becoming damaged (Fig. 20).
2. Turning the eggs for more than 2 h can result in the embryo sticking to the wall of the egg shell.
3. Transpore tape is placed on the egg shell to avoid pieces of the eggshell from falling into the egg.
4. Placing a line with a sharpie about 2–3 mm from the end of the needle will help you keep from pushing needle too far into the egg and avoid bursting the yolk.
5. Once placing the needle at a 30° angle and slightly moving inside the egg, run the tip of the needle along the egg shell to avoid contact with the yolk. Once underneath the yolk, gently remove 1–3 ml of albumin.

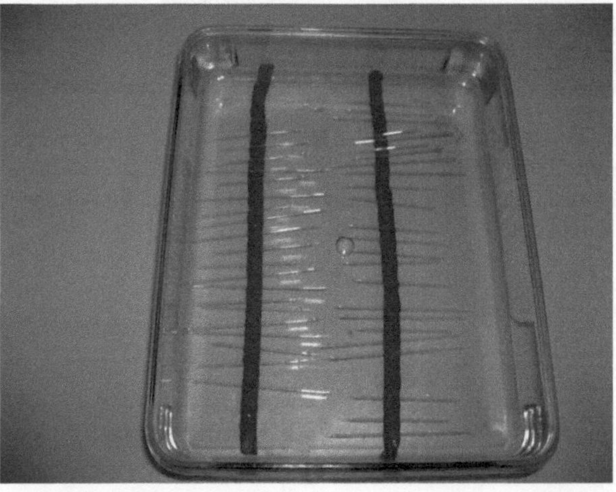

Fig. 20. Pulled pipettes in a container. Pipettes ready for microinjections are stored in a container into which strips of dental wax have been places. The pipettes are gently pressed into the dental wax to keep them from rolling around and sustaining damage.

6. To sterilize the dissecting scissors, either use an open flame or hot bead sterilizer.
7. To identify the region of the embryo, look for the halo that surrounds the embryo.
8. For successful injection into the head and optic vesicles, enter the tip of the pipette near the midbrain/hindbrain. Be very careful to enter the neural tube at an angle and do not penetrate through the opposite side of the embryo, or you will be injecting the DNA *under* the embryo. Keep the pipette at an angle that will allow you to inject DNA towards the anterior part of the neural tube. (The arrow in Fig. 17 points to the location where the tip of the pipette should enter the embryo.)
9. Placing dry electrodes on the yolk around the embryo will cause the embryo to sink concurrent with the electrodes and the electric current will pass over the embryo rather than through it. Wetting the electrodes with water or L-15 medium prior to placing next to the embryo will help with this problem.
10. It is important to make sure the tape is completely covering the window made in the egg shell. Openings will allow air to enter and dry out the embryo. A dry embryo is a dead embryo.
11. We have been able to verify green fluorescent protein (GFP) expression 3 h postelectroporation of the embryo; however, we recommend checking for GFP expression the next day.

Acknowledgments

This work was supported by the American Health Assistance Foundation and NIH grants 1R01EY019525-01 and 1R15EY020816-01.

References

1. Stern CD (2004) The chick embryo – past, present and future as a model system in developmental biology. Mech Dev 121: 1011–1012
2. De Vry J et al (2010) In vivo electroporation of the central nervous system: a non-viral approach for targeted gene delivery. Prog Neurobiol 92:227–244
3. Titomirov AV, Sukharev S, Kistanova E (1991) In vivo electroporation and stable transformation of skin cells of newborn mice by plasmid DNA. Biochem Biophys Acta 1088:131–134
4. Muramatsu T et al (1997) Comparison of three nonviral transfection methods for foreign gene expression in early chicken embryos in ovo. Biochem Biophys Res Commun 230:376–380
5. Farley EK et al (2011) Effects of in ovo electroporation on endogenous gene expression: genome-wide analysis. Neural Dev 6:17
6. Erkeland SJ et al (2006) Significance of murine retroviral mutagenesis for identification of disease genes in human acute myeloid leukemia. Cancer Res 66:622–626

7. Jaenisch R et al (1983) Expression of retroviruses during early mouse embryogenesis. Hematol Blood Transfus 28:270–274
8. Sato N et al (2002) Regulated gene expression in the chicken embryo by using replication-competent retroviral vectors. J Virol 76: 1980–1985
9. Nayak S, Herzog RW (2010) Progress and prospects: immune responses to viral vectors. Gene Ther 17:295–304
10. Papp Z, Babiuk LA, Baca-Estrada ME (1999) The effect of pre-existing adenovirus-specific immunity on immune responses induced by recombinant adenovirus expressing glycoprotein D of bovine herpesvirus type 1. Vaccine 17:933–943
11. Parr MJ et al (1998) Immune parameters affecting adenoviral vector gene therapy in the brain. J Neurovirol 4:194–203
12. Schulick AH et al (1997) Established immunity precludes adenovirus-mediated gene transfer in rat carotid arteries. Potential for immunosuppression and vector engineering to overcome barriers of immunity. J Clin Invest 99:209–219
13. McNally MM, Wahlin KJ, Canto-Soler MV (2010) Endogenous expression of ASLV viral proteins in specific pathogen free chicken embryos: relevance for the developmental biology research field. BMC Dev Biol 10. doi:10.1186/1471-213X-10-106
14. Felgner PL et al (1987) Lipofection: a highly efficient, lipid-mediated DNA-transfection procedure. Proc Natl Acad Sci USA 84: 7413–7417
15. Ohnuma S et al (2002) Lipofection strategy for the study of *Xenopus* retinal development. Methods 28:411–419
16. Panagiotou T, Fisher RJ (2011) Enhanced transport capabilities via nanotechnologies: impacting bioefficacy, controlled release strategies, and novel chaperones. J Drug Deliv. doi:10.1155/2011/902403
17. Wyber JA, Andrews J, D'Emanuele A (1997) The use of sonication for the efficient delivery of plasmid DNA into cells. Pharm Res 14: 750–756
18. Byun SJ et al (2011) Oviduct-specific enhanced green fluorescent protein expression in transgenic chickens. Biosci Biotechnol Biochem 75: 646–649
19. Heo YT et al (2011) Bone marrow cell-mediated production of transgenic chickens. Lab Invest. doi:10.1038/labinvest.2011.53
20. Koo BC et al (2010) Tetracycline-dependent expression of the human erythropoietin gene in transgenic chickens. Transgenic Res 19: 437–447
21. Matsubara Y et al (2011) Detection of the EGFP sequence in breast muscle of 3-year-old chicken after transfection using sonoporation. Anim Sci J 82:428–433
22. McGrew MJ, et al. (2010) Functional conservation between rodents and chicken of regulatory sequences driving skeletal muscle gene expression in transgenic chickens. BMC Dev Biol doi:10.1155/2011/902403
23. Motono M et al (2010) Production of transgenic chickens from purified primordial germ cells infected with a lentiviral vector. J Biosci Bioeng 109:315–321
24. Park SH et al (2010) CpG methylation modulates tissue-specific expression of a transgene in chickens. Theriogenology 74(5):805–816
25. Park SW, Wei LN (2003) Regulation of c-myc gene by nitric oxide via inactivating NF-kappa B complex in P19 mouse embryonal carcinoma cells. J Biol Chem 278:29776–29782
26. Poynter G, Huss D, Lansford R (2009) Injection of lentivirus into stage-X blastoderm for the production of transgenic quail. Cold Spring Harb Protoc 2009, pdbprot5118
27. Harada R et al (2008) Genome-wide location analysis and expression studies reveal a role for p110 CUX1 in the activation of DNA replication genes. Nucleic Acids Res 36:189–202
28. Sato M, Tanigawa M, Kikuchi N (2004) Nonviral gene transfer to surface skin of mid-gestational murine embryos by intraamniotic injection and subsequent electroporation. Mol Reprod Dev 69:268–277
29. Watanabe T et al (2007) Tet-on inducible system combined with in ovo electroporation dissects multiple roles of genes in somitogenesis of chicken embryos. Dev Biol 305: 625–636
30. Yokota Y et al (2011) Genomically integrated transgenes are stably and conditionally expressed in neural crest cell-specific lineages. Dev Biol 353:382–395
31. Baeriswyl T, Mauti O, Stoeckli ET (2008) Temporal control of gene silencing by in ovo electroporation. Methods Mol Biol 442: 231–244
32. Katahira T, Nakamura H (2003) Gene silencing in chick embryos with a vector-based small interfering RNA system. Dev Growth Differ 45:361–367
33. Pekarik V et al (2003) Screening for gene function in chicken embryo using RNAi and electroporation. Nat Biotechnol 21:93–96
34. Rao M et al (2004) In vivo comparative study of RNAi methodologies by in ovo electroporation in the chick embryo. Dev Dyn 231: 592–600

35. Luo J, Redies C (2005) Ex ovo electroporation for gene transfer into older chicken embryos. Dev Dyn 233:1470–1477

36. Sakuta H, Suzuki R, Noda M (2008) Retrovirus vector-mediated gene transfer into the chick optic vesicle by in ovo electroporation. Dev Growth Differ 50: 453–457

37. Simkin JE, McKeown SJ, Newgreen DF (2009) Focal electroporation in ovo. Dev Dyn 238: 3152–3155

38. Munoz Madero V, Ortega Perez G (2011) Electrochemotherapy for treatment of skin and soft tissue tumours. Update and definition of its role in multimodal therapy. Clin Transl Oncol 13:18–24

39. Sardesai NY, Weiner DB (2011) Electroporation delivery of DNA vaccines: prospects for success. Curr Opin Immunol 23:421–429

40. Thomson KR et al (2011) Investigation of the safety of irreversible electroporation in humans. J Vasc Interv Radiol 22:611–621

41. Vasan S et al (2011) In vivo electroporation enhances the immunogenicity of an HIV-1 DNA vaccine candidate in healthy volunteers. PLoS One. doi:10.1371

42. Wong TW et al (2011) Painless skin electroporation as a novel way for insulin delivery. Diabetes Technol Ther 13:929–935

43. Hamburger V, Hamilton HL (1952) A series of normal stages in the development of the chick embryo. Dev Dyn 195:231–272

44. Krull CE (2004) A primer on using in ovo electroporation to analyze gene function. Dev Dyn 229:433–439

Chapter 6

Targeted Microinjection of Synthetic mRNAs to Alter Retina Gene Expression in *Xenopus* Embryos

Sally A. Moody

Abstract

The individual cells of the *Xenopus* cleavage-stage embryo have been fate mapped, revealing which of these cells contribute to the retina. Using this retina fate map, one can specifically modulate levels of gene expression in retina lineages to determine the function of proteins in various aspects of early retinal development, such as formation of the eye fields and determination of specific cell fates. This chapter presents the techniques for identifying specific retina blastomere precursor cells, and injecting them with lineage tracers, mRNAs encoding wild-type and mutant constructs or morpholino antisense oligonucleotides to alter gene expression.

Key words: Retina cell fate, Morpholino, *Xenopus laevis*, Amacrine cells, Blastomeres, Fate mapping, Lineage tracing

1. Introduction

Cell lineage studies reveal what kinds of tissues descend from a single cell of an embryo. By defining which cells contribute to the various tissues and organs, one can manipulate specific precursor cells to elucidate the mechanisms that control patterning, morphogenetic movements, cell fate determination, and test the influence of a variety of genes on these developmental events. In the South African clawed frog, *Xenopus laevis*, it was possible to create complete fate maps of the early cleavage cells of the embryo, called blastomeres, nearly 30 years ago with the very simple approach of microinjecting a molecule, which could be detected either by a histochemical reaction or by fluorescence microscopy, into single, identifiable blastomeres. This was feasible because *Xenopus* embryos develop external to the mother, are very large (~1.5 mm in diameter), have a pigment pattern that indicates the dorsal–ventral and animal–vegetal axes, and tend to cleave in a very regular pattern so that

Shu-Zhen Wang (ed.), *Retinal Development: Methods and Protocols*, Methods in Molecular Biology, vol. 884, DOI 10.1007/978-1-61779-848-1_6, © Springer Science+Business Media, LLC 2012

the same blastomere can be labeled in different embryos. This is in contrast to zebrafish and mammalian embryos which do not have axial indicators or distinctly identifiable blastomeres at the cleavage stages. Thus, the ability to mark the same cell across embryos allowed complete fate maps of each of the early blastomeres from the 2- through 32-cell stages to be made (1–4).

These maps have revealed that embryonic tissues arise from specific sets of blastomeres in a highly predictable manner, which for neuronal cell types can be quantified (5, 6). Of special interest to this chapter, the cleavage blastomeres that give rise to the retina have been quantitatively mapped (7). At the 32-cell stage, one dorsal-animal blastomere gives rise to about 50% of the retinal cells, three others each produce 12–14%, and the remaining ~12% are produced in small numbers by four contralateral dorsal-animal blastomeres. This fate map can therefore be used to alter gene expression specifically in major or minor retina precursors and in cells that would not normally contribute to the retina (8). One can further combine lineage tracing with immunofluorescent detection of cell type-specific proteins to study cell fate determination. For example, we demonstrated quantitatively that different retina precursor blastomeres are biased to give rise to different neurotransmitter subtypes of amacrine cells (9, 10).

While fate maps are very useful for identifying the precursor cells for the organ of interest and describe the developmental path taken by a blastomere in the normal, intact embryo, they cannot describe the full developmental potential of a blastomere or identify the mechanisms by which its fate is determined. To test for developmental mechanisms, one needs to modify the gene expression of the precursor cell of interest. In mouse this is achieved through the use of tissue-specific promoters driving transgenes. In *Xenopus*, this can be very simply accomplished by microinjecting synthetic mRNAs or oligonucleotides into an identified blastomere of known fate to affect gene expression within a specific lineage (8, 11). One can microinject a number of different kinds of constructs, including mRNAs encoding: (1) an endogenous gene to test whether it promotes or represses target genes or tissues; (2) mutant forms to test for the function of different domains of a protein; (3) dominant-negative forms to test the effect of functional knockdown; and (4) wild-type, mutant or dominant-negative forms into ectopic locations to test whether they can convert tissues to novel fates. In addition, one can knock down gene expression by injecting antisense morpholino oligonucleotides (MOs), which can be designed to block either endogenous protein translation or mRNA splicing.

For all of these constructs, it is essential to lineage tag the blastomere that is injected so that one can determine whether the effect is confined to the descendants of the injected blastomere (i.e., is cell autonomous) or involves surrounding cells (i.e., involves cell-to-cell interactions). A lineage label is a tracer that marks not only the cell of interest but also all of its descendants. A lineage tracer must be: (1) nontoxic and nonreactive so that it does not change the developmental fate of the labeled cell; (2) small enough to diffuse quickly through the injected cell before it divides, so that all of the descendants of that cell will be labeled; (3) large enough not to pass through the gap junctions found between adjacent blastomeres (12); (4) detectable throughout development, and not be diluted by cell division or intracellular degradation; and (5) easily detectable by simple histological procedures. These requirements have been fulfilled by two classes of molecules: horseradish peroxidase (HRP) and fluorescent-dextrans (13–16). HRP is a plant enzyme that has no natural substrate in animal cells, and in frog is not recognized by the embryonic lysosomal compartment until late tadpole stages; this tracer is no longer commonly used. Dextrans are hydrophilic polysaccharides that are biologically inert, resistant to a cell's endogenous glycosidases, and can be purchased with a variety of fluorescent tags. Another type of lineage label is an mRNA encoding a tracer molecule. Because mRNAs do not diffuse as well as HRP or dextrans, they do not always mark all the descendants of the injected cell, and therefore are less accurate as lineage tracers for fate mapping studies. But, when they are mixed with a test mRNA being used to alter gene expression, they are ideal for marking those cells that express the injected test transcript. Two commonly used mRNA tracers encode β-galactosidase (β-Gal) and green fluorescent protein (GFP) (17, 18). Both proteins are derived from non-vertebrates (β-Gal from bacteria and GFP from jellyfish), can be distinguished from endogenous vertebrate proteins, are too large to diffuse through gap junctions, and have no known deleterious effects on developing vertebrate cells. Finally, while MOs also can be mixed with tracer dextrans or mRNAs, there are concerns that they may nonspecifically react with these molecules. Therefore, it is preferable to purchase MOs already tagged with a fluorescent molecule (http://www.gene-tools.com/).

2. Materials

2.1. Equipment

1. Microinjector. One of the following: PLI-100 Pico-Injector, Harvard Apparatus (http://www.harvardapparatus.com/); IM-300 Microinjector, Narashige (http://products.narishige-group.com/); or Nanoject II, Drummond Scientific (http://www.drummondsci.com/).
2. Programmable micropipette puller. A vertical puller, which makes very long, flexible tips, will suffice for injections of oocytes and 2–4 cell embryos. A horizontal puller is preferred to manufacture the short, fine tips required for injections of

later-stage embryos. Borosilicate capillary glass with an outer diameter of about 0.8–1.0 mm; glass should *not* contain a filament.

3. Dissection stereomicroscope placed on a steel plate in an area that is level and free of vibrations. Illumination should come from a fiber optic lamp, *not* by transillumination through the base of the microscope, because *Xenopus* cleavage embryos are opaque.
4. Micromanipulator with *X*, *Y*, and *Z* axes, mounted on a magnetic base secured to the steel plate.
5. Injection dish. There are several types of dishes that can be used, including: (1) a Petri dish whose base is covered with nontoxic, nondrying oil-based modeling clay (such as plasticine) in which ~1.5-mm shallow depressions have been made with a glass ball (crafted by melting the tip of a 6" Pasteur pipette); (2) a Petri dish whose base is covered with 2% agarose in which ~1.5-mm depressions have been melted with a heated glass ball tip of a 6" Pasteur pipette; or (3) a Petri dish in which a piece of polypropylene mesh (Spectrum Spectra/Mesh, 500 μm mesh size; http://www.spectrumlabs.com) has been glued to the bottom with a few drops of superglue (see Note 1).
5. Tabletop centrifuge.
6. Stage micrometer.
7. Fine sharpened forceps (e.g., Dumont #5 biologie).
8. Hair loop: place both ends of a fine hair (about 10 cm long) into the narrow tip of a 6" Pasteur pipette to form a 2–3-mm loop. Seal the hair in place with melted paraffin.
9. Cryostat with microtome blade, if intending to analyze tissue sections.

2.2. Lineage Tracers

1. 0.2 N KCl, pH 6.8, for making up fluorescent dextran tracers. Make 100 ml and adjust pH to 6.8 with 0.05 M KOH (see Note 2). Filter-sterilize it and store the solution at room temperature; solution is very stable.
2. Fluorescent dextrans: use dextrans (10,000–40,000 MW) conjugated to a fluorochrome of choice that are lysine fixable (Invitrogen/Molecular Probes) (see Notes 3 and 4). Make about 100 μl of a 0.5–1.0% solution of fluorescent dextran in the 0.2 N KCl solution (see Subheading 2.2 item 1). Vortex vigorously to dissolve. Transfer the solution into a microfiltration tube [e.g., Costar Spin-X cellulose acetate devise for aqueous solutions (Corning #8160)]. Spin at ~11,000 × *g* for 15–20 min at room temperature (see Note 5). Aliquot the flow-through in small volumes (10–20 μl) and store at –20°C for up to 6 months. Keep solution in the dark (tube wrapped in foil).

3. Tracer mRNAs: β-Gal and GFP vectors can be purchased from a number of companies (e.g., Clontech). The coding region will need to be subcloned into a *Xenopus*-appropriate expression vector (e.g., pCS2+), so that mRNA can be transcribed with upstream and downstream sequences allowing efficient translation in blastomeres after intracellular injection. Alternatively, *Xenopus* vectors already containing β-Gal or GFP can be obtained from a number of labs (search the *Xenopus* Community at http://www.xenbase.org/ or contact this author). Injection mRNA should be synthesized by in vitro transcription methods as described by kit manufacturers (e.g., Applied Biosystems mMessage MMachine kits) and dissolved in sterile, RNase-free distilled water. Stocks (usually ~1.0–3.0 μg/μl) can be stored at –80°C for about 6–9 months.

2.3. Test mRNAs and MOs

1. Test mRNAs should be synthesized in vitro, dissolved and stored as stock solutions as described for tracer mRNAs (Subheading 2.2, item 3).
2. For a working solution, the concentration of the tracer mRNA should be about 100 pg/nl or per cell. The concentration of the test mRNA should be determined based on effective doses in the literature.
3. Mix tracer and test mRNAs in the same solution so they are co-injected, using sterile, RNase-free water.
4. MOs should be diluted and stored in the dark according to manufacturer's instructions.

2.4. Embryo Production and Culture Solutions

1. HCG. Human chorionic gonadotropin made with sterile water at a concentration of 1,000 IU/ml. Should be refrigerated and used within a month.
2. Culture solutions. One of the following: (a) Steinberg's solution (1× SS): 60 mM NaCl, 0.67 mM KCl, 0.83 mM $MgSO_4$, 0.34 mM $Ca(NO_3)_2$, 4 mM Tris–HCl, 0.66 mM Tris base, pH 7.4. Autoclave the solution, store it in a refrigerator or incubator, and use within 2 months; (b) Marc's modified ringers (1× MMR): 100 mM NaCl, 2 mM KCl; 1 mM $MgSO_4$; 2 mM $CaCl_2$; 5 mM HEPES, pH 7.8; 0.1 mM EDTA, in distilled water. Filter-sterilize and store the solution at room temperature or in incubator for months; (c) Modified Barth's Solution (1× MBS): 88 mM NaCl; 1 mM KCl; 1 mM $MgSO_4$; 0.7 mM $CaCl_2$; 5 mM HEPES, pH 7.8; 2.5 mM $NaHCO_3$ in distilled water. Filter-sterilize and store the solution at room temperature or incubator for about a week (see Note 6).
3. Benzocaine. Stock solution: 10% in ethanol, store in the refrigerator. Working solution: 0.5% in 0.1× culture solution made fresh before use. Add 5 ml of benzocaine stock to 1 L of

culture solution drop wise with constant agitation to prevent precipitation.

4. Dejellying solution. Two percent cysteine hydrochloride (aqueous) pH 8.0. To adjust the pH, add 10 N NaOH drop wise while stirring. Make fresh solution each day (see Note 7).
5. Ficoll. 3–5% solution in 1× culture solution. Make fresh solution each week, filter-sterilize, and store it in refrigerator. Warm to room temperature before use (see Note 8).
6. 2% Agarose. Add 2 g of agarose to 100 ml of 1× culture solution. Autoclave to melt agarose, then keep it in a refrigerator. Before pouring plate, microwave until the solid agarose goes into solution, then pour plates.

2.5. General Solutions

1. Phosphate buffer (PB): 0.1 M, pH 7.4. Make 0.2 M stocks of monobasic (Na_2HPO_4) PB and of dibasic (NaH_2PO_4) PB in distilled water. Mix four parts monobasic PB stock, one part dibasic PB stock and five parts distilled water to reach final concentration and pH.
2. Phosphate-buffered Saline (PBS): 0.1 M PB, 0.9% NaCl. Mix 40 ml of monobasic PB stock and 10 ml of dibasic PB stock, add (and dissolve) 0.9 g of NaCl, and then bring the volume up to 100 ml with distilled water (see Note 9).
3. PB plus Tween (PBT): 0.1 M PB, pH 7.4, 0.1% Tween. Add 1 ml of Tween to 1 L of PB.
4. Tris/glycerol: one part 0.1 M Tris buffer, pH 7.5, nine parts glycerol.
5. Heavy mineral oil, sterile, RNase-free.
6. Graded series of ethanol (25, 50, 75, and 100%).

2.6. Solutions for Fixation

1. Four percent paraformaldehyde in 0.1 M PBS (for fluorescently labeled embryos): for 100 ml, add 4 g of paraformaldehyde to 40 ml of distilled H_2O. Stir constantly and heat to 60°C in a chemical fume hood with the beaker covered with foil. Do not let the temperature rise above 65°C. Add 1 N NaOH, drop wise, until the solution clears. Cool solution on ice to room temperature. Add 40 ml of monobasic 0.2 MPB stock, 10 ml of dibasic 0.2 M PB stock, and 0.9 g of NaCl. Mix thoroughly. Bring up the volume to 100 ml with distilled water, mix thoroughly, aliquot, and store at –20°C for months (see Note 10).
2. MEMFA (4% paraformaldehyde in 1× MEM salts; for β-Gal labeled or MO-injected embryos): dissolve 4 g of paraformaldehyde in 80 ml of distilled H_2O, as in Subheading 2.6 item 1. When cool, add 10 ml of stock 10× MEM salts, and bring the volume to 100 ml with dH_2O. 10× MEM salts stock: 1 M MOPS, 20 mM EGTA, 10 mM Mg SO_4. Add salts one at a

time to distilled H_2O with constant stirring. Adjust pH to 7.4 with 10 N NaOH; solution will be cloudy until it reaches the correct pH. Bring solution to final volume, filter-sterilize, and store at room temperature in foil-wrapped bottle (see Note 11).

2.7. Histochemical Solutions

1. β-Gal reaction buffer: 20 mM $K_3Fe(CN)_6$; 20 mM $K_4Fe(CN)_6 \cdot 3H_2O$; 2 mM $MgCl_2$, 1 mg/ml X-gal (5-bromo-4-chloro-3-indolyl-β-D-galactopyranoside) in 0.1 M PBT. Must be made fresh.
2. Clearing solution: (BB/BA) one part benzyl benzoate, two parts benzyl alcohol.

3. Methods

3.1. Embryo Production and Collection

1. Two major methods are used to obtain fertilized eggs: natural matings and in vitro fertilization. For natural matings, both male and female adult frogs are primed by hormone injections. Typically, males receive an injection of 100–200 IU of HCG 2 days before the experiment and again 12–14 h before the experiment. Females receive an injection of 800–1,000 IU of HCG 12–14 h before the experiment. For in vitro fertilization, only the females are injected with hormone. Details for how to inject frogs can be found in ref. (18) (see Notes 12 and 13).
2. For natural mating, place the hormone-injected male and female frogs in a 15-gallon tank filled with 8 gallons of 0.1× culture solution (see Note 6) 12 h prior to the time when fertilized eggs are desired. The bottom of the tank should contain square Petri dishes covered with a stiff plastic screen. The frogs should be left in the dark (we drape the chamber with black cloth) for the next 24 h. As eggs are laid, they drop through the plastic screen into square Petri dishes, and can be collected throughout the day. We use a specially constructed Plexiglas chamber for this purpose (Fig. 1).
3. For in vitro fertilization, the mature eggs are gently squeezed from the hormone-treated female into Petri dishes. Males are anesthetized by submersion in an ice-cold bath containing 0.5% benzocaine, sacrificed and their testes removed. The testes are minced and the released sperm are added to the eggs. Details for this method can be found in ref. 18.
4. Remove the jelly coats from fertilized eggs that have just begun to cleave by gently swirling the eggs in 4× volume of dejellying solution for about 4 min. After the jelly coats are free, immediately wash embryos four changes × 2 min in 0.1× culture solution (see Note 14).

Fig. 1. Side view of plan for a tank designed to facilitate the collection of naturally fertilized eggs. The tank is constructed of opaque Plexiglas. It is divided into two chambers. In one chamber (*left side*) the mating frogs sit on a stiff plastic grid floor (*cross hatch*) through which fertilized eggs (*small dots*) can fall. They fall into square Petri dishes that fit tightly into a drawer that occupies the space below the grid. A tightly fitting lid, with air holes drilled through, covers the top of this chamber to prevent the frogs from escaping. To collect eggs throughout the day without disturbing the frogs, one simply reaches into the chamber on the right side, pulls open the drawer and removes the Petri dishes. By replacing the dishes each time, one can collect eggs multiple times.

5. Transfer embryos to fresh 0.5× culture solution in a clean Petri dish. They can be stored at 14–20°C. The lower temperature will slow down cleavage, giving you more time to inject. However, embryos will not tolerate temperatures lower than 14°C.

3.2. Making Micropipettes

1. Glass capillary tubes are pulled into fine tips strong enough to puncture the vitelline membrane, yet fine enough to cause minimal damage to the injected cell. Using a specialized micropipette puller, adjust parameters (e.g., heat, pull time, pull strength, etc.) according to manufacturer instructions to fashion a micropipette with dimensions approximating those in Fig. 2.
2. Bevel the fine tip of the micropipette (which is fused by the heat of the pull) either with a commercial beveling devise (e.g., Narashige EG-3 grinder) or by manually breaking the tip with fine forceps. To perform the latter, hold the glass micropipette in a strip of plasticine clay attached to the top of a 35-mm Petri dish, place it under a stereomicroscope, and focus on the tip at high magnification. Snip off the very tip at about a 45° angle (Fig. 2).
3. Measure the outer diameter of the tip under a compound microscope using either a stage or eyepiece micrometer. A 10–15-μm diameter is ideal, as 32-cell retina precursor blastomeres are ~50–70 μm in diameter.

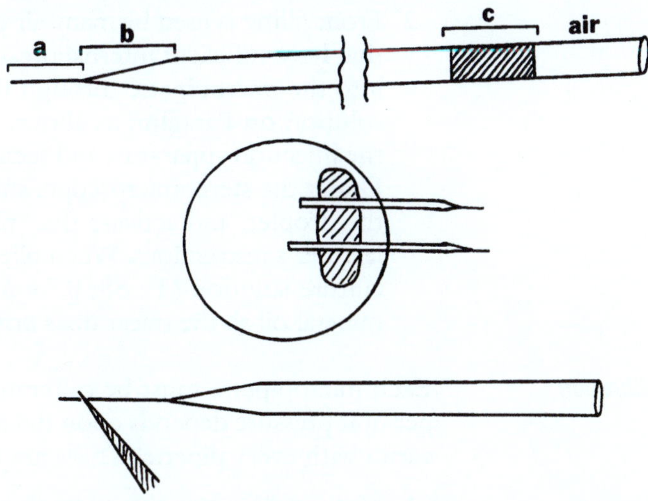

Fig. 2. *Top*: A glass micropipette should have a short primary taper (**b**) and a very fine secondary taper (**a**), each of which should be about 3–4 mm in length. (**c**) Denotes where to place tracer when backfilling the micropipette. The dead airspace behind the tracer must be filled with heavy mineral oil if using a hydraulic microinjection system. *Middle*: To manually break the tip of the micropipette, hold it in place by pressing the distal end into clay (*cross-hatched*) mounted on the lid of a 35-mm Petri dish (*circle*). *Bottom*: While viewing through a dissecting microscope, snip off the very tip of the micropipette at a 45° angle with fine forceps (*hatched triangle*).

3.3. Filling the Micropipette

The method used depends upon the type of microinjection equipment.

1. Back filling is used if the injection equipment is not capable of drawing fluid into the tip of the micropipette, e.g. hydraulic systems and pressure systems that are not designed to back draw. In this case, capillary glass should be siliconized before the micropipettes are pulled. Place a small aliquot of tracer (5–10 μl) on a clean, nonabsorbent surface (e.g. Parafilm stretched over a 35-mm Petri dish). Using either a Hamilton syringe (#701N) or one provided by the microinjection apparatus company, draw up the aliquot into the syringe. This should be done under a stereomicroscope to visually monitor that neither dust particles nor air bubbles are sucked into the needle. Insert the syringe needle into the open back end of a micropipette whose tip has already been broken. Place the tip of the syringe needle about 10 mm from the open end, and slowly deliver the aliquot. Make sure not to introduce air bubbles in the dye. For some equipment, these micropipettes can be mounted directly onto the injection apparatus. For hydraulic systems, heavy mineral oil must be delivered in the same manner, replacing the dead airspace behind the dye aliquot (Fig. 2). Oil, completely free of air bubbles, should be in a 1-cc tuberculin syringe fitted with a 26G needle.

2. Front filling is used by many air-pressure microinjection systems that have sufficient internal pressure to simply suck the solution into the micropipette through the tip. Place a small droplet of solution on Parafilm, as above. Mount the micropipette onto the injection apparatus, and secure onto the micromanipulator. Under the stereomicroscope, submerge the tip in the center of the droplet, and activate the "fill" mode, according to manufacturer's instructions. When filled, submerge the tip in a dish of culture solution (1× SS, 0.5× MMR, or 0.5× MBS) or heavy mineral oil so the tracer does not dry within the tip and clog it.

3.4. Calibration

Each micropipette must be calibrated, since the volume delivered per unit pressure depends upon the inner diameter of the tip, which varies with every pipette. There are two common methods:

1. Method #1: place the micropipette on a fine ruler, and starting at the shoulder of the primary taper (Fig. 2), mark several 1-mm lengths along the shank with a permanent marker. Fill the micropipette with about 5 μl of sterile water, as described in Subheading 3.3. Watching the meniscus of the water through the microscope, measure how many deliveries it takes to move the column of water 1 mm. Repeat a few times to assure consistency. The volume of the column of water is 1 mm $\times \Pi r^2$, where $r = ½$ the inner diameter (ID) of the capillary glass. To determine the volume of each delivery, divide the volume of the column by the number of deliveries it took to move the meniscus 1 mm. Adjust the delivery time according to manufacturer's instructions to deliver desired amount (typically 0.5–5.0 nl) (see Note 15). Expel the water and fill micropipette with mRNA (test + tracer) working solution or MO. Just before filling the micropipette, briefly (30–60 s) centrifuge the mRNA/MO solution to pellet any particulate material that might clog the tip.
2. Method #2: fill micropipette with about 5 μl of sterile water. Place a droplet of sterile heavy mineral oil on the calibration lines of a stage micrometer. Under the stereomicroscope, lower the micropipette tip into the oil, just above the calibration lines. Expel a test droplet. The oil causes the aqueous solution to form a ball at the end of the micropipette tip. Measure the diameter of this ball with the stage micrometer, and calculate volume by following equation: $4/3\Pi r^3$. Repeat several times to make sure the measurements are consistent. Adjust delivery time on the microinjection apparatus accordingly (see Note 15). Expel the water and fill the micropipette with mRNA (test + tracer) working solution or MO, as above.

3.5. Identifying Blastomeres

1. If the experimental goal is to globally express the mRNA/MO, then inject both blastomeres of the 2-cell embryo. Cleavage furrow patterns will not matter for these experiments.

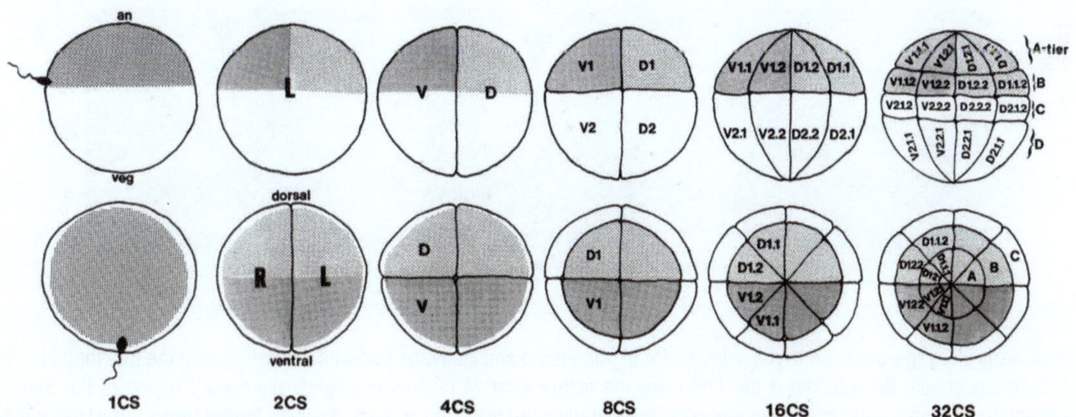

Fig. 3. Stereotypic cleavage patterns of embryos at each cell division from the 1-cell stage (1CS) to the 32-cell stage (32CS). *Top row* is a left-side view, with animal pole (an) to the top and vegetal pole (veg) to the bottom. *Bottom row* is the corresponding animal pole views of each stage. *Right* (R) and *left* (L) sides are indicated. *Shading* indicates the pigmented animal hemisphere. *Sperms* indicate their points of entry (SEP) at the 1CS. In response, at the 2CS the pigmentation becomes asymmetric due to cortical rotation (19). The dorsal side is indicated either as the region 180° opposite the SEP, or by the lightly pigmented region of the animal hemisphere if bisected by the first cleavage furrow (20, 21). Blastomere nomenclature is according to Jacobson and Hirose (25, 26). The nomenclature for 32CS blastomere tiers, as designated by Nakamura and Kishiyama (24) are indicated to the right of 32CS embryos (see Note 20) The major retina precursor blastomere at the: 8-cell stage is D1; at the 16-cell stage is D1.1 (~65%) and D1.2 (~25%); at the 32-cell stage is D1.1.1 (~50%), D1.1.2, D1.2.1, and D1.2.2 (~12% each) (7).

Injections should be placed just animal to the equator to facilitate diffusion of the mRNA/MO through the embryo (Fig. 3).

2. To deliver the mRNAs or MOs to retina precursor blastomeres, it is essential to know where the dorsal side of the embryo will be located (see Note 16). For in vitro fertilized eggs the sperm entry point (SEP) should be marked; the dorsal midline develops about 180° opposite the SEP (19). For naturally fertilized eggs the dorsal side of the embryo can be predicted very accurately (>90%) by noting the orientation of the first cleavage furrow (20, 21). At fertilization, the animal hemisphere pigmentation begins to contract towards the SEP on the ventral side, causing the dorsal equatorial region to become less pigmented (Fig. 3). If the first cleavage furrow bisects this lighter area equally between the two daughter cells, then that lighter area can be used as the indicator of the dorsal side, and the first cleavage furrow will indicate the midsagittal plane (20, 21) (Fig. 4) (see Note 17). Next, the embryos must be selected for regular cleavage furrows (see Note 18). These are found in a smaller and smaller percentage of embryos as cell divisions proceed. Embryos should be selected at each cleavage stage for their adherence to the patterns used for the published retina precursor fate maps (Figs. 3 and 4). Retina precursor blastomeres are identified in ref. 7 (see Notes 19 and 20).

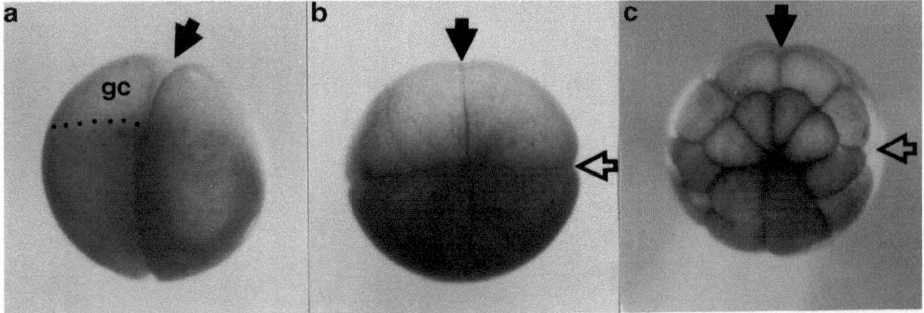

Fig. 4. Early cleavage-stage embryos selected for pigmentation and cleavage patterns consistent with the fate maps (1–4). (**a**) Embryos should be selected if the first cleavage furrow (*arrow*) bisects the lightly pigmented region of the animal hemisphere (gc, *grey crescent*). (**b**) In embryos like that depicted in (**a**) the second cleavage furrow (*open arrow*) will separate dorsal (*light*) from ventral (*dark*) blastomeres. (**c**) Stereotypic pattern of cleavages at the 32-cell stage results in identifiable blastomeres (compare to nomenclature in Fig. 3). *Black arrow* depicts the first cleavage furrow and *open arrow* depicts the second cleavage furrow.

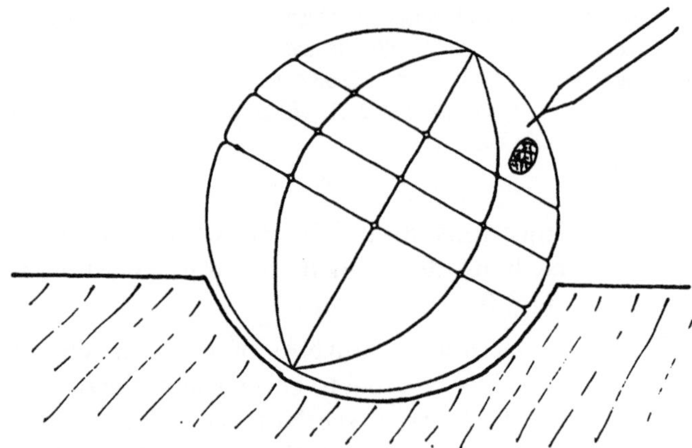

Fig. 5. Position the embryo in a well (*hatch marks*) in the injection dish so that the blastomere to be injected is facing the micropipette. The least damage will occur if the tip punctures the cell at a perpendicular angle and enters to only a short depth, avoiding the nucleus (*darkened oval*).

3.6. Microinjection

1. Place several embryos in an injection dish filled with either culture medium or Ficoll solution (see Note 8). With a hair loop or fine forceps gently angle the embryos so that the desired cell is facing the micropipette (Fig. 5).
2. Using a micromanipulator, advance the tip of the micropipette towards the blastomere to be injected. To prevent ripping the cell membrane, angle the embryo and adjust the micromanipulator so the micropipette tip will be nearly perpendicular to the center of the cell (Fig. 5).
3. When the tip touches the vitelline membrane, there will be a little resistance. Advance the micropipette with the *Z* axis control knob of the manipulator into the target blastomere

(see Note 21). Do not advance deeply into the cell to avoid damaging the nucleus (Fig. 5).

4. Deliver the tracer solution (about 1 nl) according to the equipment used (see Notes 15 and 22). Wait about 10 s before removing the micropipette from the cell to prevent tracer from leaking out or blebs from forming. Do not move the embryo for several minutes, but go on to the next embryo in the dish.
5. After all embryos in the dish are injected, transfer the group to a 35- or 60-mm Petri dish containing Ficoll in 1× SS, 0.5× MMR or 0.5× MBS.

3.7. Embryo Culture

1. About 4–5 h after injection, transfer embryos to diluted culture solution (0.5× SS, 0.1× MMR or 0.1× MBS).
2. Place no more than ten embryos in a 35-mm Petri dish (50 per 60-mm dish; 100 per 100-mm dish), and fill the dish at least 70% full of solution. These measures provide adequate oxygen diffusion.
3. Embryos injected with β-Gal mRNA can be raised on the laboratory bench at room temperature (20–22°C) or in an incubator (14–20°C). Embryos injected with fluorescent tracers should be raised in the dark at these same temperatures. Higher temperatures speed up development, but survival is better at cooler temperatures.
4. After 24 h, change medium to a lower dilution (0.1× SS, 0.05× MMR, or 0.05× MBS) and change daily. Remove any dead embryos or debris to prevent bacterial infections (see Notes 23 and 24).
5. Culture embryos until the desired stage of retina development (see Note 25).
6. Embryos labeled with fluorescent tracers can be viewed while living using epifluorescence or laser-confocal microscopy (see Note 26).

3.8. Fixation

1. If embryos are not yet swimming, they can simply be immersed in fixative. If embryos are free-swimming tadpoles, anesthetize them before fixation by cooling on ice or adding a drop of stock benzocaine to the culture solution.
2. Pick up embryos in a small volume of culture solution with a plastic transfer pipette and drop into a large volume of fixative in a vial with a tight sealing cap. Use about 40× the volume of fixative per volume of embryo.
3. Place vials on a rotator for 30–60 min at room temperature. Embryos older than stage 38 are fairly impermeable, so we snip off the tip of the tail to promote infusion of the fixative.

4. For β-Gal-labeled embryos, proceed immediately to the histochemical reaction (Subheading 3.9 step 2). For MO-injected embryos to be analyzed by ISH, dehydrate through a graded series to 100% ethanol, and store at −20°C. For fluorescent dextran- or GFP-labeled embryos, transfer to fresh PBS and store in the refrigerator for up to a week.

3.9. Whole-Mount Preparations

1. Embryos labeled with fluorescent dextrans or GFP can be mounted in depression slides, or in wells cut into an agar or Sylgard bed on a microscope slide, using either Tris–glycerol or a commercial aqueous mounting medium designed for fluorochromes. Embryos can be directly viewed under epifluorescence illumination or with laser confocal microscopy (see Note 27).
2. β-Gal-labeled embryos should be washed three times in PBT. Incubate at 37°C in reaction buffer for 10–60 min, depending upon the strength of the enzyme activity. Check color reaction frequently using a dissecting microscope. Rinse embryos three times in PBT and refix for 30–60 min on rotator to stabilize the reaction product. Rinse once in PBT. These can be viewed under epi-illumination on a dissecting or compound microscope, or cleared (see Subheading 3.9, step 3) for transillumination. If embryos will be analyzed by in situ hybridization (ISH), they should be dehydrated through a graded series to 100% ethanol stored at −20°C (see Note 28).
3. Clearing embryos: embryos in which a stable histochemical reaction product (β-Gal, ISH) has been fixed in place can be cleared to reveal the 3D patterns of the labeled cells, including those in the internal organs. After histochemical reaction and post-fixation, wash twice in PBT and dehydrate in a graded series of ethanol (30 min each in 25, 50, 75%, twice 15 min each in 100%). Transfer to BB/BA solution in a depression slide or spot plate, and view under microscope. After viewing β-Gal-labeled specimens, wash them in 100% ethanol several times (they will again become opaque), and store them in fresh 100% ethanol at −20°C (see Note 29).

3.10. Tissue Section Preparations

1. Tissue sectioning: wash embryos in PBS containing 5% sucrose overnight. One hour before embedding, wash in PBS containing 15% sucrose on a rotator. Place each embryo to be sectioned in a small volume of embedding material (e.g., TissueTek O.C.T. Compound) to remove excess PBS. Mount embryo in a mold filled with embedding medium and freeze in cryostat. When frozen, section at 10–20 μm. Pick up sections on slides (gelatin-coated or positively charged) and store the slides in freezer until ready for processing. If the tissue is fluorescent, the slides should be stored in the dark (see Notes 30 and 31).

2. Preparation of slides containing fluorescent tracers: slides should be dried on a warming plate (37°C) for 20–30 min. They can be refixed (4% paraformaldehyde in 0.1 M PBS) onto the slides for 5–10 min to prevent sections from falling off during further processing. Wash slides three times 5 min each in PBS. If desired, double-labeling with immunofluorescence techniques can be started at this point (see Note 32). Mount glass coverslips with an aqueous, nonfluorescing mounting medium, such as Tris/glycerol or a commercially available one (see Note 33). To preserve the fluorescent signal, slides should be stored in the refrigerator or freezer in the dark (see Note 31).
3. Preparation of slides containing β-Gal-labeled tissue: wash embryos 3× in PBT and incubate at 37°C in reaction buffer for 10–60 min, depending upon the strength of the enzyme activity. Rinse slides three times in PBT and refix for 10 min to stabilize the reaction product. Rinse slides once in PBT, dehydrate in a series of ethanol (1 min each in 25, 50, 75%, twice each 1 min in 100%), clear in toluene (twice 1 min each) and mount glass coverslips with a permanent medium. Allow slides to dry for at least 24 h, and then put them in a box to keep clean. These can be stored and viewed for years.

4. Notes

1. Mesh-fitted injection dishes should be soaked in several changes of distilled water for several days before use to remove any toxic contaminants of the glue. Test the dish for toxicity by growing control embryos overnight in its wells.
2. When adjusting the pH of the KCl solution, proceed slowly. The pH will overshoot and then gradually fall, so it can take about 3 h to stabilize at 6.8. The pH is important for the intracellular health of the cell to be injected, so it must be adjusted carefully. Fluorescent dextran tracers can be made in water, but they are less stable and more toxic.
3. Purchase dextrans that are conjugated to lysine. The amino acid side chains allow them to be cross-linked to intracellular proteins by paraformaldehyde.
4. Purchase dextrans that are 10,000 or 40,000 MW for maximum diffusion within the cytoplasm of the injected cell. Larger molecules might not label the entire lineage; smaller molecules may pass through gap junctions or intercellular bridges.
5. Microfiltration of dextran tracers is important to make the solution sterile and free of particles that will clog the micropipette tip.

6. SS, MMR, and MBS are virtually interchangeable. Different labs prefer one to the other, mostly due to laboratory history. MBS has a shorter shelf life because it is buffered with bicarbonate.
7. The high pH of the dejellying solution is critical for it to work properly. A lowered pH will cause the dejellying process to take too long, causing damage to the embryos.
8. Incubating embryos in Ficoll causes the vitelline membrane to collapse onto the cell surface. This is an advantage during microinjection because it exerts pressure on the puncture hole, preventing leakage of cytoplasm and your label. Do not use Ficoll if you plan to subsequently remove the vitelline membrane to perform a transplantation or dissection because the vitelline membrane will collapse so tightly against the blastomeres that its manual removal without damaging the embryo is virtually impossible. A range of 3–5% Ficoll is used by different laboratories.
9. Salt is added to fixatives and buffers for fluorescently labeled embryos to stabilize the tracer, lower background autofluorescence and improve antibody specificity if these tissues are to be labeled subsequently for protein localization.
10. Commercially purchased formaldehyde solution contains many breakdown products that either fluoresce themselves or cause tissue to autofluoresce. Therefore, fixatives to be used on fluorescently labeled embryos should be freshly prepared with paraformaldehyde. Single-use aliquots of fresh paraformaldehyde fixative for fluorochrome-labeled embryos should be frozen to deter chemical degradation. Detergents either autofluoresce or quench the fluorescence of fluorochromes, and should not be included in fixatives for fluorescently labeled embryos.
11. β-Gal and MOs injected embryos are typically assayed for changes in gene expression by ISH. MEMFA is the preferred fixative for ISH because it preserves both tracer and endogenous mRNAs.
12. In vitro fertilization is ideal for obtaining large numbers of embryos on demand, synchronized to the same stage of development. However, it requires sacrificing the male frog and the embryos do not always cleave in regular patterns that match the fate maps. Natural fertilization frequently produces regular cleavage patterns and was used for all the fate maps from the Moody lab (2–11, 20). Natural fertilization provides developmental stages of embryos spread out over a long time period, which is advantageous when complex manipulations are planned. However, frogs do not always mate successfully on a time frame convenient to the experimenter's schedule.

13. Allow female frogs to rest at least 6–8 weeks between hormone treatment to prevent stress and to allow them to replenish egg supplies.
14. Dejellying must be performed carefully. Do not dejelly prior to the appearance of the first cleavage or the dejellying solution will disable the sperm. Do not agitate the eggs, as this can cause polyspermy and very irregular cleavages; eggs should be gently swirled intermittently. Watch for signs that the jelly is falling off the eggs; the eggs will touch one another, rather than being separated by their coats. If a small amount of jelly remains, the eggs will be very sticky and nearly impossible to inject. However, leaving embryos in the dejellying solution too long will damage them.
15. A very small volume of the tracer molecule needs to be pressure-injected inside the blastomere without damaging that cell. There are several kinds of microinjection apparatus commercially available that range in complexity, but the most important feature to be considered is the typical injection volumes that you will use. Injection volumes into oocytes, fertilized eggs and 2-cell embryos can be as large as 10 nl (or sometimes larger), and thus simple equipment will suffice (e.g., the hydraulic or Drummond systems). For older stages it is best to keep injection volumes to about 1 nl per blastomere because 10 nl injections at 16- and 32-cells can result in artifactual fate changes. For example, we found that injecting >10 nl of tracer alone into some 16-cell blastomeres can drive epidermal lineages into brain lineages (22).
16. The dorsal side of the embryo can be identified by three methods that are detailed in the following references: (a) marking the SEP with a vital dye (19, 23); (b) by tipping in vitro fertilized eggs and marking one side (19, 23); or (c) by selecting embryos in which the first cleavage furrow bisects the grey crescent (20, 21).
17. One can also select embryos at the early part of the 4-cell cleavage, when the first and the second furrows at the vegetal pole can be distinguished; the first furrow should be complete and the second furrow not yet complete. If, however, you wait until the end of the 4-cell stage to select embryos, you can no longer discriminate between the first and second cleavage furrows, and the lightly pigmented cells may be dorsal ones in only about 70% of embryos (20, 21). This will render targeting retina precursor blastomeres much less accurate.
18. Observe embryos frequently until they reach the required cleavage stage to ensure that cleavage furrows are dividing the cytoplasm in a regular pattern. The stereotyped pattern used for the fate maps is illustrated in Fig. 3. If spatial localization is critical for the interpretation of the experiment, select embryos

that adhere to this ideal pattern, at least on the side of the embryo to be injected. Each cleavage cycle takes 20–30 min, depending on the temperature at which they are raised.

19. Not every blastomere in the embryo has to be "perfect". If you are targeting one specific cell, only that cell needs to cleave according to the ideal pattern.
20. Regarding blastomere nomenclature: Nakamura and Kishiyama (24) presented a very simple nomenclature for the 32-cell embryo that is simple to remember because the tiers are labeled A–D (animal to vegetal) and the rows are labeled 1–4 (dorsal midline to ventral midline) (Fig. 3). When Jacobson and Hirose (25, 26) began to map the nervous system lineages at all of the different cleavage stages, they devised a plan, similar to those used in sea urchins and ascidians, that would relate the cells to their mothers, grandmothers and descendants. Although these numbers and letters are harder to remember, there is a logic to the system that communicates lineal relationships. All cells starting with "D" are on the dorsal side of the embryo, and all cells starting with "V" are on the ventral side. A number is then added at each cleavage stage, which denotes the position of the blastomere with regards to the poles and the midlines (Fig. 3) (27).
21. If the micropipette encounters a lot of resistance at the cell surface, the tip is too blunt and will likely damage the cell. The puncture hole may be so large that cytoplasm (and your mRNA/MO) will leak out, or blebs will form at the puncture site after you remove the micropipette. Discard these embryos (see Note 15). At this point, it is best to set up a new micropipette. If the tip bends when it touches the vitelline membrane, the taper is too long. You can break the tip to a shorter length, but you need to calibrate it again. Adjust the programming of the pipette puller to forge a stronger, shorter tip.
22. If the volume or concentration of mRNA/MO is too great, the injected cell or some of its descendants will stop dividing. At the extreme, shortly after injection there will be one or two large cells in a field of smaller ones. However, damaged cells may not be noticed until a later point in development. There may be larger than normal, labeled cells incorporated into the organs of the embryo, or labeled cells that are the correct size, but spherical, rather than differentiated, in shape. This can happen to the entire clone or to only a subset of the clone. Often these cells will move to the correct regions of the embryo, but never differentiate. Another sign of damage is the accumulation of labeled cells in the spaces within the embryo, i.e., the lumens of the nervous system, gut, liver, and heart. These damaged cells probably dissociated from the rest of the embryo during gastrulation movements and accumulated

wherever space appeared. If any of these signs of damage occur, discard the embryos and inject a smaller volume or concentration of mRNA/MO in the next experiment.

23. If bacterial infections are a problem, the culture solution can be supplemented with 0.1% gentamicin. This addition will reduce the storage life of the medium to about 1 week and may lower the pH.

24. *Xenopus* embryos prior to gastrulation prefer the pH to be around 7.8. If the pH of the culture solution drops below 7.4, make up fresh solutions.

25. Fluorescent-dextran tracers and fluorescently tagged MOs are detectable immediately after injection. Tracer mRNAs require 2–3 h for adequate amounts of protein to be synthesized at detectable levels. These tracer molecules all can be detected in the descendants at least through tadpole stages (stages 45–48; (28)). Because *Xenopus* cells decrease in size by cell division up through blastula stages (during which there are basically no G phases in the cell cycle), the originally injected concentration of dextran tracer or MO remains stable. The growth of the embryo after neurulation does not appear to dilute the tracers significantly. However, once tadpoles begin to feed, dextran labeling can become granular and uneven, suggesting it is being packaged into lysosomes. Injected tracer mRNAs are probably degraded by the end of gastrulation, but both β-Gal and GFP proteins are very stable and can be detected at least through stage 45.

26. Excitation of the fluorochrome releases free radicals that can damage living cells, so view living embryos under low-light conditions for very brief periods of time.

27. Intracellular yolk platelets quench the fluorescence of tracers, so sectioning the tissue with a cryostat may be necessary. Fluorescence will be destroyed by the commonly used histological organic solvents, so sections should be mounted with an aqueous-based mounting medium, such as Tris/glycerol or one of the numerous commercially available media.

28. β-Gal activity is detected by fixing the enzyme in place with paraformaldehyde, and providing a substrate that upon being altered by enzymatic action becomes an insoluble, colored precipitate. This precipitate therefore indicates the cellular location of the enzyme, which was encoded for by the injected mRNA. Enzyme histochemical reactions provide essentially permanent specimens that can be referenced for years.

29. Clearing greatly improves the 3D visualization of the staining pattern and reveals deep members of the clones. However, β-Gal specimens should not be stored in the clearing solution because the reaction product will fade after a few days. BB/BA will

corrode any surface other than glass, especially microscope parts. Clean up spills with 100% ethanol.

30. Fluorescent tracers can be combined with other fluorescent methods, such as UV-excitable nuclear markers, fluorescent streptavidin to detect biotin-labeled compounds, and the detection of cell type-specific proteins with fluorescently tagged antibodies.

31. Fluorescent lineage tracers are not permanent, but they can be extremely hardy if the specimens are stored in the dark and refrigerated or frozen. For example, we have viewed 2-year-old dextran and GFP-labeled tissue sections stored at –80°C with no detectable diminution of the signal.

32. Anti-GFP and anti-β-Gal antibodies are commercially available if double-labeling with another protein marker is desired.

33. To improve the life span of fluorescent preparations, they must be mounted in a buffered, aqueous medium. Fluorescein absorption is especially sensitive to acidic pH, so the pH of the mounting medium must be between 7.4 and 7.6.

Acknowledgement

This work was supported by NSF grant IOS-0817902.

References

1. Dale L, Slack JMW (1987) Fate map of the 32-cell stage of *Xenopus laevis.* Development 100:279–295
2. Moody SA (1987) Fates of the blastomeres of the 16-cell stage *Xenopus* embryo. Dev Biol 119:560–578
3. Moody SA (1987) Fates of the blastomeres of the 32-cell stage *Xenopus* embryo. Dev Biol 122:300–319
4. Moody SA, Kline MJ (1990) Segregation of fate during cleavage of frog (*Xenopus laevis*) blastomeres. Anat Embryol 182:347–362
5. Moody SA (1989) Quantitative lineage analysis of the origin of frog primary motor and sensory neurons from cleavage stage blastomeres. J Neurosci 9:2919–2930
6. Huang S, Moody SA (1992) Does lineage determine the dopamine phenotype in the tadpole hypothalamus: a quantitative analysis. J Neurosci 12:1351–1362
7. Huang S, Moody SA (1993) The retinal fate of *Xenopus* cleavage stage progenitors is dependent upon blastomere position and competence: studies of normal and regulated clones. J Neurosci 13:3193–3210
8. Kenyon KL, Zaghloul N, Moody SA (2001) Transcription factors of the anterior neural plate alter cell movements of epidermal progenitors to specify a retinal fate. Dev Biol 240:77–91
9. Huang S, Moody SA (1995) Asymmetrical blastomere origin and spatial domains of dopamine and Neuropeptide Y amacrine cells in *Xenopus* tadpole retina. J Comp Neurol 360:2–13
10. Huang S, Moody SA (1997) Three types of serotonin-containing amacrine cells in the tadpole retina have distinct clonal origins. J Comp Neurol 387:42–52
11. Moore KB, Moody SA (1999) Animal-vegetal asymmetries influence the earliest steps in retinal fate commitment in *Xenopus.* Dev Biol 212:25–41
12. Guthrie S, Turin L, Warner AE (1988) Patterns of junctional communication during development of the early amphibian embryo. Development 103:769–783

13. Weisblat DA, Sawyer RT, Stent GS (1978) Cell lineage analysis by intracellular injection of a tracer enzyme. Science 202:1295–1298
14. Jacobson M (1985) Clonal analysis and cell lineages of the vertebrate nervous system. Annu Rev Neurosci 8:71–102
15. Stent GS, Weisblat DA (1985) Cell lineage in the development of invertebrate nervous systems. Annu Rev Neurosci 8:45–70
16. Gimlich RL, Braun J (1985) Improved fluorescent compounds for tracing cell lineage. Dev Biol 109:509–514
17. Chalfie M, Tu Y, Euskirchen G, Ward WW, Prasher DC (1994) Green fluorescent protein as a marker for gene expression. Science 263: 802–805
18. Sive HL, Grainger RM, Harland RM (2000) Early development of *Xenopus laevis*. A laboratory manual. Cold Spring Harbor Laboratory Press, Cold Spring Harbor, NY
19. Vincent J-P, Gerhart JC (1987) Subcortical rotation in *Xenopus* eggs: an early step in embryonic axis specification. Dev Biol 123: 526–539
20. Klein SL (1987) The first cleavage furrow demarcates the dorsal–ventral axis in *Xenopus* embryos. Dev Biol 120:299–304
21. Masho R (1990) Close correlation between the first cleavage plane and the body axis in early *Xenopus* embryos. Dev Growth Differ 32: 57–64
22. Hainski AM, Moody SA (1992) *Xenopus* maternal RNAs from a dorsal animal blastomere induce a secondary axis in host embryos. Development 116:347–355
23. Peng HB (1991) Appendix A: solutions and protocols. Methods Cell Biol 36:657–662
24. Nakamura O, Kishiyama K (1971) Prospective fates of blastomeres at the 32-cell stage of *Xenopus laevis* embryos. Proc Jpn Acad 47: 407–412
25. Hirose G, Jacobson M (1979) Clonal organization of the central nervous system of the frog. I. Clones stemming from individual blastomeres of the 16-cell and earlier stages. Dev Biol 71:191–202
26. Jacobson M, Hirose G (1981) Clonal organization of the central nervous system of the frog. II. Clones stemming from individual blastomeres of the 32- and 64-cell stages. J Neurosci 1:271–284
27. Sullivan SA, Moore KB, Moody SA (1999) Early events in blastomere fate determination. In: Moody SA (ed) Cell lineage and cell fate determination. Academic, New York, pp 297–321
28. Nieuwkoop PD, Faber J (1967) Normal table of *Xenopus laevis* (Daudin). Elsevier-North Holland Publishing Co., Amsterdam

Part II

Tracing Cell Fate

Chapter 7

Testing Retina Fate Commitment in *Xenopus* by Blastomere Deletion, Transplantation, and Explant Culture

Sally A. Moody

Abstract

The lineages of individual cells of the *Xenopus* cleavage-stage embryo have been fate-mapped to reveal the subset of blastomeres that are the major and minor precursors of the retina. Using this retina fate map, one can test the commitment of each of these cells to various retinal cell fates by manipulating the environment in which they develop. This chapter presents the techniques for identifying specific retina blastomere precursor cells, deleting them to test whether they are required for producing specific kinds of retinal cells, transplanting them to novel embryonic locations in host embryos to test whether they are committed to produce specific kinds of retinal cells, and growing them in explant culture to determine if their ability to produce specific kinds of retinal cells is autonomous.

Key words: Retina cell fate, *Xenopus laevis*, Blastomeres, Fate mapping, Lineage tracing, Fate bias, Fate commitment

1. Introduction

The retina derives from the central region of the anterior neural plate called the eye field, which becomes specified to this fate via local signaling that induces the expression of retina-specific transcription factors, such as *Rx1*, *Pax6*, and *Six3* (reviewed in ref. 1). Fate mapping studies in *Xenopus* identified the subset of cleavage-stage blastomeres that contribute to the eye field, and subsequently to the retina (2). As described in ref. (3), the *Xenopus* embryo is ideal for fate mapping studies because its cleavage blastomeres are easy to identify and their lineages are highly reproducible. These maps revealed that the retina derives primarily from the blastomeres that occupy the animal-dorsal quadrant of the embryo (Fig. 1). This fate map can be used to manipulate gene expression in major or minor retinal precursor blastomeres to test for the function of these genes in retinal fate commitment (1, 4–9), as described in

Shu-Zhen Wang (ed.), *Retinal Development: Methods and Protocols*, Methods in Molecular Biology, vol. 884,
DOI 10.1007/978-1-61779-848-1_7, © Springer Science+Business Media, LLC 2012

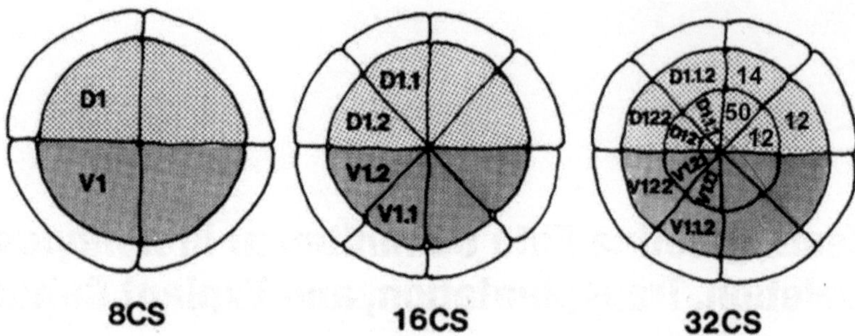

Fig. 1. Diagrams of cleavage embryos depicting the animal pole blastomeres at the 8-, 16- and 32-cell stages (CS). Nomenclature for each blastomere, according the Jacobson and Hirose (18, 19, 20) is given on the *left side* of each embryo. At the 32-CS, the major retina precursor blastomeres are D1.1.1, D1.1.2, D1.2.1, and D1.2.2; the percentage of retina cells that descend from each of these blastomeres (2) is given on the *right side*. Blastomere V1.2.1 and the contralateral dorsal animal blastomeres can contribute very small percentages of retina cells.

detail in ref. (3). However, to test the commitment of a blastomere to autonomously express a retinal fate requires exposing that cell to novel embryonic environments.

There are at least three ways to test the commitment of a blastomere to produce retinal cells: deletion, transplantation and explant culture. These experimental approaches can test whether a cell is committed to a retinal fate by comparing its normal repertoire of descendants with those expressed after each experimental manipulation. The term "commitment" refers to whether a cell's fate is fixed (committed) regardless of the surrounding cellular environment, or can be influenced by external factors provided by its neighbors. A cell that is committed to its fate will produce its normal descendants regardless of whether it is grown in other regions of the embryo or in culture. Alternatively, a cell may easily change fate to produce descendants appropriate to any environment it occupies. A cell also may demonstrate an intermediate, "biased" state, in which it produces some, but not all, of its normal progeny when grown in a novel environment.

In this chapter, I will present three ways to manipulate blastomeres in order to gain information about their state of commitment to a retinal fate. A first approach is to delete the cell of interest. This will not reveal the state of commitment of the deleted cell, of course, but will test whether its presence is necessary for the normal retinal development of the remaining cells, and whether these neighbors can change fate to reconstitute the entire retinal repertoire of cells. By counting the number of lineage-labeled cells in the retina after deletion of the major 32-cell precursor blastomere, for example, we showed that usually a normal-sized retina was produced. While dorsal neighbors made fewer retinal cells, ventral neighbors produced more, indicating that blastomere position in the embryo dictates its overall contribution to the retina (2).

It has yet to be determined whether the production of specific subtypes of retinal cells, which descend in a biased manner from the different retina precursor blastomeres (10, 11), are affected by this manipulation. A second approach is to transplant a retina precursor blastomere into to a novel region of the embryo to test whether it maintains its original set of descendants, or alters its progeny in accord with its new environment. We have used this approach to determine that ventral animal blastomeres transplanted to the retinogenic site of the cleavage embryo can contribute to retina as well as the original cell, but vegetal blastomeres are not competent to make retina (2). Using gene expression manipulations, as described in ref. (3), combined with blastomere transplantation we showed that vegetal blastomeres are not able to make retinal cells because they contain endo-mesoderm determinants and are unable to respond to head-inducing factors (4, 9). A third approach is to prevent the cell from communicating with its normal neighbors by removing it to explant culture, which will test whether it produces retinal cells autonomously. We have used this approach to show that dorsal animal blastomeres have an autonomous ability to produce neural and dorsal mesodermal tissues (12), but have not yet used it to study specific retinal fates. An advantage of this approach is that one can supplement the medium with signaling factors or antagonists to determine which cell-to-cell communication pathway might be involved. For each of these approaches one can analyze gene expression and the production of specific types of retinal cells using all of the standard approaches. Together, these manipulations provide simple, yet powerful, ways to demonstrate the fundamental developmental mechanisms that regulate how the embryo creates specific retinal lineages.

2. Materials

2.1. Equipment

1. For microinjection: microinjection apparatus, programmable micropipette puller, dissection stereomicroscope (at least ×50 magnification) placed on a steel plate in an area that is level and free of vibrations, equipped with a fiber optic lamp and a micromanipulator mounted on a magnetic base. See ref. 3 for details.
2. A second dissection stereomicroscope for performing microsurgery, equipped with a fiber optic lamp.
3. Injection dish: described in detail in ref. (3).
4. Microsurgery dish: 60-mm plastic Petri dish coated with 1 mm thickness of 2% agarose dissolved in culture solution. After the agarose has hardened, melt about 10–12 shallow depressions in the agarose in a circle. To do this, melt the tip of a 6″ Pasteur pipette into a ball with a Bunsen burner. Warm it briefly in the

flame and lightly touch the heated ball onto the agarose surface. Make several dishes in advance, and store in the refrigerator wrapped in plastic wrap so they do not dehydrate.

5. Explant culture dish: lightly coat each well of a 24-well plate with 2% agarose dissolved in culture solution. After the agarose has hardened, melt a shallow well in the center of each well using a heated Pasteur pipette, as above (Subheading 2.1, item 4).
6. Tabletop centrifuge to spin lineage tracers, as described in ref. (3).
7. Stage micrometer to calibrate micropipette injections, as described in ref. (3).
8. Fine sharpened forceps (e.g., Dumont #5 biologie). Two sets are advisable, in case one is damaged during the course of an experiment. Each forceps set should consist of one that you have honed to a square tip and one that you have sharpened to a very fine tip (see Note 1).
9. Hair loop: this can be crafted by placing both ends of a fine hair (about 10 cm long) into the narrow tip of a 6″ Pasteur pipette to form a 2–3-mm loop. Seal the hair in place with melted paraffin. Sterilize before use by dipping in 70% ethanol and air-drying.
10. Six-inch glass Pasteur pipettes, autoclaved.

2.2. Lineage Tracers

1. Fluorescent dextrans, as described in ref. (3).
2. Tracer mRNAs (β-Gal, GFP), as described in ref. (3).

2.3. Embryo Production and Culture Solutions

1. HCG. Human chorionic gonadotropin made with sterile water at a concentration of 1,000 IU/ml. Should be refrigerated and used within a month.
2. Benzocaine. Stock solution: 10% in ethanol, store in the refrigerator. Working solution: 0.5% benzocaine in 0.1× culture solution (see Subheading 2.3, item 4) made fresh before use. To make working solution, add 5 ml of benzocaine stock to 1 L of culture solution, drop wise with constant agitation (see Note 2).
3. Dejellying solution. Two percent cysteine hydrochloride (aqueous), pH 8.0. Adjust pH by adding 10 N NaOH solution drop wise. Should be made fresh just before use (see Note 3).
4. Culture solution. One of the following: (a) *Steinberg's solution* (1× SS): 60 mM NaCl, 0.67 mM KCl, 0.83 mM $MgSO_4$, 0.34 mM $Ca(NO_3)_2$, 4 mM Tris–HCl, 0.66 mM Tris base, in distilled water, pH 7.4. Autoclave, store in incubator for months. (b) *Marc's Modified Ringers* (1× MMR): 100 mM NaCl, 2 mM KCl; 1 mM $MgSO_4$; 2 mM $CaCl_2$; 5 mM HEPES, pH 7.8; 0.1 mM EDTA, in distilled water, pH 7.4.

Filter-sterilize and store in incubator for months. (c) *Modified Barth's Solution* (1× MBS): 88 mM NaCl; 1 mM KCl; 1 mM $MgSO_4$; 0.7 mM $CaCl_2$; 5 mM HEPES, pH 7.8; 2.5 mM $NaHCO_3$ in distilled water, pH 7.4. Filter-sterilize and store in incubator for about a week (see Note 4).

3. Methods

3.1. Embryo Production and Collection

1. Two major methods are used to obtain fertilized eggs: natural matings and in vitro fertilization. For both methods, adult frogs are primed by hormone injections. Typically, males receive an injection of 100–200 IU of HCG 2 days before the experiment and again 12–14 h before the experiment. Females receive an injection of 800–1,000 IU of HCG 12–14 h before the experiment. Details for how to inject frogs can be found in ref. (13). Methods for setting up a natural mating or an in vitro fertilization are described in detail in ref. (3) (see Note 5).
2. Remove the jelly coats from fertilized eggs that have just begun to cleave (2–4 cells) by gently swirling the eggs in 4× volume of dejellying solution for about 4 min. After the jelly coats are free, immediately wash embryos 4×2 min in diluted culture solution (0.5× SS, 0.1× MMR, or 0.1× MBS) (see Note 6).
3. Transfer embryos to fresh diluted culture solution in a clean Petri dish. They can be stored at 14–20°C. The lower temperature will slow down cleavage, giving you more time to inject and perform the microsurgeries. Embryos will not tolerate temperatures lower than 14°C.

3.2. Microinjections (See Ref. 3 for Details)

1. Pull glass capillary tubes into fine tips strong enough to puncture the vitelline membrane, yet fine enough to cause minimal damage to the injected cell. Bevel the tip to make it patent.
2. Fill the micropipette with lineage tracer. The method used depends upon the type of microinjection equipment. Calibrate delivery according to chosen tracer and microinjection apparatus used.
3. Set up the microinjection station as described in ref. (3). Fill injection dish with 1× embryo culture medium *without* Ficoll (see Note 7).
4. For transplantation experiments, label the entire embryo with lineage tracer by injecting both blastomeres of the 2-cell embryo in the animal pole region.
5. For deletion experiments, inject a single, identified retina precursor blastomere at the desired stage (8-, 16-, or 32-cells; Fig. 1). The dorsal side must be identified at the 2-cell stage and

subsequent cleavages monitored, as described in ref. (3), to ensure correct identification of the retina precursor blastomeres.

6. For explant experiments, the cell to be placed in culture does not need to be lineage labeled, unless you intend to recombine the cell with other cells/tissues. In this latter case, label the entire embryo with lineage tracer as in Subheading 3.2, step 4.

3.3. Blastomere Deletion

1. Microinject lineage tracer into a single blastomere that is to remain in the experimental embryo. This will be the blastomere whose change in fate you wish to test after deleting a neighboring cell.
2. Transfer embryo to a dissection dish filled with diluted culture solution (0.5× SS, 0.1× MMR, or 0.1× MBS). Using a hair loop, gently push the embryo into one of the shallow depressions, animal pole up, and position it so that you can see the transparent vitelline membrane separated by a clear space (perivitelline space) above the surface of the animal pole of the embryo.
3. Using a square-tipped forceps in your subdominant hand (left hand if you are right handed), grasp the vitelline membrane above the perivitelline space. Using the fine-tipped forceps, grasp the membrane close to the first forceps tip, and gently pull in opposite directions to peel the membrane away (see Note 8). You can tell that the membrane has been removed because the embryo will flatten (see Note 9). It is useful to grasp the membrane over the cell you intend to delete, because you may damage the underlying cell when first grasping the vitelline membrane.
4. Orient the embryo with the cell to be deleted (the "victim" cell) facing up; make sure the embryo is in the shallow depression in the agarose. Holding the squared-tipped forceps in an open position, gently place them over the embryo to stabilize it and hold it in place. Using the fine-tipped forceps, grab the middle of the "victim" cell and gently pull. Use the squared-tipped forceps to hold down neighboring cells so they do not also pull free with the "victim" cell. This works very well for 16-cell and older blastomeres (see Note 10). For 8-cell blastomeres, operate towards the end of the cell cycle (i.e., when you can start to see the beginning of the next cleavage furrow) because cytoplasmic bridges are likely to be small and the cells are less adhesive. Place closed tines of the forceps between cells and gently open them to separate the neighboring blastomeres, snip any visible cytoplasmic bridges, and dissect away the "victim" cell piecemeal (see Note 11).
5. Remove any cellular debris with a sterile, glass Pasteur pipette so surgery is clean (Fig. 2) (see Note 12).
6. Repeat the process in the same dish at the other depressions melted into the agarose. I usually perform about ten deletions per 60-mm microsurgery dish.

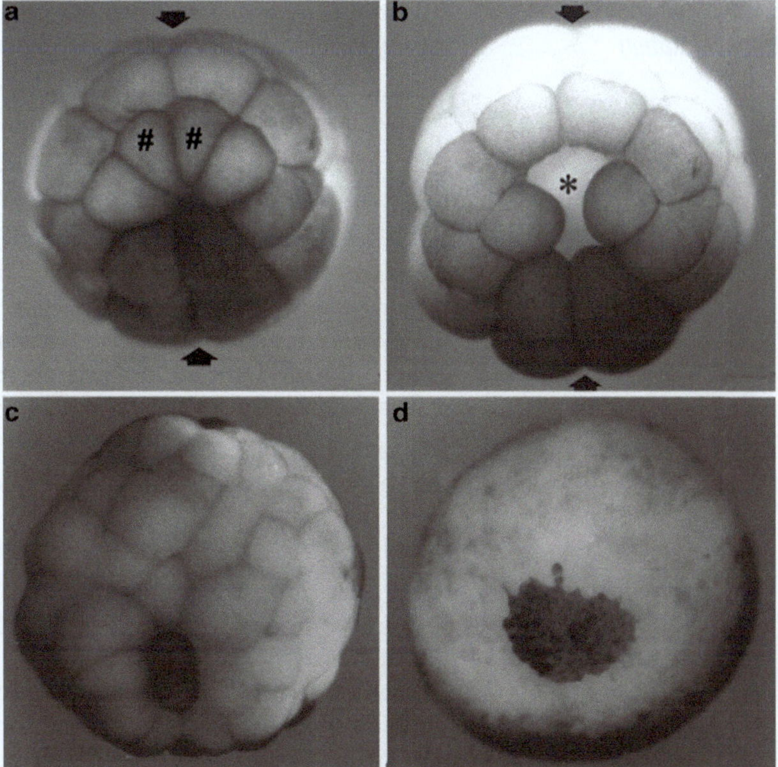

Fig. 2. Examples of blastomere deletion and transplantation. (**a**) A 32-cell embryo, still within the vitelline membrane, oriented as in Fig. 1. *Arrows* denote the midline, dorsal is to the *top*. The two major retina progenitor blastomeres (D1.1.1 *left* and *right*) are noted by *hash*. (**b**) The same embryo shortly after deletion of the two D1.1.1 blastomeres; the resulting space, cleared of cellular debris is noted by *asterisk*. The embryo is flattened, compared to (**a**), due to removal of the vitelline membrane. (**c**) Vegetal view of a 64-cell embryo in which a 32-cell retina precursor blastomere (*dark cell*) was transplanted to a ventral, vegetal position. Vegetal cells do not contain melanin pigment and appear *white*. Note that the transplanted cell has divided once. (**d**) The same embryo a few hours later showing that the descendant cells of the transplanted blastomere (*darkly pigmented patch*) have neatly integrated into their new location.

7. After the last operated embryo heals (15–30 min), remove about half of the culture solution (and any remaining debris) with a sterile, glass Pasteur pipette, and replace with fresh diluted culture solution (0.5× SS, 0.1× MMR, or 0.1× MBS) containing 0.1% gentamicin (see Note 13).
8. Culture at 14–20°C overnight next to a dish containing sibling embryos of the same stage (see Note 14).
9. The next day, transfer embryos to fresh culture medium plus gentamicin in a clean agarose-coated Petri dish using a sterile, glass Pasteur pipette. Culture until siblings reaching the desired stage for analysis (see Note 15).

3.4. Blastomere Transplantation

1. Inject lineage tracer into both cells at the 2-cell stage to create totally labeled embryos that will donate a blastomere.
2. Set aside same-staged embryo siblings to act as host recipients, but do not inject these with a lineage tracer.
3. Place several host embryos in the circle of depressions melted into the microsurgery dish. Place an equal number of injected donor embryos in the center of the circle.
4. Remove the vitelline membrane from one host embryo, making the initial grasp above the cell that you will delete to make room for the transplanted donor cell (in case you damage the cell during removal of the vitelline membrane). Delete this cell as described above (Subheading 3.3).
5. Remove the vitelline membrane from one donor embryo, making the initial grasp distant from the cell that you wish to transplant, so that you do not damage it during removal of the membrane.
6. To dissect out the donor cell, grab one neighboring cell with the square-tipped forceps in your subdominant hand and use this cell as a "handle" so you do not actually touch the cell to be transplanted. With the fine-tipped forceps, gently pull the remaining neighboring cells away from the embryo. The "handle" cell is likely to leak and fall apart during this process; this is fine because you will only use it to move the donor cell over to the gap you created in the host.
7. Check that the hole from the deleted cell made in the host embryo has not healed during donor cell dissection. Open up the hole by placing closed forceps tines in the hole and gently opening them to stretch the space apart. Grab the "handle" cell to move the donor cell in place over the hole, and drop it into place. Alternatively, you can use a sterile hair loop to float the donor cell over to the host. You may need to gently push it into position with the side of your forceps or the hair loop. If pieces of the "handle" cell are still attached, dissect them free (Fig. 2).
8. Remove any cellular debris with a sterile, glass Pasteur pipette (see Note 12).
9. Repeat the process in the same dish at the other depressions melted into the agarose. I usually perform about ten transplantations per dish.
10. After the last embryo heals (~60 min), remove about half of the culture solution (and any remaining debris) with a sterile, glass Pasteur pipette, and replace with fresh diluted culture solution (0.5× SS, 0.1× MMR or 0.1× MBS) containing 0.1%

gentamicin. Make sure the transplanted cell has healed in place and is dividing (Fig. 2) (see Note 16).

11. Culture at 14–20°C overnight next to a dish containing sibling embryos of the same stage (see Note 14).
12. The next day, transfer embryos to fresh culture medium plus gentamicin in a clean agarose-coated Petri dish using a sterile glass Pasteur pipette. Culture until siblings reach desired stage of analysis (see Note 15).

3.5. Blastomere Explant Culture

1. Fill each well in the explant culture dish with diluted culture solution (0.5× SS, 0.1× MMR, or 0.1× MBS) containing 0.1% gentamicin (see Note 13).
2. Place an unlabeled embryo in a depression in a microsurgery dish, remove vitelline membrane, and dissect free the blastomere of interest, as described for donor cell transplantation (Subheading 3.4, steps 4–6). For retina precursors, it is very convenient to dissect free both midline D1.1 cells as a pair at the 16-cell stage. For better survival you can combine the same blastomere from more than one embryo into a single explant (see Note 17).
3. Pick up the blastomere or blastomere pair, with a sterile, glass Pasteur pipette, avoiding air bubbles and excessive pressure to prevent these very fragile cells from exploding. Place the tip of the Pasteur pipette under the surface of the culture solution in a well of the explant culture dish, and gently expel them. Use a hair loop to maneuver the blastomere(s) into the shallow depression. If you combine several blastomeres in one explant, push them together in a mound so they adhere together (see Note 18).
4. After about an hour, remove debris surrounding the healed explant with a sterile, glass Pasteur pipette.
5. Culture at 14–20°C overnight next to a dish containing sibling embryos of the same stage (see Note 14).
6. Transfer explants to fresh culture medium in a clean agarose-coated Petri dish using a sterile, glass Pasteur pipette. Culture until siblings reach desired stage of analysis (see Notes 18 and 19).

3.6. Cell Fate Analysis

1. Embryos or explants can be fixed and processed for whole mount or tissue section analysis of lineage label, gene expression, and immune-detection of protein expression, as described in ref. 3. The method of fixation and tissue processing will depend upon the lineage label used and the analysis to be performed.

4. Notes

1. The tips of forceps can be finely honed to desired shapes, and easily repaired if bent by gently stroking them across a piece of Alumina abrasive film (Thomas Scientific #6775E-38 (course; 12 μm), -46 (medium, 3 μm), -54 (fine, 0.3 μm)). Do this under the microscope for precision. The course film can be used for major repair and making square tips, the medium film is used for fine sharpening and the fine film is used for a final polish.
2. Benzocaine stock solution will form a white precipitate on the surface unless it is added drop wise with agitation. Use a dedicated container for anesthesia because benzocaine precipitate clings to the surfaces of the container.
3. The pH of the dejellying solution is critical for it to work properly. A lower pH will cause the dejellying process to take too long or be incomplete, and a higher pH will damage the embryos.
4. SS, MMR, and MBS are virtually interchangeable. Different labs prefer one to the other, mostly due to laboratory history. MBS has a shorter shelf life because it is buffered with bicarbonate.
5. In vitro fertilization is ideal for obtaining large numbers of embryos on demand, synchronized to the same stage of development. However, it requires sacrificing the male frog and the embryos do not always cleave in regular patterns that match the fate maps. Natural fertilization frequently produces regular cleavage patterns and was used for all the fate maps from the Moody lab (14–17). Natural fertilization provides developmental stages of embryos spread out over a long time period, which is advantageous when complex manipulations are planned. However, frogs do not always mate successfully on a time frame convenient to the experimenter's schedule.
6. Dejellying must be performed carefully. Do not dejelly prior to the appearance of the first cleavage or the dejellying solution will disable the sperm. Do not agitate the eggs, as this can cause polyspermy and very irregular cleavages; eggs should be gently swirled at intervals. Watch for signs that the jelly is falling off the eggs; the eggs will touch one another, rather than being separated by their coats. If a small amount of jelly remains, the eggs very sticky and nearly impossible to inject. However, leaving embryos in the solution too long will damage them.
7. Incubating embryos in Ficoll causes the vitelline membrane to collapse onto the cell surface. This is an advantage during

microinjection because it exerts pressure on the puncture hole, preventing leakage of cytoplasm. However, to perform the microsurgeries described in this chapter the vitelline membrane needs to be manually removed. The Ficoll incubation causes the vitelline membrane to collapse so tightly against the blastomeres that its manual removal without damaging the embryo is extremely difficult.

8. This procedure requires some practice and manual dexterity, and should be mastered before embarking on any of the more complicated microsurgeries. Since embryos removed from the vitelline membrane stick to Petri dish plastic, perform all manipulations on the smooth, slippery, inert surface of the 2% agarose bed.
9. Embryos normally are firmly held together by the vitelline membrane. Once this is removed, they relax into a flattened spheroid, which is accentuated when they are grown on a flat agarose bed. Placing them in small, rounded depressions melted into the agarose bed holds them in place during microsurgery and helps them maintain a more rounded shape.
10. Blastomere deletions are performed in culture media containing a lowered concentration of cations, compared to the injection media, in order to lessen the strength of cell-to-cell adhesions. Do not use calcium-/magnesium-free media because the cells become too fragile and leaky to manipulate. In some batches of eggs, the "victim" cell will lift out of the embryo cleanly with a single tug. In others, the "victim" cell will rip open and leak cytoplasm; this is not a problem as you can remove chunks of the "victim" cell with both forceps until all the pieces are removed. The ideal resulting embryo will have a single cell cleanly removed, with no leaking neighbors or oozing debris that could interfere with further development (Fig. 2b). If cells fall apart without tugging on them, increase the cation concentration; if they are impossible to separate, lower it.
11. There is a high mortality rate with 8-cell blastomere deletions, so be prepared to perform large numbers in order to get 10–20 successful cases.
12. Cellular debris contains proteases that will damage the unprotected embryos, and can foster harmful bacterial growth.
13. Without the jelly and vitelline protective membranes, embryos can succumb to bacterial infection. Solutions containing gentamicin have a short storage life, and thus should be made up fresh.
14. If you increased the cation concentration to promote healing, lower it to 0.5× SS (0.1× MMR, 0.1× MBS) for the overnight incubation to promote normal gastrulation movements.

Unmanipulated, sibling embryos serve to inform whether the microsurgery has delayed developmental progression.

15. Embryos without a vitelline membrane stick enthusiastically to plastic, so transfer them only with a glass pipette. Do not let them touch an air bubble or the surface of the solution; they explode at the air–water interface.
16. Transplanted embryos take longer to heal than deletion embryos, so the dish should not be moved until integration of the donor cell is visually confirmed. In an hour, it should have divided once or twice (Fig. 2c). Transplantation is not a simple procedure, and mortality is quite high. After a lot of practice, a 20–30% survival rate is outstanding.
17. We have had trouble successfully culturing single blastomeres older than the 16-cell stage with this method, but 16-cell or younger blastomeres or small groups (2–6 cells) of the same 32-cell blastomere grow quite well (12, 18).
18. Blastomeres avidly stick to plastic, so glass pipettes must be used. Transfer must be done with very little sucking on the cell or it will tear apart in the shear forces of the pipette. Avoid air bubbles in the pipette and make sure the tip is below the surface of the solution when you expel the cell; blastomeres explode if they touch the air–water interface.
19. These explants survive quite well even if some of the cells disintegrate. When changing the medium, explore the turbid mass at the bottom of the depression with a hair loop to determine whether there is a buried, healthy mass of cells that can be transferred to a new dish containing fresh culture solution.

Acknowledgment

This work was supported by NSF grant IOS-0817902.

References

1. Moore KB, Mood K, Daar IO, Moody SA (2004) Morphogenetic movements underlying eye field formation require interactions between the FGF and ephrinB1 signaling pathways. Dev Cell 6:55–67
2. Huang S, Moody SA (1993) The retinal fate of *Xenopus* cleavage stage progenitors is dependent upon blastomere position and competence: studies of normal and regulated clones. J Neurosci 13:3193–3210
3. Moody SA (2013) Targeted microinjection of synthetic mRNAs to alter retina gene expression in *Xenopus* embryos. In: Retinal Development: Methods and Protocols (S.Z. Wang, ed), Methods in Molecular Biology vol 884
4. Moore KB, Moody SA (1999) Animal-vegetal asymmetries influence the earliest steps in retinal fate commitment in *Xenopus*. Dev Biol 212:25–41
5. Kenyon KL, Zaghloul N, Moody SA (2001) Transcription factors of the anterior neural plate alter cell movements of epidermal progenitors to specify a retinal fate. Dev Biol 240: 77–91
6. Lee H-S, Bong Y-S, Moore KB, Soria K, Moody SA, Daar IO (2006) Dishevelled mediates

ephrinB1 signaling in the eye field via the planar cell polarity pathway. Nat Cell Biol 8:55–63

7. Zaghloul NA, Moody SA (2007) Changes in Rx1 and Pax6 activity at eye field stages differentially alter the production of amacrine neurotransmitter subtypes in *Xenopus*. Mol Vision 13:86–95. http://www.molvis.org/molvis/v13/a10/
8. Zaghloul NA, Moody SA (2007) Alterations of *rx1* and *pax6* expression levels at neural plate stages differentially affect the production of retinal cell types and maintenance of retinal stem cell qualities. Dev Biol 306:222–240
9. Yan B, Moody SA (2007) The competence of *Xenopus* blastomeres to produce neural and retinal progeny is repressed by two endo-mesoderm promoting pathways. Dev Biol 305:103–119
10. Huang S, Moody SA (1995) Asymmetrical blastomere origin and spatial domains of dopamine and neuropeptide Y amacrine cells in *Xenopus* tadpole retina. J Comp Neurol 360:2–13
11. Huang S, Moody SA (1997) Three types of serotonin-containing amacrine cells in the tadpole retina have distinct clonal origins. J Comp Neurol 387:42–52
12. Gallagher BC, Hainski AM, Moody SA (1991) Autonomous differentiation of dorsal axial structures from an animal cap cleavage stage blastomere in *Xenopus*. Development 112: 1103–1114
13. Sive HL, Grainger RM, Harland RM (2000) Early development of *Xenopus laevis*. A laboratory manual. Cold Spring Harbor Laboratory Press, Cold Spring Harbor, NY
14. Moody SA (1987) Fates of the blastomeres of the 16-cell stage *Xenopus* embryo. Dev Biol 119:560–578
15. Moody SA (1987) Fates of the blastomeres of the 32-cell stage *Xenopus* embryo. Dev Biol 122:300–319
16. Moody SA, Kline MJ (1990) Segregation of fate during cleavage of frog (*Xenopus laevis*) blastomeres. Anat Embryol 182:347–362
17. Moody SA (1989) Quantitative lineage analysis of the origin of frog primary motor and sensory neurons from cleavage stage blastomeres. J Neurosci 9:2919–2930
18. Hainski AM, Moody SA (1996) An activin-like signal activates a dorsal-specifying RNA between the 8- and 16-cell stages of *Xenopus*. Dev Genet 19:210–221
19. Hirose G, Jacobson M (1979) Clonal organization of the central nervous system of the frog. I. Clones stemming from individual blastomeres of the 16-cell and earlier stages. Dev Biol 71:191–202
20. Jacobson M, Hirose G (1981) Clonal organization of the central nervous system of the frog. II. Clones stemming from individual blastomeres of the 32- and 64-cell stages. J Neurosci 1:271–284

Chapter 8

Application of Cre-loxP Recombination for Lineage Tracing of Adult Zebrafish Retinal Stem Cells

Rajesh Ramachandran, Aaron Reifler, Jin Wan, and Daniel Goldman

Abstract

The Cre-loxP recombination system is widely used as a genetic tool to achieve conditional gene expression and for lineage tracing. Though extensively used in mice, this technology has only recently been applied to zebrafish. Here we describe Cre-loxP methodology for conditional expression of transgenes in zebrafish and their use in lineage tracing Müller glia as they undergo cellular reprogramming and proliferation to repair damaged retinal circuitry following mechanical injury. This methodology can be used for conditional gene expression and lineage tracing at any stage of development and in any cell type.

Key words: Zebrafish, Cre, Recombination, Lineage tracing, Retina, Regeneration, Stem cells, Müller glia, Conditional gene expression, Tubulin, Tamoxifen

1. Introduction

Effective design of a system that will label progenitor cells and remain persistent through their differentiation and maturation requires three key elements. First, there must be some way of targeting or identifying a particular progenitor population. Second, there must be a means of inducing a signal within the cells of that population that is specific to those cells, so as not to follow too broad a group, and also transient, so as to specify genesis both spatially and temporally. Third, and most important if one wants to trace the lineage of a particular progenitor, the reporter mechanism must be stable, innocuous and non-diffusible, so that cells initially labeled will maintain the label indefinitely. Conditional gene expression based on genetic recombination appears to fit these criteria as long as one has identified promoters that allow temporal and spatial control of gene expression.

The Cre-loxP recombination system is well suited for conditional gene expression and lineage tracing by stimulating targeted deletion,

Shu-Zhen Wang (ed.), *Retinal Development: Methods and Protocols*, Methods in Molecular Biology, vol. 884, DOI 10.1007/978-1-61779-848-1_8, © Springer Science+Business Media, LLC 2012

insertion, inversion, and exchange of chromosomal DNA (1, 2). Cre is a recombinase initially identified from bacteriophage P1 and is a member of the integrase family of recombinases that recognizes a specific 34-bp DNA sequence called loxP. The loxP sequence contains an 8-bp core (for providing orientation) flanked by 13 bp of complementary, palindromic DNA. Because loxP sites are absent from vertebrate genomes they can be exploited for Cre-mediated targeted recombination using transgenic approaches. Further improvements to the Cre-loxP system include the generation of a ligand-dependent chimeric Cre recombinase where Cre is fused to the mutant ligand-binding domain of the human estrogen receptor (CreER^{T2}) (3). CreER^{T2} is efficiently activated by the synthetic estrogen receptor ligand 4-hydroxytamoxifen (4-OHT) and provides improved temporal control over Cre-mediated recombination.

We utilized the CreER^{T2}-loxP system to investigate the multipotent character of Müller glia-derived progenitors during retina regeneration (4). To accomplish this we developed transgenic fish harboring a truncated *α1tubulin* promoter (*1016tuba1a*) driving CreER^{T2} expression that is activated specifically in dedifferentiated Müller glia in the adult retina upon injury (4, 5). To permanently label the dedifferentiated Müller glia and follow their fate as they repair the damaged retina, we developed another transgenic fish harboring a recombination reporter that contained the promiscuous *β-actin* promoter driving the expression of the mCherry fluorescent protein whose DNA sequence was flanked by loxP sites and followed by an out-of-frame enhanced green fluorescent protein (EGFP) expression cassette. After retinal injury, Müller glia at the injury site expressed CreER^{T2} which was activated by tamoxifen or 4-OHT exposure and induced reporter recombination, resulting in *mCherry* deletion and consequently brought *egfp* into the reading frame. Thus the permanent expression of GFP could be followed through reprogramming, proliferation, and redifferentiation into mature neural cell types.

In this chapter, we discuss the detailed procedures for using the CreER^{T2}-loxP system for tracing the lineage of transiently induced retinal stem cells after an injury. This system can be adapted, with appropriate promoters and recombined transgenes, to study the fate of other cells and for conditional expression of any gene product to gain insight into its function.

2. Materials

2.1. Transgenic Fish Lines

1. *1016tuba1a:CreER^{T2}* transgenic line harbors the 1,016 bp *tuba1a* promoter driving CreER^{T2} expression and the 906-bp *tuba1a* promoter driving cyan fluorescent protein (CFP) expression to facilitate identification of transgenic fish (see Note 1).

2. *β-actin2:loxP-mCherry-loxP-EGFP* transgenic line harbors 3,851 bp of 5′ flanking DNA, exon 1, and intron 1 of the zebrafish β-*actin2* gene fused in frame with *loxP* flanked *mCherry* and followed by an out-of-frame *egfp* sequence (see Note 2). The mCherry reporter allows identification of transgenic fish.

2.2. Retinal Injury and Dissection

1. Tricaine methanesulfonate: For 10× stock solution, dissolve 200 mg tricaine methanesulfonate in 90 ml deionized water. Mix 0.14 g Tris base in 2 ml of water and add drop-wise to tricaine solution until pH reaches ~7. Make up the volume to 100 ml with deionized water.
2. 30G needle.
3. Stainless steel forceps-5F.
4. Surgical blade.
5. 70% Alcohol.
6. Sterile plastic Petri dish.
7. Surgical grade fine scissors.
8. Sponge bed for anesthetized fish.

2.3. Recombination

1. 10 mM 4-hydroxytamoxifen (4-OHT): 3.875 mg of 4-hydroxy tamoxifen is dissolved in 1 ml of ethanol. Store at −20°C (see Note 3).
2. 10 mg/ml Tamoxifen: 5 mg of Tamoxifen is dissolved in 20 μl of ethanol and then 480 μl of sunflower oil is added.
3. 10-μl Hamilton syringe: sterilize with alcohol and rinse with sterile water before use.
4. Fluorescence microscope: we use a Zeiss Axiophot fluorescence microscope equipped with digital camera or Olympus FluoView FV1000 confocal imaging system.
5. Dissection microscope: we use a Stemi DV4 Zeiss dissecting microscope.

2.4. Tissue Cryoprotection and Sectioning

1. 1 M phosphate buffer (10× PB): dissolve 13 g of $NaH_2PO_4 \cdot H_2O$ and 103 g of $Na_2HPO_4 \cdot 7H_2O$ in 350 ml of deionized water. Adjust pH to 7.4 with 2 N HCl (if required). Adjust to a final volume of 500 ml with deionized water (see Note 4).
2. 10× Phosphate buffered saline (10× PBS): dissolve 2.76 g of $NaH_2PO_4 \cdot H_2O$, 11.36 g of Na_2HPO_4, 87.6 g of NaCl and 1.87 g of KCl in 800 ml of deionized water. Adjust to 1 L with deionized water (see Note 5).
3. 4% Paraformaldehyde (PFA): dissolve 4 g of PFA in 80 ml of 1× PB and heat at 65°C for approximately 1 h until completely dissolved. Cool to room temperature and adjust pH to 7.4

with 2 N HCl and bring to a final volume of 100 ml with 1× PB (see Note 6).

4. 5% Sucrose (wt/vol): dissolve 5 g of sucrose in 100 ml of 1× PB (see Note 7).
5. 20% Sucrose (wt/vol): dissolve 20 g of sucrose in 100 ml of 1× PB (see Note 7).
6. OCT cryostat embedding medium (Tissue Tek).
7. Superfrost slides (Fisher Scientific).

2.5. Immunodetection

1. 20 mM Bromodeoxy uridine (BrdU): dissolve 303.5 mg of BrdU in 50 ml of 5% DMSO. Store at 4°C (see Note 8).
2. Primary and secondary antibodies: Rabbit anti-GFP polyclonal antibody (Invitrogen); Rat anti-BrdU monoclonal antibody (Abcam); Mouse anti-Zpr1 (Double-cone photo receptor cell) monoclonal antibody (Zebrafish International Resource Center); Mouse anti-glutamine synthetase (Muller glia cell) monoclonal antibody (Chemicon/Millipore); Mouse anti-HuC/D (Amacrine cell) monoclonal antibody (Invitrogen); Mouse anti-Zn5 (differentiating ganglion cell) monoclonal antibody (Zebrafish International Resource Center); Goat anti-PKCβ-1 (Bipolar cell) polyclonal antibody (Santa Cruz Biotechnology); Donkey anti rabbit IgG-alexa488 conjugate (Invitrogen); Donkey anti mouse IgG-alexa555 conjugate (Invitrogen); Donkey anti goat-Cy3 conjugate (Invitrogen); and Donkey anti rat-AMCA conjugate (Jackson Immunoresearch, USA) (see Note 9).
3. Blocking solution: 3% donkey serum with 0.1% Triton X-100 in 1× PBS.
4. 10 mg/ml DAPI nuclear stain (100×): dissolve 10 mg of DAPI into 1 ml of deionized water.
5. CoverWell incubation chamber (RPI).
6. 2 N HCl: in a 50-ml falcon tube containing 41.7 ml of deionized water slowly add 8.3 ml of concentrated (12 N) HCl (see Note 10).
7. Sodium borate: dissolve 3.81 g of sodium borate in 100 ml of deionized water. Adjust to pH 8.0 with 2 N HCl.

3. Methods

3.1. Establishment of Transgenic Zebrafish

1. Inject approximately 1 nl of a mixture of transgene plasmid (20 μg/ml) and in vitro transcribed Tol2 transposase mRNA (20 μg/ml) into fertilized zebrafish embryos at the 1–2-cell stage using a micropipette and micromanipulator and visualized using a dissecting

microscope, to create transgenic fish of *1016tuba1a:CreER*^{T2} and β*-actin2:loxP-mCherry-loxP-EGFP* using Tol2 transposase-mediated integration of the transgenes (6).

2. Raise injected fish that exhibit transgene expression as embryos to adults, breed with wild-type fish, and their embryos are screened for transgene expression, which indicates germ-line integration of the transgene. We generally screen 200–300 embryos before concluding a fish is negative.

3. Raise a number of different lines of *1016tuba1a:CreER*^{T2} and β*-actin2:loxP-mCherry-loxP-EGFP* transgenic fish to adults and then breed with each other to test for basal Cre activity and strong reporter activity after 4-OHT or tamoxifen treatment (see Subheadings 3.2 and 3.3 below). When working with novel lines it is best to examine them for basal recombination and for 4-OHT or tamoxifen-induced recombination. DNA integration sites have a dramatic affect on transgene expression levels and selecting optimal lines is important for unambiguous interpretation of experimental results (see Note 11).

3.2. Retinal Injury

1. Select healthy adult double transgenic fish that harbor both *1016tuba1a:Cre-ER*^{T2} and β*-actin2:loxP-mCherry-loxP-EGFP* transgenes (see Note 11).

2. Anesthetize the fish in 50 ml of 0.02% Tricaine Methanesulfonate in a glass beaker until the animal loses its balance and is unresponsive to touch (see Note 12).

3. Place the anesthetized fish on a moist sponge bed under the dissecting scope. Gently pull the right eye from its socket using fine forceps exposing the back of the eye and stab four times (once in each retinal quadrant) through the sclera with a 30 G needle. Take care to insert the needle only up to the length of the bevel to avoid any undesirable damage to the eye (see Note 13).

3.3. Recombination

1. At the time of retinal injury, inject 0.5–1 μl of 4-OHT (10 μM in 50% ethanol) into the eye of anesthetized *1016tuba1a:Cre-ER*^{T2};β*-actin2:loxP-mCherry-loxP-EGF* fish by attaching a Hamilton syringe to a 30 G needle. The injury and 4-OHT injection are performed by inserting the needle to the length of the bevel as described in Subheading 3.2 (see Note 14).

2. Alternatively, recombination is effected through intraperitoneal injection of 10 μl of tamoxifen (10 mg/ml) into anesthetized fish every day for 4 days starting on the day of retinal injury.

3. Alternatively, recombination can be stimulated by immersing fish in fish water containing 1 μM 4-OHT for at least 12 h at 28°C (this is most convenient for assaying recombination in embryos and young fry) (Fig. 1).

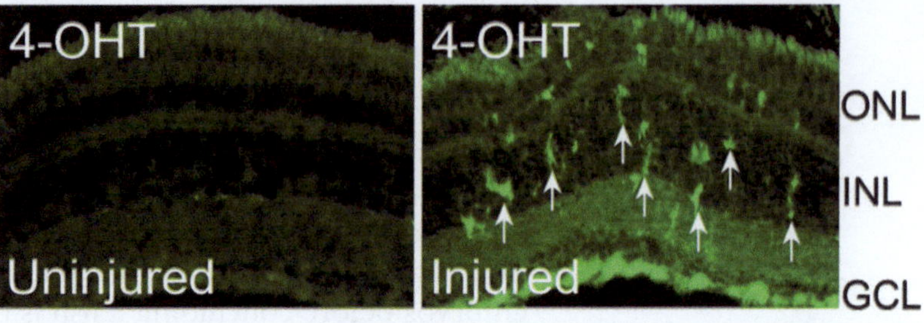

Fig. 1. 4-Hydroxytamoxifen (4-OHT) and injury-dependent recombination in the retina. Adult *1016 tuba1a:Cre-ER*^{T2};β-*actin2:loxP-mCherry-loxP-EGFP* transgenic fish were anesthetized and retinas were left uninjured or injured by a needle poke. Twenty hours later fish were immersed in 4-OHT for 3 days with twice daily changes to fresh water and intermittent feeding. Fish were then returned to their tanks. Retinas were harvested 1 week later and retinal sections stained with anti-GFP antibody.

3.4. BrdU Labeling

1. On day 4 post injury, anesthetize the fish in 0.02% tricaine methanesulfonate and give a 25 μl intraperitoneal injection of BrdU (20 mM) using an insulin syringe; 3 h later, euthanize the fish and harvest the retina as described in Subheading 3.5 (see Note 15).
2. Alternatively, proliferating cells can be labeled by immersing fish in fish water containing BrdU to a final concentration of 5 mM for at least 12 h at 28°C (see Note 15).

3.5. Whole Eye and Retina Dissection

1. Harvest eyes from treated zebrafish over the course of several weeks at multiple time points. If retinas are dissected from the eye it is best to dark adapt the fish for at least 3 h prior to surgery.
2. Overdose fish with 0.2% tricaine methanesulfonate before dissecting the eye.
3. Sever the extraorbital muscles surrounding the eye using fine scissors. Cut the optic nerve behind the eye before pulling the eyeball out of its socket with a forceps.
4. Place the eye in a Petri dish containing 4% PFA [step 6(a)] or PBS [step 6(b)].
5. Make an incision in the cornea near the lens using a surgical blade and then, using a forceps, gently remove the lens.
6. (a) For preparation of tissues to be cryoprotected and sectioned, immediately transfer the remaining eye to a 1.5-ml microcentrifuge tube containing 4% PFA at 4°C (see Note 16). (b) If the retinal tissue is to be used for RNA extraction, the dissection should be carried out in sterile PBS. Separate the retina from the sclera with fine forceps. Tease the retina, which is pale yellow in color, away from the pigment epithelium and store in Trizol reagent (Invitrogen) (see Note 17).

3.6. Tissue Sectioning

1. Fix dissected eyes in 4% PFA overnight at 4°C with gentle shaking on a rocking platform or rotator (see Note 18).
2. Cryoprotect samples in graded sucrose solutions. Prepare 5 and 20% stock sucrose solutions in 1× PB. Drain PFA from the retinas contained in the 1.5-ml microcentrifuge tube and replace with 5% sucrose at room temperature for 30 min. This is followed by three steps of increasing concentration of sucrose by mixing 5 and 20% sucrose at a ratio 2:1, 1:1, and 1:2 for 45 min each at room temperature.
3. Perform one final cryoprotection in 20% sucrose overnight at 4°C. Eyes are then mixed with a 2:1 solution of 20% sucrose and OCT for 30 min before freezing (see Note 19).
4. Place cryoprotected eyes into shallow molds containing OCT. Molds of about 1-ml volume can be made with aluminum foil wrapped about electroporation cuvettes and filled with approximately 1 ml of OCT. Orientate eye so the lens hole is facing up with a dorsal side facing one wall of the mold. This wall can be identified by a paper flag (which is used to label details of the experiment) inserted into the OCT at room temperature (see Note 20).
5. Cool 50 ml of methyl butane in a stainless steel beaker on top of several pieces of dry ice for about 10–15 min (see Note 21). Using a forceps or tongs, slowly lay the OCT-filled mold containing the cryoprotected eye onto the surface of the dry ice-cooled methyl butane. Hold for 2–3 min until the entire transparent OCT freezes into an opaque milky white block with the eye buried within. This block can be stored several months in −80°C freezer in a 50-ml Falcon centrifuge tube (see Note 22).
6. Remove the frozen OCT containing the cryoprotected eye from the mold and mount in a cryostat at −20°C. Section tissue at 8–10 μm and collect on Superfrost slides. Air-dry the slides overnight at room temperature and store at −20°C for up to 3 weeks.

3.7. Immunostaining

1. Dry slides with tissue sections at 37°C for 1 h before proceeding to immunostaining (see Note 23). The dried sections are incubated with 1× PBS (1 ml per slide) for 10 min at room temperature.
2. Immerse slides in blocking solution for 30 min at room temperature (see Note 24).
3. Dilute primary antibody (1:1,000 for GFP, 1:500 for all other antibodies) in 1% donkey serum, 0.1% triton X-100, 1× PBS, place on slides and cover with a CoverWell incubation chamber (see Note 25). Place slides in a humid chamber (air-tight

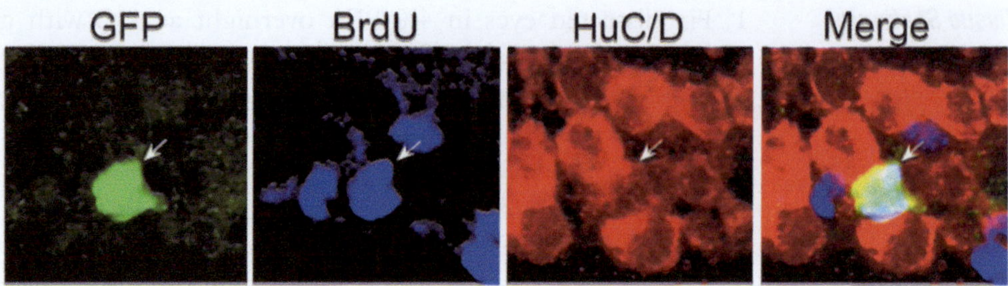

Fig. 2. Lineage tracing shows HuC/D+ amacrine cells are regenerated from Müller glia in the injured retina. Adult *1016 tuba1a:Cre-ER^{T2};β-actin2:loxP-mCherry-loxP-EGFP* transgenic fish were anesthetized and retinas were injured by a needle poke and simultaneously received an injection of 4-OHT. At 4 days postinjury fish received a single intraperitoneal injection of BrdU and 3 weeks later sacrificed. Retinal sections were processed for immunofluorescence to detect: (1) the recombination marker GFP; (2) the proliferation marker BrdU; and (3) the amacrine cell marker HuC/D.

container with wet paper towels and a track of glass/plastic rods on which slides are placed) overnight at 4°C.

4. Remove the primary antibody and wash slides three times, 10 min each in 1× PBS containing 1% donkey serum and 0.1% triton X-100. Dilute secondary antibody (1:1,000 for all antibodies except anti-rat AMCA antibody which is 1:250 dilution) in 1% donkey serum, 0.1% triton X-100, 1× PBS (see Note 26). Apply the secondary antibody mixture to the slides and incubate at room temperature in a humid chamber for 2 h. Remove the secondary antibody and wash three times in 1× PBS containing 0.1% triton X-100 (no donkey serum this time) for 10 min each. If the AMCA secondary is not used, the slides can be washed once in 1× PBS and nuclei stained with DAPI (10 ng/ml) for 2 min. Wash twice in deionized water and air-dry for cover slipping (see Note 27).

5. If sections are to be stained for BrdU immunofluorescence, overlay the slide with 1 ml of 4% PFA for 2 min at room temperature (see Note 28). Wash in 1× PBS and perform epitope retrieval by placing slides in a coplin jar containing 2 N HCl prewarmed to 37°C and incubate at 37°C for 20 min (see Note 29). Remove slides from 2 N HCl and overlay with 1 ml of 100 mM sodium borate, pH 8.0 for 10 min. Repeat sodium borate incubation for an additional 10 min. Wash in 1× PBS for 5 min (see Note 30). Dilute the primary anti-BrdU antibody 1:500 and follow steps 3 and 4 above.

6. Observe slides under a fluorescence microscope for cell type-specific or BrdU immunofluorescence (Fig. 2).

4. Notes

1. The *1016tuba1a* promoter provides retinal progenitor specificity to CreER^{T2} expression which is crucial to tracing the lineage of progenitors during regeneration. Because the CreER^{T2} transgene was not fused to a fluorescent reporter, we chose to insert a second transgene harboring the 906-bp *tuba1a* promoter driving CFP expression into the first intron of the *1016tuba1a:CreER*^{T2} construct to facilitate the identification of transgenic fish (4). Alternatively, PCR could be used to screen for transgenic fish. There are also a number of different Cre fish lines available (7–16).
2. We used the constitutively active and promiscuous *β-actin2* promoter. This can serve as a universal reporter for recombination in any cell type. However, other ubiquitous or cell type-specific promoters can be used depending on the goal (see refs. 7–16). *β-actin2:loxP-mCherry-loxP-EGFP* transgenic fish use EGFP to report recombination, but this could be any gene of interest to study gene function.
3. Tamoxifen and 4-OHT are to be handled with extreme care. They can act as estrogen analogs possibly causing undesirable effects in humans. Wear gloves and a lab coat when handling these compounds and follow OSEH guidelines for disposal.
4. 10× PB is stable at room temperature for several weeks. For prolonged storage it is best to autoclave the solution.
5. For long-term storage keep 10× PBS at 4°C.
6. 4% PFA needs to be made with extreme care as the fumes can cause irritation to the eyes and nose. Best to work in a fume hood when dissolving PFA. Care should be taken to avoid contact with skin. PFA is most soluble when the solution is basic. Add a few drops of 10 N NaOH to the PFA, 1× PB mixture to facilitate dissolving the PFA. Be sure to adjust pH back to 7.4 after cooling to room temperature. This solution can be stored in aliquots frozen at −20°C for several months. The solution remains stable at 4°C for 1 week.
7. Sucrose solutions can be stored at 4°C for several weeks and care should be taken not to contaminate the stock as bacteria/fungal molds can grow easily in them.
8. BrdU is a nucleotide analog. Wear gloves and lab coat when working with BrdU. Dispose of this solution according to OSEH recommendations.
9. It is desirable not to freeze primary or secondary antibodies. However, if they need to be stored for a long period of time, they can be diluted into 50% glycerol and stored in small aliquots at −20°C.

10. Caution: wear gloves, lab coat, and goggles when working with HCl.
11. Embryos can be checked for recombination by immersing in 4-OHT (1 μM) for 12 h and recombination (EGFP expression) assayed by fluorescence microscopy. It is also important to check untreated embryos for basal recombination. For adults we generally assay for recombination in 1-year-old fish +/– retinal injury. Recombination can also be verified by RT-PCR using retinal RNA and primers flanking the loxP sites.
12. Tricaine methanesulfonate should be stored in 4°C to avoid any contamination or degradation. Care should be taken not to overdose the fish as it may cause death. Tricaine methanesulfonate can be toxic to humans if exposed in large quantities. Precaution should be taken not to expose directly to skin. Wear gloves and lab coat when working with tricaine methanesulfonate.
13. While injuring the retina, care should be taken not to cut the blood vessel or optic nerve emanating from the back of the eye. A small amount of bleeding, due to needle poke, is common and does not severely affect regeneration. Surgical tools are sterilized in 70% alcohol. Retinal injury can also be induced with bright light and cytotoxic agents (17–19).
14. We have found similar results injecting 4-OHT at 2 and 4 days post injury as at the time of injury. For inducing recombination by immersion in 4-OHT containing water we use a final concentration of 1 μM and protect the solution from light by wrapping in tin foil or placing in a 28°C incubator. Immersion in 4-OHT was less effective than intravitreal injection at inducing recombination in the retina of adult fish but worked fine for inducing recombination in the olfactory pits and other surface structures. Immersion in 4-OHT works fine for embryos and small fry. Recombination can be assayed by fluorescence microscopy and by RT-PCR using retinal RNA and primers flanking the loxP sites.
15. Approximately, 3 h of BrdU exposure is sufficient to label proliferating cells; however, longer exposures can be done if the experiment demands such a regime. If immersing fish in BrdU water is necessary, we recommend at least 10 ml of water/fish with water change every 12 h. We have had difficulty keeping fish alive when immersed in a solution containing both BrdU and 4-OHT. If both of these compounds need to be delivered simultaneously we recommend that they are delivered by different routes (i.e. eye or intraperitoneal injection of at least one of the compounds).
16. Care should be taken not to damage the retina while removing the lens. This can be avoided by a quick and gentle incision, which can be widened by scissors to help ease extraction of the lens.

17. Rapid dissection and immersion of retinas in Trizol ensures RNA stability. Care should be taken to remove the pigment epithelium completely. Removal of pigment epithelium is facilitated by putting fish in complete darkness (a dark cabinet) for 3 or more hours prior to dissection. Dissections are also performed in low light conditions to facilitate removal of the pigment epithelium.
18. Excessive fixing in 4% PFA can increase autofluorescence and result in poor immunostaining. Fixing in 4% PFA can be reduced to 2 h at room temperature, with gentle shaking if required.
19. Gradual cryoprotection by increasing concentration of sucrose is desirable for maintaining retinal anatomy. The incubation in 20% sucrose and OCT is crucial for preventing separation of the retinal section from the OCT during sectioning.
20. Orienting the eye ball in a dorsal–ventral fashion and cutting from either the ventral or dorsal end of the eye ensure that all tissue sections harboring an injury site will contain normal tissue flanking the site of injury.
21. Methyl butane should be cooled to dry ice temperature before immersing the OCT-filled tissue block. The temperature can be checked my putting a small piece of dry ice into the methyl butane; if no bubbling is observed it has reached the right temperature.
22. Retina blocks should be stored in an air-tight container as excessive drying can cause rigidity to the block causing difficulty in sectioning and damage to the tissue.
23. Slides must be dried before antibody staining. The condensation of water on sections immediately after removal from freezer can cause peeling of sections.
24. Blocking in donkey serum reduces background fluorescence. If required, blocking can be done at 4°C for 2 h.
25. In general we use manufacturer's instructions as a guide for antibody dilutions; however, these dilutions may need to be determined empirically to obtain optimum results. Care should be taken not to pour solutions directly on to the sections as this may cause peeling from the glass surface.
26. Secondary antibodies should be centrifuged before use to remove particulate matter that may increase background.
27. Air-drying is done in the dark to preserve fluorescence. This step is important since water droplets can affect mounting and cause freeze damage when stored at −20°C.
28. Secondary PFA fixing is important before acid treatment. Skipping this step can cause drastic reduction in the fluorescence signal.
29. Prolonged exposure (beyond 20 min) of tissue samples with 2 N HCl can cause damage to the nuclear architecture.

This will be reflected by indistinct nuclear boundaries in DAPI-stained sections.

30. Sodium borate ensures neutralization of the sections. Final wash in 1× PBS prepares the sections for smooth transition to the antibody. Blocking in 3% donkey serum is unnecessary in this step.

Acknowledgments

This work was supported by NIH grant NEI RO1 EY018132.

References

1. Branda CS, Dymecki SM (2004) Talking about a revolution: the impact of site-specific recombinases on genetic analyses in mice. Dev Cell 6:7–28
2. Nagy A (2000) Cre recombinase: the universal reagent for genome tailoring. Genesis 26:99–109
3. Indra AK et al (1999) Temporally-controlled site-specific mutagenesis in the basal layer of the epidermis: comparison of the recombinase activity of the tamoxifen-inducible Cre-ER(T) and Cre-ER(T2) recombinases. Nucleic Acids Res 27:4324–4327
4. Ramachandran R, Reifler A, Parent JM, Goldman D (2010) Conditional gene expression and lineage tracing of tuba1a expressing cells during zebrafish development and retina regeneration. J Comp Neurol 518:4196–4212
5. Fausett BV, Goldman D (2006) A role for alpha1 tubulin-expressing Muller glia in regeneration of the injured zebrafish retina. J Neurosci 26:6303–6313
6. Urasaki A, Morvan G, Kawakami K (2006) Functional dissection of the Tol2 transposable element identified the minimal cis-sequence and a highly repetitive sequence in the subterminal region essential for transposition. Genetics 174:639–649
7. Collins RT, Linker C, Lewis J (2010) MAZe: a tool for mosaic analysis of gene function in zebrafish. Nat Methods 7:219–223
8. Feng H et al (2007) Heat-shock induction of T-cell lymphoma/leukaemia in conditional Cre/lox-regulated transgenic zebrafish. Br J Haematol 138:169–175
9. Hans S, Kaslin J, Freudenreich D, Brand M (2009) Temporally-controlled site-specific recombination in zebrafish. PLoS One 4:e4640
10. Liu X, Li Z, Emelyanov A, Parinov S, Gong Z (2008) Generation of oocyte-specifically expressed Cre transgenic zebrafish for female germline excision of loxP-flanked transgene. Dev Dyn 237:2955–2962
11. Wang L et al (2008) Functional characterization of Lmo2-Cre transgenic zebrafish. Dev Dyn 237:2139–2146
12. Mosimann C et al (2011) Ubiquitous transgene expression and Cre-based recombination driven by the ubiquitin promoter in zebrafish. Development 138:169–177
13. Le X et al (2007) Heat shock-inducible Cre/Lox approaches to induce diverse types of tumors and hyperplasia in transgenic zebrafish. Proc Natl Acad Sci USA 104:9410–9415
14. Wang Y, Rovira M, Yusuff S, Parsons MJ (2011) Genetic inducible fate mapping in larval zebrafish reveals origins of adult insulin-producing beta-cells. Development 138: 609–617
15. Hans S et al (2011) Generation of a non-leaky heat shock-inducible Cre line for conditional Cre/lox strategies in zebrafish. Dev Dyn 240: 108–115
16. Jopling C et al (2010) Zebrafish heart regeneration occurs by cardiomyocyte dedifferentiation and proliferation. Nature 464:606–609
17. Montgomery JE, Parsons MJ, Hyde DR (2010) A novel model of retinal ablation demonstrates that the extent of rod cell death regulates the origin of the regenerated zebrafish rod photoreceptors. J Comp Neurol 518:800–814
18. Kassen SC et al (2007) Time course analysis of gene expression during light-induced photoreceptor cell death and regeneration in albino zebrafish. Dev Neurobiol 67:1009–1031
19. Fimbel SM, Montgomery JE, Burket CT, Hyde DR (2007) Regeneration of inner retinal neurons after intravitreal injection of ouabain in zebrafish. J Neurosci 27:1712–1724

Chapter 9

Fate Tracing of *neurogenin2*-Expressing Cells in the Mouse Retina Using CreER™: LacZ

Wenxin Ma and Shu-Zhen Wang

Abstract

Delineating the final fate of progenitor cells that transiently express a regulatory gene may shed light on how the gene participates in regulating retinal development. We describe the steps in tracing final fates of progenitor cells that once transiently express *neurogenin2* (*ngn2*) during mouse retinal development with the binary, conditional Ngn2-CreER™—LacZ reporter system. Ngn2-CreER™ mice (Zirlinger et al. Proc Natl Acad Sci USA 99:8084–8089, 2002), in which *ngn2* promoter drives the expression of Cre-estrogen receptor CreER™ (Littlewood et al. Nuc Acid Res 23:1686–1690, 1995; Hayashi and McMahon Dev Biol 244:305–318, 2002), are crossed with Rosa26-LoxP-LacZ reporter mice (Soriano Nat Genet 21:70–71, 1999), in which the expression of *lacZ* requires the removal of "stop" by Cre recombinase (Wagner et al. Transgenic Res 10:545–553, 2001). 4-hydroxytamoxifen (4-OHT), a synthetic ligand with high affinity for ER™, is administered to double transgenic embryos and/or neonatal mice. Binding of 4-OHT to Cre-ER™ activates Cre recombinase, which then catalyzes the removal of the "stop" sequence from the LoxP-LacZ transgene, leading to *lacZ* expression in cells that express *ngn2*. Retinal tissues are fixed at different time points after 4-OHT treatment and analyzed for LacZ activities by colorimetric reaction. Double-labeling with a cell type-specific marker can be used to define the identity of a LacZ$^+$ cell. Combining persisted *lacZ* expression through the life of the cell and the short half-life (0.5–2 h) of 4-OHT (Danielian et al. Curr Biol 8:1323–1326, 1998), this system offers the opportunity to track the final fates of cells that have expressed *ngn2* during the brief presence of 4-OHT administered during retinal development.

Key words: Retinal neurons, Progenitor cells, Differentiation, Cell fate tracing, Cre-LoxP, Cre-ER™, ROSA26

1. Introduction

One problem often encountered in studying genes that regulate cell fate specification is their elusive association with the cell type that they regulate, because of their expression often being confined to progenitor/precursor cells and becoming no longer detectable once the cells have differentiated sufficiently to exhibit their terminal

Shu-Zhen Wang (ed.), *Retinal Development: Methods and Protocols*, Methods in Molecular Biology, vol. 884, DOI 10.1007/978-1-61779-848-1_9, © Springer Science+Business Media, LLC 2012

identities. With the development of Cre-mediated conditional DNA recombination (1–3), a bipartite genetic-system based lineage-tracing emerged and has been used to produce valuable information on the final fates of progenitor cells that transiently express a gene of interest. We have used the binary, conditional Ngn2-CreER™—LacZ reporter system to map the fate of *ngn2*-expressing cells in the developing mouse retina (4). In Ngn2-CreER™ mice, generated by Zirlinger et al. (5), the coding sequence of *ngn2* is replaced with CreER™. CreER™ is a fusion protein of Cre recombinase with the hormone-binding domain of a mutant mouse estrogen receptor ER™ (6, 7), which fails to bind the naturally released ligand 17b-estradiol at normal concentrations but retains relatively high affinity for the synthetic ligand 4-hydroxytamoxifen (4-OHT) (6–8). For LacZ reporter mice we used ROSA26, which was created by transgenesis with a construct of LoxP-stop cassette/Neo-LoxP-LacZ (9). In ROSA26 mice, *lacZ* is not expressed and its expression requires Cre-mediated removal of the "stop" sequence from the DNA cassette (2).

Because Ngn2-CreER™ homozygotes often die soon after birth, heterozygous Ngn2-CreER™ $^{+/-}$ mice are crossed with homozygous ROSA26-LacZ reporter mice. 4-OHT is administered to double transgenic embryos (through pregnant mice) and to neonates to activate Cre recombinase in *ngn2*-expressing cells. The activated Cre recombinase removes the "stop" sequence, a cell-heritable DNA rearrangement event, from the LoxP-stop cassette/Neo-LoxP-LacZ, leading to the persistent activation of the reporter LacZ in the progeny of cells transiently expressing *ngn2*. Since the 4-OHT has a half-life of only 0.5–2 h in vivo (5, 10), the activation of CreER™ is further restricted to a relatively narrow time window after the ligand is injected. Thus, the final expression pattern of the *lacZ* will exclusively identify the progeny of cells expressing *Ngn2-CreER™* during this time window, and will not include cells that express *Ngn2* at either earlier or later times (5).

2. Materials

Prepare all solutions using ultrapure water and analytical grade reagents. Prepare and store all reagents at room temperature (unless indicated otherwise). Adhere to policies and regulations when disposing waste materials. We do not add sodium azide to solutions prepared in our own laboratory.

2.1. Mice

Ngn2-CreER™ mice: provided by Dr. David Anderson (Caltech); Rosa26-LoxP-LacZ reporter mice: obtained from the Jackson Laboratory (Stock #003310, 129s-Gt(ROSA)26Sor) (see Note 1).

2.2. Reagents for Genotyping

1. Digestion buffer. Add 5 ml of 5 M NaCL, 2.5 mL of 1 M Tris–Cl (pH 8.0), 50 mL of 0.5 M EDTA (pH 8.0), 5 mg of pancreatic RNase, 12.5 mL of 10% SDS, and adjust volume to 250 mL with H_2O. Store at 4°C.
2. Proteinase K: 20 mg/mL in PBS. Store at −20°C.
3. 10 M ammonium acetate (NH_4Ac).
4. Phenol chloroform (pH 8.0): Phenol:chloroform:isoamyl alcohol mixture(25:24:1) (see Note 2).
5. Ethanol: 100% and 70% ethanol.
6. 5 M betaine.
7. 10× TE. Add 100 mL of 1 M Tris–HCl, pH 7.5, 20 mL of 0.5 M EDTA, pH 8.0, 880 mL of H_2O, and mix. TE: mix 100 mL of 10× TE with 900 mL of H_2O. Aliquot into 15-mL and 50-mL tubes.
8. 0.5 M EDTA, pH 8.0. Add 186.1 g of disodium ethylenediamine tetraacetate·$2H_2O$ to 800 mL of H_2O, stir vigorously, and adjust the pH to 8.0 with NaOH. EDTA will slowly dissolve as the pH approaches 8.0. Adjust the volume to 1 L with H_2O and sterilize the solution by autoclaving.
9. TAE buffer (50×). Add 121 g of Tris base into 250 mL of H_2O, stir to dissolve, add 28.6 mL of 1 N acetic acid, 50 mL of 0.5 M EDTA (pH 8.0), and adjust the volume to 500 mL with H_2O.
10. Loading buffer (6×) (Sigma).
11. Ethidium bromide (Sigma).
12. 1% Agarose gels. Add 0.75 g of agarose to 75 mL of 1× TAE buffer and microwave for 2–3 min. Cool the melted agarose for 5 min at room temperature. Add 5 μL of ethidium bromide, mix by swirling, and pour into a mid-size gel casting tray with comb. Let gel solidify at room temperature for about 20–30 min (see Note 3).
13. HotStartTaq Plus Master Mix Kit (1000) (Qiagen) (see Note 4).
14. 1 kb DNA ladder mix.
15. Oligonucleotides. Ngn2-CreER™: Cre: 5′ atccgaaaagaaaacgttga 3′ and 5′ atccaggttacggatatagt 3′ (see Note 5); ngn2: 5′ ctgagactctggagttgaag 3′ and 5′ ctagatacagtccctggcg 3′ (see Note 6). ROSA26: 5′ gcgaagagtttgtcctcaacc 3′, 5′ aaagtcgctctgagttgttat 3′, and 5′ ggagcgggagaaatggatgtg 3′ (see Note 7). Internal control GAPDH (glyceraldehyde-3-phosphate dehydrogenase): 5′ catgaccacagtccatgccat 3′ and 5′ cctcttgctcagtgtcctt 3′ (see Note 8).

2.3. Materials for Tissue Preparation

1. 4-OHT solution (10 mg/mL). Dissolve 25 mg of 4-OHT with 500 μL ethanol first, then mix dissolved 4-OHT with

2 mL of autoclaved sun flower seed oil (Sigma, Cat#: S5007) to 10 mg/mL, and sonicate for approximately 20 min. Aliquot out, wrap with foil, and store in –20°C. Thaw the solution completely before using (see Note 9).

2. Specimen molds (Tissue-Tek).
3. Superfrost/Plus glass slides (Fisher Scientific).
4. 22 × 50 mm cover glass.
5. X-Gal solution (5-bromo-4-chloro-3-indolyl-beta-D-galactopyranoside). Add 10 mL of dimethyl sulfoxide (DMSO) to 500 mg of X-gal to make 50 mg/mL solution. Aliquot out, wrap with foil, and store at –20°C.
6. X-gal rinsing buffer: 0.2 M sodium phosphate, pH 7.3, 2 mM magnesium chloride, 0.02% NP-40, and 0.01% sodium deoxycholate.
7. X-gal staining buffer: X-gal rinsing buffer plus 5 mM potassium ferricyanide, 5 mM potassium ferrocyanide. Wrap the container with foil and store in 4°C. Before staining, add X-gal to a final concentration of 1 mg/mL (see Note 10).
8. ImmEdge Pen (Vector, cat#H-4000).
9. PBS (Phosphate Buffered Saline). Dissolve the following in 800 mL of H_2O: 8 g of NaCl, 0.2 g of KCl, 1.44 g of Na_2HPO_4, 0.24 g of KH_2PO_4. Adjust pH to 7.4 and the final volume to 1 L with H_2O. Sterilize by autoclaving.
10. 0.2 M phosphate buffer (PB) (pH 7.4). Dissolve 6.4 g NaH_2PO_4 and 21.8 g Na_2HPO_4 into 1 L of H_2O.
11. PBST: 0.1% (V/V) Tween-20 in PBS.
12. 4% Paraformaldehyde solution. Dilute 16% paraformaldehyde solution (EMS) with 0.2 MPB (see Note 11).
13. 1% Triton-PBS: 1% (V/V) of Triton-X100 in PBS. 0.2% Triton-PBS: 0.2% (V/V) of Triton-X100 in PBS.
14. 20% Sucrose: 20% (W/V) sucrose in PBS.
15. OCT:Sucrose solution. Mix 1 volume of OCT with 2 volumes of 20% Sucrose (see Note 12).
16. Maleic Acide buffer (100 mM Maleic Acide, 150 mM NaCl, pH 7.5): Dissolve 11.61 g of maleic Acid and 8.76 g of NaCl into 900 mL of H_2O. Adjust pH to 7.4 with 7.5 g solid NaOH, plus 30–50 drops of 10 N NaOH. Bring the volume to 1 L with H_2O, and autoclave.
17. Blocking reagent. Stock solution (10%): dissolve blocking reagent (Roche) with maleic acide buffer for a 10% (w/v) concentration, autoclave, aliquot into 50-mL tubes, and store in –20°C. Working solution (2%): diluted with maleic acide buffer.
18. Mounting solution: Vector Mount AQ.

2.4. Equipment and Instruments

1. 37°C oven.
2. 55–60°C water bath.
3. Dissecting microscope.
4. Dissection instruments: curved and straight fine forceps, microsurgery scissors.
5. Microwave.
6. PCR machine.
7. Electrophoresis unit.
8. Microcentrifuge for 1.5-mL tubes.
9. Sonicator.
10. Syringes with a 29/30 gauge needle.
11. Cryostat.
12. Research microscope with camera.
13. UV light/camera system for taking pictures of DNA on agarose gel.

3. Methods

3.1. Genotyping

1. Collect 0.5–1 cm of mouse tail and place it in a 1.5-mL microcentrifuge tube (see Note 13).
2. Add 200 μL of the digestion buffer and 20 μL of Proteinase K solution.
3. Mix well and incubate in a 55–60°C water bath for 3 h or overnight.
4. Take samples out and leave it at room temperature for 30 min to equilibrate to room temperature.
5. Add 200 μL phenol/chloroform, mix gently by inverting the tube several times, followed by centrifugation ≥ 12,000 g (12,000 rpm) for 10 min (see Note 14).
6. Transfer 100 μL of the top aqueous phase (containing DNA) to a new 1.5-mL tube (see Note 15).
7. To the DNA add 100 μL of phenol/chloroform to extract one more time.
8. Centrifuge at ≥ 12,000 g (12,000 rpm) for 10 min. Transfer ~90 μL of the top aqueous phase (containing DNA) to a new 1.5-mL tube.
9. Add 1/10th volume of 10 M ammonium acetate.
10. Add 2.5 volume of 100% ethanol, swirl gently, and leave on ice for 30 min, then centrifuge ≥ 12,000 g (12,000 rpm) for 12 min.
11. Discard the supernatant and wash pellet two times with 0.5 mL of 70% ethanol with centrifugation ≥ 12,000 g (12,000 rpm) for 5 min each (see Note 16).

12. Air-dry the DNA for 3–5 min and dissolve the DNA in 60–100 μL of H_2O (or TE) (see Notes 17 and 18).
13. Set up PCR (final volume 20 μL). For Cre/ngn2/GAPDH: add 10 μL of 2× HotStartTaq Plus Master Mix, 7 μL of H_2O, 1 μL of each of the two oligonucleotides, and 1 μL of genome DNA. Mix. Modification for PCR of LacZ: 6 μL H_2O and 3 μL oligonucleotide mix.
14. PCR condition: 95°C for 5 min, 56°C/60°C for 1 min, 72°C for 1 min, 34 cycles of "95°C for 15 s, 56°C/60°C for 1 min, 72°C for 1 min" 72°C for 5 min, and 12°C for holding. Modification for PCR of LacZ: annealing at 65°C (see Note 19).
15. Analyze PCR products with agarose gel electrophoresis and taking pictures under UV light (see Note 20).

3.2. Mouse Breeding and Sample Collection

All experimental procedures involving the use of animals must be approved by institutional animal use and care committee.

1. Breed Ngn2-CreER™ mice and select the heterozygous after genotyping to establish a colony.
2. Verify Rosa26-LoxP-LacZ (+/+) reporter mice by PCR genotyping and breed to establish a colony (see Note 21).
3. Cross heterozygous Ngn2-CreER™ mice with Rosa26-LoxP-LacZ homologous mice to generate embryos that are Ngn2-CreER™ (+/−)/Rosa26-LoxP-LacZ (+/+) (Fig. 1). The pregnant mice and the neonatal mice are used for experiments (see Note 22).
4. Time pregnancies. The age of embryos is determined by designating as embryonic day 0.5 (E0.5) when the vaginal plug is observed at midday (see Note 23).
5. Inject 200–500 μL of 4-OHT (10 mg/mL) intraperitoneally (5 mg/20 g weight) with 29 gauge needle attached to a syringe into pregnant mice with embryos at a desirable age, from E10 to E20, or neonatal mice between postnatal day 1 (P1) to day 14 (P14). Include multiple (3–5) mice for each developmental stage (see Note 24).
6. At P17–P20, sacrifice the mice with CO_2. Collect tail snips for genotyping (see Note 25).
7. Enucleate the eyes with a pair of curved forceps and place the eyes in a dish with ice-cold PBS.
8. Remove the cornea, the lens, and the vitreous. Fix the eye cup with ice-cold 4% paraformaldehyde for 30 min (see Note 26).
9. Rinse the eye cup twice with PBS, followed by two washes with X-gal rinsing buffer for 10 min each. Add X-gal staining buffer and incubate overnight with gentle shaking in dark (see Note 27).

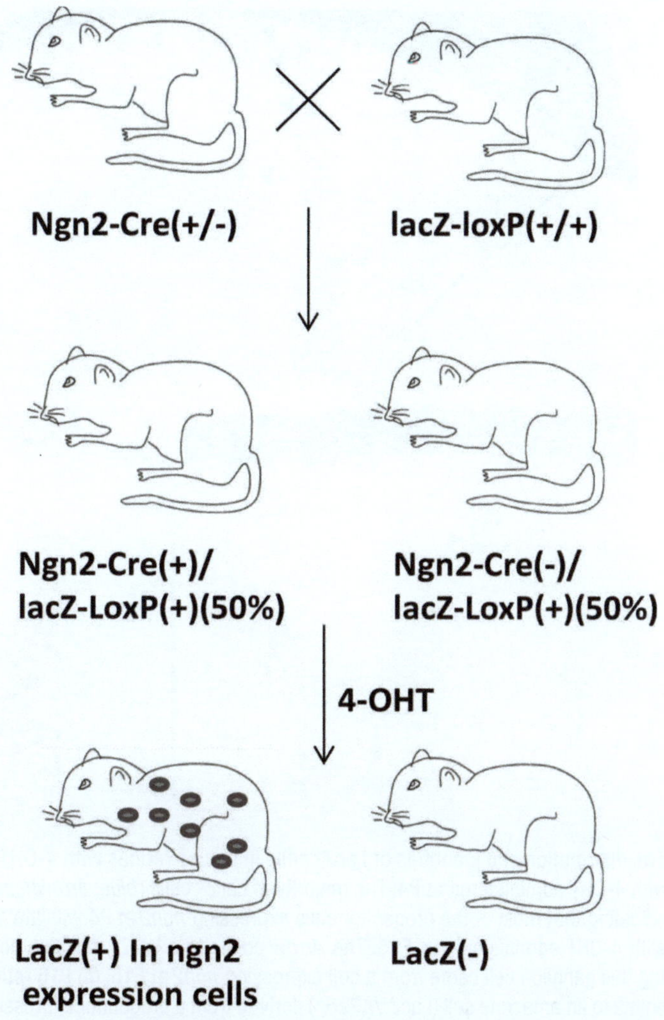

Fig. 1. The overall experimental scheme. LacZ is only expressed [LacZ(+)] in cells that once expressed *ngn2* in the Ngn2-Cre(+)/lacZ-LoxP(+) double transgenic mice.

10. After three washes with PBS, post fix the sample for 30 min with 4% paraformaldehyde, followed by three washes with PBS.
11. Embedd the samples with OCT: sucrose (1:2) in specimen molds, and freeze the sample with liquid nitrogen or dry ice. Store the frozen samples at −80°C (see Note 28).
12. Section the sample with a cryostat, and collect 5–12-μm retinal cryosections onto Superfrost/Plus glass slides.
13. Look for blue colored (LacZ$^+$) cells under a microscope (Fig. 2).
14. Score the number of LacZ$^+$ cells, the total number in a retinal section as well as in each individual retinal nuclear layer

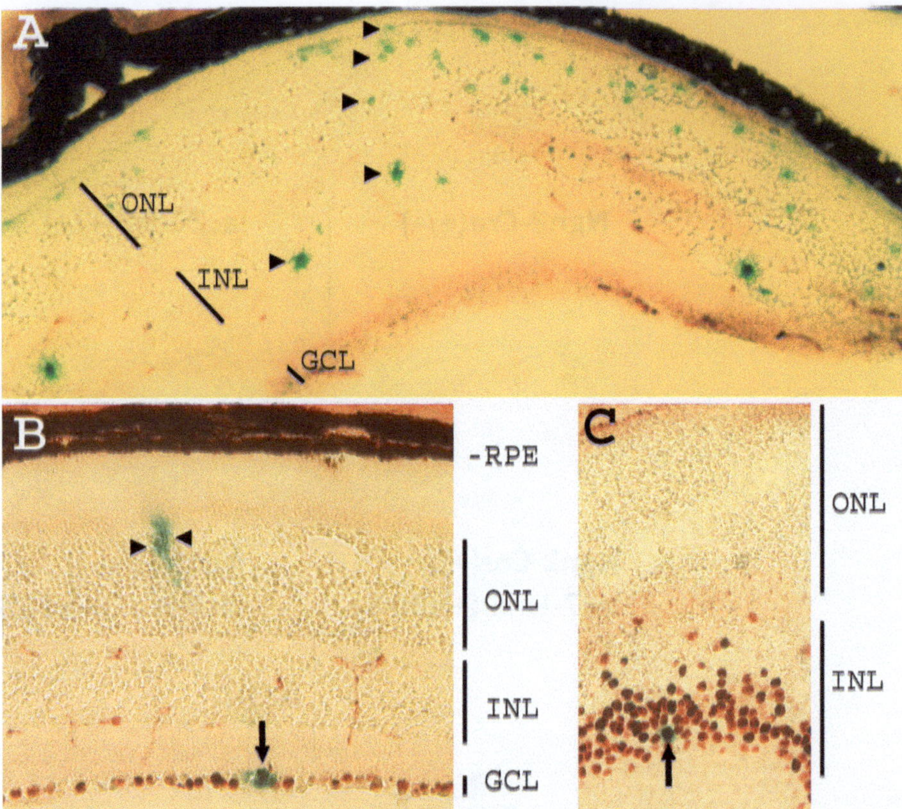

Fig. 2. The distribution and identifies of LacZ⁺ cells in mouse retinas with 4-OHT administered at different time. (**a**) P18 retina with 4-OHT administered at P4. The majority of LacZ⁺ cells (*blue*; *arrowheads*) are localized within the outer nuclear layer, indicating that most of the progenitor cells expressing *ngn2* at P4 will late differentiate into photoreceptors. (**b**) P17 retina with 4-OHT administered at E15. The *Arrow* points to a LacZ⁺ cell also positive for a ganglion cell marker Brn3A, indicating this ganglion cell came from a cell expressing *ngn2* at E15. (**c**) P16 retina with 4-OHT administered at E18. The *Arrow* points to an amacrine cell (LacZ⁺/AP2α⁺) derived from a progenitor expressing *ngn2* at E18. *ONL* outer nuclear layer, *INL* inner nuclear layer, *GCL* ganglion cell layer.

(i.e., the outer nuclear layer, the inner nuclear layer, and the ganglion cell layer).

15. If needed, continue with immunohistochemistry staining to confirm and/or determine the identities of the LacZ⁺ cells. An example image of such double-labeling is shown in Fig. 2.

4. Notes

1. Several reporter lines are available. The choice of a particular mouse line depends on the detection method an investigator plans to use.
2. Commercial sources often provide phenol/chloroform solutions of different pH. Remember to use the pH 8.0 one for DNA purification, as the pH 4.5 one is for RNA extraction.

3. 1% agarose gel works well for separating DNA fragments of >500 bp. If the size of PCR products is smaller than 500 bp, use a 2% or 3% agarose gel.
4. Investigators may choose any commercially available PCR reaction mixture. If a laboratory wishes to prepare its own, following this formula: 2 μL of 10× PCR Buffer (usually comes with Taq polymerase), 2 μL of 2 mM dNTP, 1 μL of 25 mM $MgCL_2$ (final concentration to 1.5–2.5 mM), 1 μL of DMSO, 4 μL of 5 M betaine, 2 μL of oligoucleotids, 1 μL of Taq polymerase (20,000 U/mL), and 7 μL of H_2O (to a total of 20 μL).
5. Cre oligoucleotides: Cre Expression Vector PTN75 CRE (NCBI: AB542060.1), sequence site: 1,871–2,500, a total length of 630 bp.
6. Ngn2 oligoucleotides: mouse *ngn2*, NCBI: NM_009718, sequence site: 328–1,106, a total length of 779 bp.
7. Information on the primers and PCR conditions to verify the Rosa26-loxP-LacZ reporter genotypes is provided by Jackson laboratory. (http://jaxmice.jax.org/protocolsdb/f?p=116:2:1946125058970626::NO:2:P2_MASTER_PROTOCOL_ID,P2_JRS_CODE:4615,003309).
8. GAPDH (glyceraldehyde-3-phosphate dehydrogenase) oligoucleotides: NCBI: NM_008084, sequence site: 566–1,098, a total length of 533 bp.
9. 4-OHT does not dissolve in sun flower seed oil readily. The cloudy suspension needs to be sonicated in a sonicator bath or with sonicator probe until the cloud disappears.
10. X-gal staining buffer is light sensitive. Protect it from the light with foil and store 4°C. Before staining, add X-gal to 1 mg/mL final concentration.
11. 4% paraformaldehyde: dilute 10 mL of 16% paraformaldehyde solution with 30 mL of 0.2 M PB, mix well, and leave it on ice for 30 min before using.
12. OCT:Sucrose (1:2). In a 500-mL clear container, combine one bottle of OCT with two bottles of 20% sucrose solution (using the OCT bottle), and mix well. Allow the air bubbles to dissipate before using.
13. The mice tail snips can be stored at −20°C.
14. Don't vertex the tube after phenol/chloroform is added to avoid shearing of genomic DNA. Instead, mix the content by inverting the tube several times. The phenol/chloroform is toxic. The worker should wear the glove and masks, operate in a fume hood, and dispose waste properly.

15. After centrifugation, hold the tube at the angle similar to its position in the centrifuge to avoid mixing of its contents. Slowly take out 100 μL of the top transparent, aqueous phase with a 200-μL micropipette into a new 1.5-mL microcentrifuge tube.
16. Care should be taken when discarding the 70% ethanol in the tube after centrifugation. We recommend slow aspiration to avoid accidently losing the DNA pellet.
17. Do not overdry DNA pellet. Overdried DNA pellet is difficult to dissolve. The DNA can be dissolved in H_2O if used immediately; otherwise dissolve the DNA in TE.
18. This method of genomic DNA isolation from mouse tail snips works well in our hand. However, investigators may choose to use a commercially available genome DNA isolation kit.
19. Choose annealing temperature by considering both the T_m (melting temperature) of the oligonucleotides and the results of trial tests. Usually, annealing temperature is around 4–6°C below the T_m of the nucleotides. The T_m of the oligonucleotides described in this chapter is ~60°C, but results from out trail tests with gradient temperature showed that 56°C is optimal for ngn2 and GAPDH, and 60°C for Cre, and 65°C for LacZ.
20. Anticipated Ngn2-CreER™ DNA genotyping results: heterozygote, PCR positive for Cre and for Ngn2; homozygote, PCR positive for Cre only; WT, PCR positive for Ngn2 only. Anticipated ROSA-LacZ DNA genotyping results: heterozygote, PCR products of 340 bp and 650 bp; homozygote, 340 bp only; WT, 650 bp only. Labeling the mice according to the PCR results with ear marker or ear wholes punch for the colonies' amplification.
21. The 129 s-Gt(ROSA)26Sor mice from Jackson Lab are from cryopreseved embryos and have at least 1 129 s-Gt(ROSA)26Sor transgenic (http://jaxmice.jax.org/strain/003310.html). Upon receiving them, genotyping is needed to identify the positive ones, which can be amplified to get the more 129 s-Gt(ROSA)26Sor(+/−) mice. Crossbreeding these 129 s-Gt(ROSA)26Sor(+/−) mice is carried out to generate a sufficient number of homozygous reporter mice.
22. Crossing heterozygous Ngn2-CreER™ mice will generate ¼ homozygous, ½ heterozygous, and ¼ of wild type.
23. The age of pregnancies is important for data analysis and interpretation. We check the mice two times a day to determine

whether the vaginal plug is present because the vaginal plug may remain only for 16–48 h.

24. Working 4-OHT solution is prepared fresh each time. 4-OHT is absorbed by abdominal omentum majus and distributed to the entire body. Studies show that 4-OHT-mediated Cre recombination is dose dependent (7). Thus, sufficient amount of 4-OHT (no less than 3 mg/20 g body weight) should be injected into each animal to activate *lacZ* expression. Otherwise, X-gal staining may be too weak to be detected. A pregnant mouse receives 500 μL of 4-OHT (10 mg/mL) and a neonatal mouse (P1–P14) receives 100–200 μL of 4-OHT.
25. We sacrifice all 4-OHT-treated mice at P17–P20, when the retina has structurally and functionally matured.
26. Before fixing, place the enucleated eye balls immediately into a 35-mm dish with ice-cold PBS. Punch a hole on the cornea with 29/30 gauge needle attached to a syringe and cut the cornea out behind the limbus with microsurgery scissors under a dissecting microscope. Remove the lens and vitreous, and transfer the eye cup (retina + RPE) immediately into a small vial with fixation solution.
27. X-gal is light sensitive. Stock solution and samples in staining solution should be covered with foil. If staining is not satisfactory, replenish with fresh X-gal staining solution and continue the incubation for color development. Include positive control to rule out that negative results are due to improper X-gal detection. The postfix after staining helps to maintain the blue color and tissue histology during subsequent processes.
28. Samples are frozen with OCT:Sucrose (1:2) using either dry ice or liquid nitrogen. In case of liquid nitrogen, freeze the samples slowly to avoid cracking of the block and for easy cryosectioning.

Acknowledgments

The authors thank Dr. David Anderson for providing the Ngn2-CreER™ mice. This work was supported by NIH/NEI grant EY11640 and an unrestricted grant to UAB Department of Ophthalmology from Research to Prevent Blindness.

References

1. Nagy A (2000) Cre recombinase: the universal reagent for genome tailoring. Genesis 26:99–109
2. Sauer B (1998) Inducible gene targeting in mice using the Cre/*lox* system. Methods 14:381–392
3. Tsien J, Chen D, Gerber D, Tom C, Mercer E, Anderson D, Mayford M, Kandel E, Tonegawa S (1996) Sub region and cell type-restricted gene knockout in mouse brain. Cell 87:1317–1326
4. Ma W, Wang SZ (2006) The final fates of neurogenin2-expressing cells include all major neuron types in mouse retina. Mol Cell Neurosci 31:463–469
5. Zirlinger M, Lo L, McMahon J, McMahon AP, Anderson DJ (2002) Transient expression of the bHLH factor neurogenin-2 marks a subpopulation of neural crest cells biased for a sensory but not a neuronal fate. Proc Natl Acad Sci USA 99:8084–8089
6. Littlewood TD, Hancock DC, Danielian PS, Parker MG, Evan GI (1995) A modified oestrogen receptor ligand-binding domain as an improved switch for the regulation of heterologous proteins. Nucleic Acids Res 23:1686–1690
7. Hayashi S, McMahon AP (2002) Efficient recombination in diverse tissues by a tamoxifen-inducible form of Cre: a tool for temporally regulated gene activation/inactivation in the mouse. Dev Biol 244:305–318
8. Schwenk F, Kühn R, Angrand P-O, Rajewsky K, Stewart AF (1998) Temporally and spatially regulated somatic mutagenesis in mice. Nucleic Acids Res 26:1427–1432
9. Soriano P (1999) Generalized LacZ expression with the ROSA26 Cre reporter strain. Nat Genet 21:70–71
10. Danielian PS, Muccino D, Rowitch DH, Michael SK, McMahon AP (1998) Modification of gene activity in mouse embryos in utero by a tamoxifen-inducible form of Cre recombinase. Curr Biol 8:1323–1326

Part III

In Vitro Systems

Chapter 10

In Vitro Explant Culture and Related Protocols for the Study of Mouse Retinal Development

Kangxin Jin and Mengqing Xiang

Abstract

The mouse retina is composed of many cell types and subtypes with distinct morphology and function; how these cells are differentiated from the multipotent progenitors is still largely unknown. Retinal in vitro explant culture has proven to be a useful tool to study the molecular and cellular mechanisms underlying retinal development. Here, we provide detailed descriptions about how to prepare retroviruses, dissect retinal cups, perform in vitro explant culture, and collect explant samples.

Key words: Retina, Explant culture, Development, Differentiation, Retrovirus, Electroporation

1. Introduction

The mouse retina is a very delicate neural tissue composed of three cellular layers and seven classes of cells. Most of the classes can be divided into subgroups based on morphological and functional differences (1, 2). For instance, the class of amacrine cells has more than 27 subgroups, each of which has distinct morphology and function. During retinogenesis, all classes of cells are derived from multipotent progenitors, and the differentiation processes are under tight control of both intrinsic and extrinsic factors, such as growth factors, membrane receptors, and most importantly transcription factors (3, 4). Loss- and gain-of-function studies have shown that a large number of factors are crucial for retinal development, but the list is far from complete (3–9).

Retinal in vitro explant culture (RIVEC) is a very important tool for both loss- and gain-of-function studies during retinal development. Homozygous mutation (or knockout) of a gene might be lethal at embryonic or early postnatal stages; yet, the differentiation of most retinal cell types has not started or been incomplete by

Shu-Zhen Wang (ed.), *Retinal Development: Methods and Protocols*, Methods in Molecular Biology, vol. 884, DOI 10.1007/978-1-61779-848-1_10, © Springer Science+Business Media, LLC 2012

these early stages. Under such circumstances, RIVEC provides an important tool to investigate the differentiation defects in the mutant. As an example, *Math3*$^{-/-}$;*NeuroD*$^{-/-}$ double-mutant mice die soon after birthday; RIVEC was used to investigate the double-mutant effect on retinal development (6). The loss-of-function studies can also be achieved by RNAi technologies, with RNAi target sequence(s) cloned into plasmids or viral vectors (10). RIVEC is even more widely used in the gain-of-function studies. One or more genes, wild type, mutated, or modified, can be cloned into plasmids and driven by different promoters. The genes in plasmids can be directly delivered by the electroporation method or packaged into retroviruses or lentiviruses. In our lab, RIVEC has been successfully used to study the functions of Foxn4 (5), Brn3b (11), Nr4a2 (12), and Ebf factors (13). The RIVEC can be also applied to other studies, for instance, to test promoters/enhancers or other DNA regulatory sequences.

Any method has its own advantages and drawbacks. The major benefit of the RIVEC is that it mimics the in vivo environment and can maintain a reasonably intact tissue structure. The drawback is that retinal ganglion cells (RGCs) usually die within several days of culture due to the fact that long-term survival of RGCs requires neurotrophic factors retrogradely transported from their brain targets. Compared to other approaches such as in utero injection (14, 15), one obvious advantage of RIVEC is that different factors can be easily added to the medium or explant tissues at any time. Prior to doing any experiment, the gains and losses of each approach should be carefully weighed.

Practically speaking, there are some general principles that should be considered when adapting the RIVEC. First, the time window for generating each cell type must be considered. For example, to study whether overexpression of a gene can promote the RGC or horizontal cell (HC) fate, it would be wise to choose E13 or earlier retinas to carry out the RIVEC, since after E14, the multipotent progenitors capable of adopting an RGC or HC fate may become too few to show a detectable effect. Second, the culture period should be as short as possible, since there are more and more cells committing apoptosis as the culture progresses, especially RGCs; the retinal structure boundaries become obscure too with time. Generally, you need to design your experiment according to the developmental processes of the cell types under study.

2. Materials

Prepare all solutions with ultrapure water filtered by Millipore filters. Unless otherwise specified, all materials should be sterile or autoclaved.

2.1. Instruments (Keep Clean, Unnecessary to Be Sterile)

1. Zeiss Stemi SV6 dissecting microscope (for tissue dissection).
2. Schott ACE I light source (for tissue dissection).
3. BTX Electro Square Porator ECM 830 (for electroporation).
4. Beckman L8-70M ultracentrifuge and SW-28 rotor (for retrovirus preparation).

2.2. Dissecting Tools

These tools can be acquired from Fine Science Tools Inc. or other similar sources.

1. 10-cm Straight forceps.
2. 10.5-cm Fine iris scissors.
3. Dumont #7 forceps with large radius curved shanks.
4. Dumont #5 or 55 forceps (for fine dissection).
5. Extra delicate mini-Vannas style iris spring scissors.
6. Micro-curette, diameter 1 mm (for electroporation).

2.3. Tissue Dissection

1. 70% Ethyl alcohol.
2. DMEM medium (Invitrogen).

2.4. Electroporation

1. Nepagene Micro Electroporation Chamber Model CUY532.
2. Electroporation solution containing 1× PBS (pH 7.4) or Hanks buffer (see Note 1), 0.5–3 μg/μl of your plasmid (see Note 2).

2.5. Retrovirus Preparation and Transfection

1. 100 mm × 20 mm and 150 mm × 25 mm Cell Culture Dish (Corning).
2. 0.25% Trypsin–EDTA (Invitrogen).
3. pBMN-GFP vector and/or derivatives with your gene insertions (Gentaur Molecular Products) (see Note 3).
4. Phoenix Eco packaging cell line (Gentaur Molecular Products) (see Note 4).
5. OPTI-MEM I medium (Invitrogen) (see Note 5).
6. Lipofectamine (Invitrogen).
7. 1,000× Puromycin (5–10 mg/ml in H_2O) (see Note 6).
8. Culture medium: DMEM medium,10% FBS, 1× penicillin/streptomycin/glutamine.
9. Screening medium: Culture medium with 1× puromycin (see Note 6).
10. 10× Polybrene (50 μg/ml) (see Note 7).
11. 250-ml Filter System (0.45 μm, from Corning) (see Note 8).
12. Beckman Ultra Clear Centrifuge Tubes (25 mm × 89 mm) (see Note 9).

2.6. Explant Culture

1. Falcon 6-well cell culture plate.
2. Millipore Millicell-CM Low Height Culture Plate Inserts (0.4-μm pore size) (see Note 10).
3. Explant culture medium (see Note 11): 42.5% DMEM, 42.5% F-12 Nutrient Mixture, 15% Fetal Bovine Serum, 1× penicillin/streptomycin/L-glutamine (Invitrogen), 5 mM Forskolin (optional), and 1× serotonin/transferrin/insulin (optional, Invitrogen).

2.7. Tissue Collection and Treatment (Unnecessary to Be Sterile)

1× PBS, pH 7.4 (sterile).
30% sucrose in 1× PBS (sterile).
Tissue-Tek O.C.T. Compound.
Tissue-Tek Cryomold Standard (25 mm × 20 mm × 5 mm).
Fresh 4% paraformaldehyde (PFA) (see Note 12), prepared as following:

(a) Heat 90 ml ddH$_2$O to 50–70°C.
(b) Add 30 μl 5 N NaOH.
(c) Add 4 g PFA and shake by hand until dissolved (see Note 13).
(d) Cool down immediately on ice (see Note 14).
(e) Add 30 μl 5 N HCl.
(f) Add 10 ml 10× PBS (pH 7.4).
(g) Use filter paper to remove the undissolved particles (optional).
(h) Store at 4°C and use it on the same day (see Note 15).

3. Methods

3.1. Retrovirus Preparation (Optional, Only if Retroviruses Are Needed)

1. Culture the Phoenix Eco cells in a 100-mm dish or split them into multiple dishes if you have multiple plasmids, and wait until the cells are 40–60% confluent (see Note 16).
2. Prepare the following solutions (see Note 17):
   (a) A: Dilute 10 μg of pBMN-GFP (or derivatives) into 1 ml of OPTI-MEM I, and mix well.
   (b) B: Dilute 50 μl of lipofectamine into 1 ml of OPTI-MEM I, and mix well.
   (c) Combine A and B, add 1 ml of OPTI-MEM I, mix gently, and incubate for 30 min to allow DNA–lipofectamine complexes to form at room temperature.
3. Discard the culture medium and wash cells with OPTI-MEM I medium.
4. Overlay the DNA–lipofectamine complexes onto the washed cells, and put the cells back into a cell culture incubator at 37°C for 5 h.

5. Remove the DNA–lipofectamine complexes, add 10 ml of Culture Medium, and incubate the cells for 1 day.
6. Remove the Culture Medium, add 10 ml of Screening Medium, and continue culturing until cells grow to confluence. Change Screening Medium every 2 days.
7. Transfer the cells into a 150-mm culture dish and add 20–25 ml of Screening Medium. Refresh Screening Medium every 2 days until cells grow to confluence (see Note 18).
8. Split the cells into three 150-mm culture dishes and continue culturing with the Screening Medium until cells grow to confluence.
9. Remove the Screening Medium and wash once with 1× PBS to get rid of the residual puromycin; add 13 ml of Culture Medium and continue culturing.
10. After culturing for 24 h, collect the medium (containing the viruses) into a 50-ml tube and store at 4°C; add 13 ml of Culture Medium, and collect again after 24 h (see Note 19).
11. Combine the collected medium, remove the cell debris through a 0.45-μm filter, transfer the medium into ultracentrifuge tubes, place the tubes into a SW-28 rotor, and centrifuge at 21,000 rpm (~79,000 × g force) for 3 h at 4°C using the Beckman L8-70M Ultracentrifuge (see Note 20).
12. Pour off the supernatant and aspirate the last drop from the lip of the tube. Seal the tube with parafilm and shake on ice for 1–2 h (see Note 21).
13. Aliquot 10–20 μl to each Eppendorf tube and store at −80°C (see Note 22).

3.2. Retinal Cup Preparation

1. Embryonic stage:
   (a) Expose the mother to carbon dioxide inhalation in a sealed chamber to euthanize the mouse. It usually takes 1–3 min (see Note 23).
   (b) Clean its abdominal area with 70% ethanol.
   (c) Cut the abdomen open with scissors. Now you should be able to see the embryos. Take the embryos out of wombs with the Dumont #5 (or #55) forceps very carefully, and put them into cold DMEM medium in the Petri dish (see Note 24).
   (d) Carefully isolate the eyeball with the Dumont #5 (or #55) forceps, and transfer it to a new Petri dish with fresh cold DMEM.
   (e) Under a Zeiss dissecting microscope, carefully remove the sclera, choroid, and other structures wrapping the retina, and then remove the lens using the Dumont #5 (or #55) forceps. Now you have a retinal cup.

2. Neonatal stage:
   (a) Euthanize the pup by decapitation (see Note 23).
   (b) Clean the area around the eyes with 70% ethanol.
   (c) Take out the eyeball with the Dumont #7 forceps and place it into cold DMEM medium (see Note 25).
   (d) Dissect under a Zeiss dissecting microscope. First, punch a hole on the iris–retina midline with forceps, rip a small fissure along the line with two pairs of forceps, and use scissors to cut open the sclera along the iris–retina midline (see Note 26 and Fig. 1a); and then remove the remaining sclera and choroid using the Dumont #5 (or #55) forceps; remove the lens last.

3.3. Electroporation (Optional Step, Only for Plasmid Transfection)

1. Add about 30 μl of electroporation solution (containing the plasmids) to Nepagene Micro Electroporation Chamber (see Note 27); transfer the retina into the chamber using forceps, and adjust the orientation of the retina so that the RGC layer faces the positive electrode (see Note 28, Fig. 1c).
2. Electroporation conditions: 10–12 V, 50 ms duration, 950 ms interval, five pulses (see Note 29). Push the start button to electroporate (see Note 30). Then, transfer the retina back into cold DMEM medium using the micro-curette (see Note 31).

3.4. Explant Culture

1. Now you have retinal cups from either Subheading 3.2 or 3.3 in cold DMEM. Make four incisions from the margin of the eye cup half way through toward the bottom at 0, 3, 6, and 9 o'clock positions. The retina looks like a four-petal flower after the incisions (see Fig. 1d).
2. Use a pipette to transfer up to three to four retinas onto a Millipore Millicell Cell Culture Insert; use forceps to adjust retinal orientation and position so that the RGC layer faces up and all four petals spread outwards. Carefully remove the medium along the boundary with a 200-μl pipette (see Note 32, Fig. 1e).
3. Place the insert into a 6-well plate with 1 ml of culture medium in each well (see Note 33). And transfer the plate into a cell culture incubator (37°C, 5% CO_2).
4. Virus infection (optional step, for retrovirus only):
   (a) After 5 h of culturing, infect retinas with virus in the following steps.
   (b) Add 1 μl of 10× polybrene (see Note 7) to every 10 μl of concentrated virus and mix well.
   (c) Add 2–3 μl of the mixture to each retina. Repeat once after 10 min.
   (d) Continue culturing in the incubator.
5. Change the culture medium every 1–2 days until collection of the explant tissue.

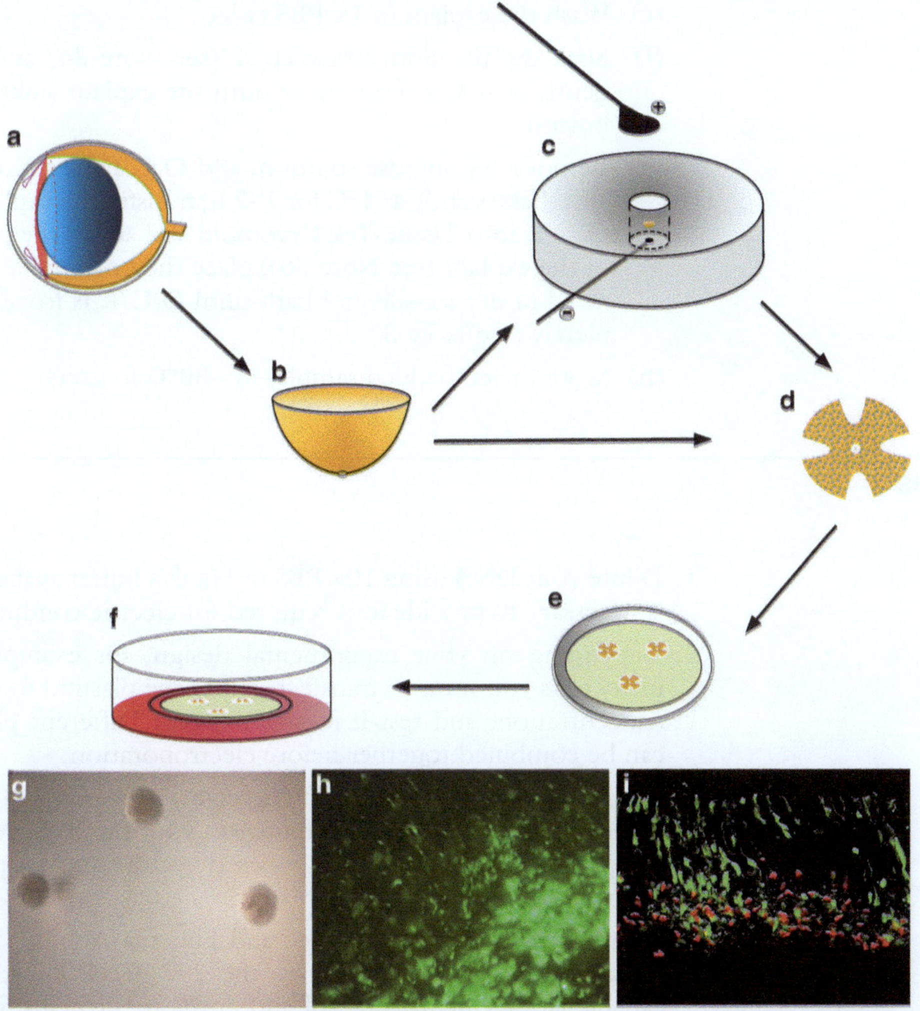

Fig. 1. Illustration of the retinal in vitro explant culture (RIVEC) protocol. (**a**) Structure of the mouse eye. The *orange color* represents the retina. When trying to separate postnatal stage retinas, use scissors to cut along the *dashed red line* to get rid of cornea, sclera, and other tissues. (**b**) A complete retinal cup after removing other tissues. (**c**) Electroporation in the Nepagene Micro Electroporation Chamber. (**d**) The retina looks like a four-petal flower after incisions. (**e**) Retinal explants were transferred onto a culture plate insert. (**f**) The insert was placed into one well of a 6-well plate with culture medium. (**g**) Retinal explants growing on the insert filter. (**h**) A microscopic image showing numerous GFP-positive cells in an electroporated explant. (**i**) A confocal microscopic image of an explant section showing cells immunoreactive for GFP (*green*) and Brn3a (*red*).

3.5. Sample Collection and Treatment for Immunohistochemistry

1. Depending on your experimental design, collect retinal explant samples following 2–14 days of culture.
2. Prepare fresh 4% PFA (see Note 12).
3. Collecting samples:
   (a) Remove the culture medium.
   (b) Wash with 1× PBS once.
   (c) Carefully wash off the explant (see Fig. 1g) with 1× PBS from the insert along the boundary.
   (d) Fix the explant in fresh 4% PFA for 10–15 min on ice.

(e) Wash the explant in 1× PBS twice.

(f) Soak the tissue in 30% sucrose (see Note 34) and shake gently at 4°C overnight or until the explant sinks to the bottom.

(g) Remove the sucrose solution, add O.C.T. (see Note 35), and shake gently at 4°C for 1–2 h; transfer the explant and O.C.T. into Tissue-Tek Cryomold and adjust the position of the explant (see Note 36); place the bottom half of the mold in dry ice–alcohol bath until O.C.T. is frozen completely (see Note 37).

(h) Store the embedded samples in –80°C freezers.

4. Notes

1. Dilute your DNA using 10× PBS or Hank's buffer as the buffer is necessary to provide ions required for electric conduction.
2. Depending on your experimental design, for example how many cells you want to transfect, dilute the plasmid to desired concentration, and test if it is optimized. Different plasmids can be combined together before electroporation.
3. pBMN-GFP is an MMLV-based retroviral expression vector commercially available. Depending on your purpose, you may use pBMN-Z or other MMLV-based vectors and adapt your protocol accordingly.
4. Phoenix cell lines, Phoenix-Eco and Phoenix-Ampho, are retrovirus producer lines based on the 293 T-cell line that are capable of producing gag-pol and envelope proteins for ecotropic and amphotropic viruses.
5. Opti-MEM® I Reduced Serum Medium is modified from Eagle's Minimum Essential Medium, and optimized to improve the transfection efficiency.
6. pBMN-GFP vector contains the puromycin-resistant gene that can be used to screen transfected cells. Phoenix cells without the resistant gene will be killed and float in the culture medium.
7. Polybrene is a small, positively charged molecule that binds to cell surfaces and neutralizes surface charge. It increases retrovirus gene transfer efficiency by enhancing receptor-independent virus adsorption on target cell membranes (16). Attention: Target cells could be killed by high concentration of polybrene. This is especially a known problem with some B- and T-cell lines. Lower the concentration of polybrene if necessary.
8. Filters with 0.45-μm pore size are recommended since smaller pores increase the chance of damaging the retrovirus surface proteins when they pass through the pores.

9. Those tubes are able to hold around 30 ml of liquid.
10. For E12 or earlier stages, the retina can be suspended in the culture medium for a few days. The major drawback is tissue hypoxia and cell death. The problem can be partially solved by using culture inserts that allow for studying three-dimensional explant structures.
11. Forskolin is commonly used to raise the levels of cyclic AMP (cAMP) and thus promotes cell survival. Serotonin/transferrin/insulin are also used to promote cell survival. All these factors are optional.
12. PFA is a suspected carcinogen. All operations should be carried out under the ventilation hood.
13. There are always some unsolvable white small particles, which usually do not affect your experiment. Those particles can be filtered out in the following steps.
14. PFA will be decomposed to formaldehyde at high temperature.
15. Some labs tend to store concentrated PFA (8–20%) at -20°C, and dilute it when needed. We prefer 4% PFA freshly prepared to guarantee consistent results, especially for antibodies that are highly sensitive to fixation conditions.
16. Pass the Phoenix Eco cells at a split ratio of 1:4–5 when they reach 70–80% confluence. They should reach 40–60% confluence before transfection; however, some people suggest that the cells should reach 70–80% confluence to obtain higher transfection efficiency.
17. Diluting plasmid and lipofectamine separately helps to mix and distribute DNA–lipofectamine complexes uniformly in the transfection mixture.
18. You should be able to see the dead cells floating in the medium. Under an inverted fluorescence microscope, you should clearly see the GFP-positive green cells adhering to the surface of the plate.
19. Due to the exhaustion of nutrition and pH value change, the medium should become yellowish instead of pink.
20. The speed of SW-28 at 21,000 rpm is equivalent to around 79,000 × *g* force, in case you use other ultracentrifuge equipment.
21. After the last drop aspirated from the lip of the tube, there is still some residual supernatant left, usually in the range of 100–150 μl. The virus concentration will be diluted if there is more supernatant left.
22. Aliquot according to the volume you need later. Avoid repeated freeze–thaw cycles which decrease the titer of the virus dramatically.
23. Depending on the protocol approved by your institute, the mice can be euthanized by other methods.

24. Generally, a C57BL6/J mother will have 5–9 embryos and a CD1 mother will have 10–17. To avoid tissue degradation caused by the lack of oxygen and other factors, dissect out retinas as quickly as you can.
25. Tip: Rip and push down the eyelids with two fingers, and the eyeball will protrude out; use forceps to clamp the bottom of the eyeball and pull it out.
26. The sclera becomes more and more tenacious after P0. Scissor cutting helps to keep better integrity of the retina.
27. The small chamber can hold about 30 μl liquid; however, the solution needs to be replaced after several cycles of electroporation, since the fluid becomes very viscous with broken tissues. Prepare more solution in advance.
28. The negative electrode is at the bottom of the chamber, so the plasmid migrates from the bottom to the top. Try not to bring extra fluid into the chamber when transferring the retinas. The retina looks like "sticking" to the tip of the forceps while being transferred.
29. Early-stage retina is very fragile and higher voltage tends to damage the retina.
30. You should hear one "beep" sound for each pulse of electroporation; there are also numerous bubbles forming in the buffer. After five pulses, you need to push the "start" button again to reset to the original state.
31. If the retina sticks to the Micro-curette, use a pipette to wash it down carefully.
32. The trick is that when you use the pipette to "suck" the fluid, as a result, the "petals" spread naturally.
33. Do not use more than 1 ml of medium; otherwise, the medium would seep through the filter and "flood" the explant, leading to hypoxia.
34. Prepare the sucrose solution in 1× PBS. Sucrose helps to keep better tissue morphology in later steps.
35. Some labs prefer to freeze the tissues in a mixture of O.C.T. and 30% sucrose at a 50:50 ratio. Remember to label the blocks properly using a permanent marker pen.
36. Tissues should be oriented in the block appropriately for sectioning (cross sections, longitudinal sections, etc.); label the orientation as well if necessary, for example, to check the cone cells in the future.
37. Add 90% or higher concentration of ethyl alcohol to the dry ice, which makes the freezing process much faster. The alcohol can be reused repeatedly for this purpose. Alternatively, if dry

ice is not available, freeze the mold in liquid nitrogen. Place the bottom half of the mold into liquid nitrogen until O.C.T. is completely frozen.

Acknowledgment

This protocol is adapted from a number of procedures of different labs, and we would like to specifically thank Dr. Constance Cepko's lab at the Harvard Medical School for their tremendous original work. We thank Kamana Misra for critical reading of the manuscript. This work was supported by the National Institutes of Health grants EY012020 and EY020849 (to M.X.).

References

1. Masland RH (2001) The fundamental plan of the retina. Nat Neurosci 4:877–886
2. Masland RH (2001) Neuronal diversity in the retina. Curr Opin Neurobiol 11:431–436
3. Cepko CL (1999) The roles of intrinsic and extrinsic cues and bHLH genes in the determination of retinal cell fates. Curr Opin Neurobiol 9:37–46
4. Livesey FJ, Cepko CL (2001) Vertebrate neural cell-fate determination: lessons from the retina. Nat Rev Neurosci 2:109–118
5. Li S, Mo Z, Yang X, Price SM, Shen MM, Xiang M (2004) Foxn4 controls the genesis of amacrine and horizontal cells by retinal progenitors. Neuron 43:795–807
6. Inoue T, Hojo M, Bessho Y, Tano Y, Lee JE, Kageyama R (2002) Math3 and NeuroD regulate amacrine cell fate specification in the retina. Development 129:831–842
7. Furukawa T, Morrow EM, Li T, Davis FC, Cepko CL (1999) Retinopathy and attenuated circadian entrainment in *Crx*-deficient mice. Nat Genet 23:466–470
8. Fujitani Y, Fujitani S, Luo H, Qiu F, Burlison J, Long Q, Kawaguchi Y, Edlund H, Macdonald RJ, Furukawa T, Fujikado T, Magnuson MA, Xiang M, Wright CV (2006) Ptf1a determines horizontal and amacrine cell fates during mouse retinal development. Development 133:4439–4450
9. Wang SW, Kim BS, Ding K, Wang H, Sun D, Johnson RL, Klein WH, Gan L (2001) Requirement for *math5* in the development of retinal ganglion cells. Genes Dev 15:24–29
10. Matsuda T, Cepko CL (2004) Electroporation and RNA interference in the rodent retina in vivo and in vitro. Proc Natl Acad Sci USA 101:16–22
11. Qiu F, Jiang H, Xiang M (2008) A comprehensive negative regulatory program controlled by Brn3b to ensure ganglion cell specification from multipotential retinal precursors. J Neurosci 28:3392–3403
12. Jiang H, Xiang M (2009) Subtype specification of GABAergic amacrine cells by the orphan nuclear receptor Nr4a2/Nurr1. J Neurosci 29:10449–10459
13. Jin K, Jiang H, Mo Z, Xiang M (2010) Early B-cell factors are required for specifying multiple retinal cell types and subtypes from postmitotic precursors. J Neurosci 30:11902–11916
14. Petros TJ, Rebsam A, Mason CA (2009) In utero and ex vivo electroporation for gene expression in mouse retinal ganglion cells. J Vis Exp. (31):e1333, doi:10.3791/1333
15. Garcia-Frigola C, Carreres MI, Vegar C, Herrera E (2007) Gene delivery into mouse retinal ganglion cells by in utero electroporation. BMC Dev Biol 7:103
16. Davis HE, Morgan JR, Yarmush ML (2002) Polybrene increases retrovirus gene transfer efficiency by enhancing receptor-independent virus adsorption on target cell membranes. Biophys Chem 97:159–172

Chapter 11

In Vitro Biochemical Assays to Monitor Rhodopsin Function

Joshua Sammons and Alecia K. Gross

Abstract

Rhodopsin is the dim-light photoreceptor responsible for initiation of the visual transduction cascade. In the dark its activity is very low, while light activation catalyzes the activation of its G-protein transducin. The first step in resetting rhodopsin and the phototransduction cascade involves the phosphorylation of light-active rhodopsin by rhodopsin kinase. Here, we describe assays to monitor the function of rhodopsin or rhodopsin mutants.

Key words: Rhodopsin, Phototransduction, Transducin assay, Rhodopsin kinase assay, Rod outer segment preparation

1. Introduction

Mutations causing blinding diseases such as congenital stationary night blindness, retinitis pigmentosa, and Leber congenital amaurosis are being found in genes encoding phototransduction proteins with increasing numbers (1–3). This has led to the need for functionally testing the activity of the mutant proteins to help uncover the possible pathophysiology seen in patients.

The phototransduction cycle in vertebrate vision begins in the with the absorption of a photon of light by rhodopsin, a G-protein coupled receptor composed of the apoprotein opsin covalently bound to its chromophore 11-*cis* retinal. This photon absorption induces a conformational change in the protein thereby activating rhodopsin (R*). R* in turn activates its G-protein transducin (G_t) by catalyzing the exchange of GDP for GTP on its alpha subunit ($G_{t\alpha}$). This leads to the eventual hyperpolarization of the rod cell that is sensed by downstream retinal neurons. Each protein within the phototransduction cascade is returned to its inactive state resetting the cell's polarity. This is accomplished by a series of proteins,

Shu-Zhen Wang (ed.), *Retinal Development: Methods and Protocols*, Methods in Molecular Biology, vol. 884, DOI 10.1007/978-1-61779-848-1_11, © Springer Science+Business Media, LLC 2012

including the phosphorylation of rhodopsin by rhodopsin kinase (RK), providing a high-affinity binding site for arrestin to block further signaling.

Over the past several decades, biochemical assays have been developed to test the ability of rhodopsin to activate and become inactivated in vitro. Here, we include protocols for two assays to monitor rhodopsin activity: the transducin assay, a filter-binding assay used to monitor rhodopsin's ability to catalyze the light-dependent activation of transducin, and the RK assay, which monitors the ability of RK to add radiolabeled ^{32}P from ATPγ^{32}P onto the C-terminus of rhodopsin (4–7).

2. Materials

2.1. Biological Materials

1. 150 Bovine Retinas (InVision BioResources).
2. COS cells (ATCC CRL-1650 or equivalent).

2.2. Apparatuses

2.2.1. Centrifuges, Rotors, and Bottles

1. Beckman L8-M Ultracentrifuge or equivalent.
2. Tabletop ultracentrifuge.
3. Beckman JA-19 rotor and bottles or equivalent.
4. Beckman SW-28 rotor and tubes or equivalent.
5. Beckman 50.2 Ti rotor and tubes or equivalent.

2.2.2. Syringes and Pipettors

1. Standard air-displacement pipets and serological disposable pipets.
2. 60- and 10-ml syringe.
3. 18-G needle.

2.2.3. Tissue Culture

1. Cell/tissue culture hood.
2. Humidified cell/tissue culture incubator (37°C, 5% CO_2).
3. 150-mm cell/tissue culture dishes.
4. Cell scraper (Fisher Scientific 3011).

2.2.4. UV/Vis Spectrophotometer (Cary 50 or Equivalent)

1. 0.45-μm nitrocellulose filters (Millipore HAWP02500).

2.2.5. Vaccuum Manifold (Millipore XX2702550)

2.2.6. Radioisotope Detection

1. [^{32}P] H_3PO_4 (10 mCi/ml, ICN Pharmaceuticals).
2. Scintillation counter (Beckman Coulter LS6500 or equivalent).
3. Scintillation vials (Beckman Poly Q vial or equivalent).
4. Scintillation fluid (Atlantic Nuclear).

2.2.7. Isoelectric Focusing

1. GE Multiphor II electrophoresis system (GE Healthcare, 18-1018-06) or equivalent.
2. Pharmalyte pH 2.5–5 (GE Healthcare, 17-0451-01).
3. Pharmalyte pH 5–8 (GE Healthcare, 17-0453-01).

2.2.8. Diethylaminoethyl Cellulose Column

1. Diethylaminoethyl (DEAE) cellulose anion exchanger (Whatman GE Healthcare, DE52).
2. Kontes Flex-Column Chromatography Column (VWR 420400-2510) or equivalent with stop valve (VWR 100133-888).
3. Dialysis. Spectra/por membrane tubing (flat width 2.3 mm, diameter 14.6 mm, MWCO 6-8000) or equivalent and Spectra/por closure clamps or equivalent.

2.3. Buffers and Reagents

All buffers to be filtered are filtered through 0.45-μm nitrocellulose membranes for protein and large particle removal. Just before filtration, add 1 M DTT to a final concentration of 1 mM, unless otherwise specified. Immediately after filtration, add several micrograms of solid PMSF to prevent protein aggregation and to inhibit proteolysis, respectively.

2.3.1. Buffers for ROS Preparation

1. 10× Buffer A: 20 mM MOPS pH 7.4, 1 mM $CaCl_2$.
2. 1× Buffer A: prepare from 10× Buffer A by adding 1 part 10× Buffer A to nine parts ddH_2O. Add 2 mM DTT, filter, and add solid PMSF.
3. 34% Sucrose (1 L): 390.2 g of sucrose, 100 ml of 10× Buffer A. Fill to approximately 980 ml with ddH_2O, filter through 0.45-μm filter, and adjust density to 1.15 ± 0.001 with ddH_2O or sucrose.
4. 30% Sucrose (100 ml): 11.76 ml of 1× Buffer A, 88.24 ml of 34% sucrose. Adjust density to 1.130 ± 0.001 with 1× Buffer A or 34% sucrose.
5. 26% Sucrose (100 ml): 23.53 ml of 1× Buffer A, 76.47 ml of 34% sucrose. Adjust density to 1.110 ± 0.001 with 1× Buffer A or 34% sucrose.
6. 24% Sucrose (100 ml): 29.41 ml of 1× Buffer A, 70.59 ml of 34% sucrose. Adjust density to 1.109 ± 0.001 with 1× Buffer A or 34% sucrose.

2.3.2. Buffers for COS Cell Transfection

1. Media A: 1 L of DMEM, 10 ml of 100× pen-strep, 10 ml of 100× L-glutamine, and 100 ml of 10× FBS.
2. Media B: 1 L of DMEM, 10 ml of 100× pen-strep, and 10 ml of 100× L-glutamine.
3. DEAE-dextran: 5 μM DEAE-dextran in Media B. The solution should be filter-sterilized in a sterile hood.

4. Mevalonic acid lactone (MAL) cocktail: 4 mM MAL in Media A.
5. 15 ml Rhodopsin DNA cocktail for 150-mm plate: 6 μg DNA in SV40 origin of replication-containing plasmid (such as pMT2), 0.5 μM DEAE-dextran, 100 mM Tris–HCl, pH 7.3, and 33.3 μM chloroquine in Media B.
6. 15 ml Rhodopsin-chloroquine cocktail for 150-mm plate: 30 μM chloroquine in Media A.
7. 10 ml RK DNA cocktail for a 150-mm plate: 100 mM Tris–HCl, pH 7.3, 30 μg DNA, 0.5 μM DEAE dextran, and 100 μM chloroquine in Media B.
8. 10 ml RK chloroquine cocktail for 150-mm plate: 4 mM MAL, 100 μM chloroquine in Media A.

2.3.3. Buffers for Rhodopsin Purification by 1D4 Sepharose Column

1. 10 ml 2% DM: 200 mg DM in 1× PBS (pH 7.0).
2. 6 ml Wash A: 0.1% DM, 2 mM $NaPO_4$, and 150 mM NaCl.
3. 6 ml Wash B: 0.1% DM and 2 mM $NaPO_4$.
4. 1 ml Elution: 0.1% DM, 2 mM $NaPO_4$, and 160 mM 1D4 peptide.

2.3.4. Buffers for Transducin Purification

1. 200 ml Buffer C: 2 mM Tris–HCl pH 7.4, 8 mM NaCl, 1 mM DTT, and PMSF.
2. 500 ml Buffer D: 5 mM Tris–HCl pH 7.4, 50 μM EDTA, 1 mM DTT, and PMSF.
3. 500 ml 2× Buffer E: 10 mM Tris–HCl, pH 7.4, 2 mM $MgCl_2$, 2 mM DTT, and PMSF. The DTT and PMSF will degrade over time so these should be added just before use.
4. 50 ml Buffer E: 10 mM Tris–HCl, pH 7.4, 2 mM $MgCl_2$, 1 mM DTT, and PMSF. This can be made from the 2× stock.
5. 70 ml Buffer J: 1× Buffer E, 7 mM NaCl, 1 mM DTT, and PMSF.
6. 50 ml Buffer K: 1× Buffer E, 25 mM NaCl, 1 mM DTT, and PMSF.
7. 3 L Dialysis buffer: 1× Buffer E and 50% glycerol.

2.3.5. Buffers for Transducin Assay

1. 250 ml 10× transducin (G_T) buffer: 100 mM Tris–HCl, pH 7.4, 1 M NaCl, 500 mM $MgCl_2$, 20 mM EDTA. This can be diluted 1:10 to make 1× G_T Buffer, but is also needed is small portions in the reaction mixture.
2. 200 ml 1× G_T buffer: 10 mM Tris–HCl, pH 7.4, 100 mM NaCl, 50 mM $MgCl_2$, and 2 mM EDTA. This is used to wash the reaction through the filters.
3. 50 μl 100 mM DTT.
4. 100 μl GTPγS* (50 μl): 75 μM GTPγS, 0.2 μCi/μl GTPγ^{35}S.

2.3.6. Buffers for Rhodopsin Kinase Assays

1. Phosphate-free Krebs buffer: 100 mM Hepes pH 7.4, 120 mM NaCl, 5 mM KCl, 1 mM $MgSO_4$, 1 mM $CaCl_2$, and 10 mM glucose.
2. Urea buffer: 20 mM Tris–HCl, pH 7.4, 6 M urea, 5 mM EDTA.
3. Resuspension buffer: 80 mM Tris–HCl, pH 8.0, 10 mM EDTA, 4 mM $MgCl_2$, 2 mM $CaCl_2$, and 0.5 mg/ml protease inhibitors (Boehringer Mannheim).
4. 50 ml 10× RK buffer: 750 mM Bis-Tris propane (BTP), 10 mM $Mg(OAc)_2$, 10 mM DTT.
5. DNaseI (Boehringer Mannheim).
6. RK extract.
7. 100 μl ATP*: 1.85 mM ATP, 5 mCi/ml ATPγ^{32}P.
8. Homogenization buffer: 25 mM Hepes (pH 7.5), 100 mM EDTA, 50 mM NaF, 5 mM adenosine, protease inhibitors (Boehringer Mannheim), 1 mM sodium vanadate (make fresh), 100 nM okadaic acid, and 15 μM fenvalerate.
9. Regeneration buffer: 10 mM Hepes (pH 7.5), 1 mM $MgCl_2$, 0.1 mM EDTA, 2% BSA, 50 mM NaF, 5 mM adenosine, protease inhibitors, 1 mM sodium vanadate (make fresh), 100 nM okadaic acid, 15 μM fenvalerate, and 100 μM 11 *cis*-retinal.
10. Solubilization buffer: 10 mM Hepes (pH 7.5), 1 mM $MgCl_2$, 10 mM NaCl, 0.1 mM EDTA, 1% dodecyl-maltoside, and 1 mM DTT.

2.3.7. Buffers for PDE Assay

1. 5× MOPS buffer: 100 mM MOPS (pH 8.0), 750 mM KCl, and 10 mM $MgCl_2$.
2. 5× pH assay buffer: 100 mM MOPS (pH 8.0), 750 mM KCl, 10 mM $MgCl_2$, 10 mM cGMP, and 0.5–20 nM PDE.

2.4. Nucleotides

1. 100 mM cGMP (Sigma-Aldrich G6129; sodium salt; MW 367.2).
2. 100 mM GTP (Sigma-Aldrich G3776; sodium salt; 100 mM).
3. 150 μM GTPγS (Sigma-Aldrich G6834; tetralithium salt; MW 363).
4. 100 mM ATP(Sigma-Aldrich G6129; sodium salt; MW 507.2).
5. GTPγ^{35}S (Perkin Elmer NEG030H; 10 mM tricine buffer pH 7.6, 10 mM DTT; 1250 Ci/mmol, and 12.5 mCi/ml).
6. ATPα^{32}P (Perkin Elmer BLU003H; 50 mM tricine buffer pH 7.6; 3,000 Ci/mmol, 10 mCi/ml).
7. ATPγ^{32}P (Perkin Elmer BLU003H; 10 mM tricine buffer pH 7.6; 3,000 Ci/mmol, 10 mCi/ml).

3. Methods

3.1. Protein Purification

3.1.1. Rod Outer Segment Preparation

Commercially available bovine retinas can be used to purify many of the proteins found within the retina. A thorough protocol for preparation of bovine rod outer segment (ROS) has been previously described by Papermaster and Dryer (8). This ROS prep can provide rhodopsin, transducin, and PDE depending on which is desired. Bovine ROS obtained from any commercial vender should be wrapped in foil to prevent rhodopsin activation. When performing the protocol to obtain rhodopsin, all steps must be performed in the dark under dim red light to prevent premature rhodopsin activation. Purification of transducin or PDE6 can be performed in the light. After obtaining the ROS, additional steps must be taken to purify the protein of interest (see Note 1).

1. Homogenize ROS.

   Begin by thawing three vials of bovine retinas (150 retinas) on ice at 4°C. Put two 1-L beakers on ice. Add ~180 ml chilled 40% sucrose to one of the beaker, several PMSF flakes, and 500 μM DTT. Add the bovine retinas and rinse each vial out with 5 ml of chilled 24% sucrose. Using a 60-ml syringe, suck/transfer bovine retinas and sucrose into the second beaker making sure to minimize amount of bubbles formed. Repeat until retinas have been homogenized in each beaker three times (a total of six times).

2. Centrifuge ROS.

   Split homogenized retinas into two JA-19 bottles and fill and balance them with 34% sucrose. Centrifuge the bottles at 4,600 rpm (12,900 × *g*) for 20 min 4°C. The ROS will be an orange pellet floating at the top of the bottle. Unwanted retinal material will be the black pellet at the bottom of the bottle. Pour ROS and sucrose into a clean beaker and dilute to ~800 ml with ice-cold 1× ROS buffer. Mix well. Add DTT and PMSF. Split ROS into four JA-19 bottles and centrifuge 9,900 rpm (27,950 × *g*) for 45 min 4°C to pellet the ROS.

3. Isolate ROS with sucrose gradient.

   Add 2 ml of 34% sucrose to each of six SW-28 tubes. With a transfer pipet, slowly add 6 ml of 34% sucrose to the top of the solution. Be sure not to disturb the gradient being formed or the final ROS band will be diffuse. Gently add 7.5 ml of 30% sucrose on top of the 34% sucrose. Slowly add 7.5 ml of 26% sucrose on top of the 30% sucrose layer. Decant supernatant from ROS and leave ROS pellet on ice. Gently resuspend each pellet in 4 ml of 24% sucrose and rinse each bottle with 2–3 ml of 24% sucrose. Add equal amounts of ROS on top of each gradient. Ensure that the ROS gradient has filled each tube to within 6–8 mm of top. Gently place tubes into SW-28 buckets and balance each pair of buckets with 24% sucrose. Secure each bucket in the rotor and centrifuge 27,000 rpm (136,000 × *g*) for 2.5–3 h 4°C using a soft acceleration and deceleration setting to ensure the gradient is not disturbed.

4. Obtain ROS from sucrose gradient.

   Attach an 18-G needle to a 10-ml syringe and place the end of the needle at the base of the ROS-containing lipid layer between 26 and 30% sucrose. Broken ROS will reside at the interface between 30 and 34% sucrose. Pool ROS into a 50-ml tube and add several PMSF flakes and DTT to 1 mM concentration. The concentration of rhodopsin in the ROS membranes can be determined using UV/visible spectroscopy by solubilizing the cells in 10% CHAPS in 1× PBS, spinning down the non-solubilized membranes, and measuring the absorbance using the extinction coefficient for rhodopsin at 500 nm of 42,700 (9).

3.1.2. Heterologous Expression of Rhodopsin

Heterologous COS cell transfection is an effective method for assaying wild-type and mutant rhodopsins. COS cells (10, 11) can be transfected with the DNA for a mutant rhodopsin and be ready for harvesting within 72 h. This protocol calls for the transfection of five 150-mm Petri dishes but can be scaled up or down as desired (see Note 2).

1. Transfect COS cells.

   COS cells grown in Media A supplemented with FBS, L-glutamine, and pen/strep should be grown to 90% confluency. Wash cells with 10 ml of Media B lacking FBS, and add 15 ml of rhodopsin DNA cocktail to each dish. Incubate the dishes at 37°C for 6 h to allow the DNA to migrate into the cells. After aspiration of the DNA cocktail, the cells are incubated in 15 ml of rhodopsin chloroquine cocktail overnight at 37°C to prevent excessive cell death. The next morning, the chloroquine cocktail is aspirated and 15 ml of media is added to the cells to encourage cell growth and protein expression. Seventy-two hours after transfection, rhodopsin expression should be at a maximum and the cells should be harvested.

2. Harvest COS cells.

   While rhodopsin expression is at a maximum, COS cells can be harvested by first washing away the Media A with 10 ml of 1× PBS (pH 7.0). Though a pH of 7.0 is not critical for harvesting of the cells, it does play an important role in rhodopsin purification. Aspirate PBS and add 2 ml of 1× PBS (pH 7.0) to cells. Scrap the cells off the plate with a cell scraper and transfer the cells to a 15-ml conical tube. A cell scrapper is used because it causes little harm to the cells. Spin cells down at 3,000 × *g* and wash once with 1× PBS (pH 7.0). Repeat. Purify or freeze cells at −20°C for up to several weeks.

3.1.3. Heterologous Expression of Rhodopsin Kinase

Like rhodopsin, RK can be purified from native tissue (12) or heterologously expressed (13). As RK degrades quite rapidly after purification, here we briefly describe the heterologous expression. After harvesting and sonication of the cells, the lysate is used to phosphorylate rhodopsin or frozen at –20°C for use in several days.

1. Transfect COS cells.

   Transfect cells at 60% confluency as follows. Wash cells with 10 ml of Media B twice to remove any serum. Add 10 ml of RK DNA cocktail and incubate at 37°C for 6 h to allow integration of the DNA. Aspirate cocktail, wash with 5 ml of Media B, and add 6 ml of DMSO. The cells should be incubated at RT for 3 min. Aspirate DMSO and add 10 ml of RK chloroquine cocktail. Incubate at 37°C for 2 h. Aspirate cocktail, wash with 10 ml of Media B two times, then wash with 10 ml of MAL cocktail, and incubate at 37°C overnight in 25 ml of MAL cocktail. Aspirate MAL cocktail, add 15 ml of Media A and incubate at 37°C for 24 h.

2. Harvest COS cells.

   RK-transfected cells should be harvested 48 h post-transfection for optimal expression. Wash harvested cells with 10 ml of 1× PBS (pH 7.0). Aspirate PBS and add 2 ml of 1× PBS (pH 7.0) to cells. Transfer cells to a 15-ml conical tube after scrapping the cells off the plate with a cell scraper. Spin cells down at 3,000 × *g* and wash once with 1× PBS (pH 7.0). Repeat.

3.1.4. Rhodopsin Purification

There are many published methods on rhodopsin purification, ranging from concanavalin A affinity chromatography to purification by surfactant overload (14, 15). One of the more commonly used methods, and the method described in this section, is the 1D4 sepharose immunoaffinity column purification (10). This method can purify wild-type rhodopsin as well as rhodopsin mutants containing an intact C-terminal 1D4 epitope (ETSQVAPA). Truncation, mutation, or addition of extra amino acids prevents proper binding of the protein to the beads (see Note 3).

1. Reconstitution of rhodopsin.

   For rhodopsin purification from COS cells, collect transfected cells from five 150-cm plates in a 15-ml tube and resuspend cells with 5 ml of 20 μM 11-*cis* retinal in 1× PBS (pH 7.0). The cells are incubated on a rocker at 4°C for at least 1 h to reconstitute rhodopsin. Lyse cells by adding 5 ml of 2% DM in 1× PBS (pH 7.0) containing PMSF to inhibit proteolysis and incubating on rocker at 4°C for another hour.

   Rhodopsin obtained from a dark ROS prep will already be reconstituted and thus ROS equaling 20 μg rhodopsin should be lysed in an appropriate volume of 1% DM.

2. Binding rhodopsin to the column.

   Place 150 μl 1D4 bead slurry in a 15-ml conical tube and wash three times with 1 ml of 0.1% DM in 1× PBS. Centrifuge cell lysate for 10 min at 5,000 × *g* to pellet cell debris. Add cell supernatant to 1D4 beads in the 15-ml tube and incubate on a rocker at 4°C for at least 2 h. In the meantime, prepare the column by stuffing a 1-ml syringe to 100 μl with glass wool. After 2 h, centrifuge beads (now bound with rhodopsin) at ≤3,000 × *g* and save supernatant in a different 15-ml tube. Add the beads to the syringe to make column.

3. Wash and elute rhodopsin.

   Wash the column five times with Wash 1 and then five more times with Wash 2. Place a piece of parafilm on bottom of the syringe, add 100 μl of 1:10 dilution of 1D4 peptide, and place a piece of parafilm on top of syringe. Addition of the 1D4 peptide, which has a higher affinity to the 1D4 antibody, to the column elutes rhodopsin by competing for antibody binding. Incubate at room temperature for at least 30 min. Remove parafilm and centrifuge column at 3,000 × *g* for 2 min in a clean 15-ml conical tube to collect rhodopsin elution. Add an additional 100 μl of 1:10 dilution 1D4 again and repeat final steps. Measure spectrum of each elution.

3.2. Transducin Purification

3.2.1. Prepare ROS to Obtain Transducin

Dilute ROS obtained from ROS prep to 200 ml with Buffer A. Transfer into a JA-19 bottle and centrifuge at 15,000 rpm (42,550 × *g*) at 4°C for 15 min. Decant supernatant and resuspend pellet in 200 ml of Buffer C. Centrifuge 18,000 rpm (51,100 × *g*) at 4°C for 15 min. Decant supernatant again and resuspend pellet in 200 ml of Buffer D. Centrifuge 18,000 rpm (51,100 × *g*) at 4°C for 15 min.

3.2.2. Reconstitute Transducin

Decant supernatant and resuspend in100 ml of Buffer D with 40 μM GTP (40 μl of 100 mM GTP will work). Incubate on ice for 30 min to allow GTP to bind transducin. Pour G_T into four 50.2 Ti tubes and centrifuge 44,000 rpm (301,580 × *g*) at 4°C for 15 min.

3.2.3. Setup Column and Bind Transducin

Add 10 ml of 1× Buffer E to 4 ml of DEAE cellulose beads and load onto column. Wash beads with 50 ml of 1× Buffer E at a flow rate of 500 μl/min. When Buffer E reaches ~0.5 cm from top of beads, gently add G_T and keep flow rate at 500 μl/min (see Note 4).

3.2.4. Wash Column and Elute Transducin

Wash column with 35 ml of Buffer E and then with 70 ml of Buffer J. Begin collecting flow through in 500-μl aliquots after proximally 35 ml of Buffer J has flowed through the column. Add 50 ml of Buffer K to start eluting the transducin. Elution of transducin will happen fairly quickly. Measure the absorbance of each aliquot over the range of 200–500 nm to determine which aliquots to pool together.

3.2.5. Concentrate Transducin

Pool aliquots together, being aware that the more aliquots pooled the more diluted the overall sample. Obtain approximately 10 inches per sample spectra/por membrane tubing and wet inside and out with ddH_2O. Tie off one end of the tubing and clamp a spectra/por closure just above the knot. Ensure the tubing contained no leaks and remove all ddH_2O. Add pooled sample and place second clamp just above sample volume. Tie a knot above the clamp and place in dialysis buffer to dialyze for 12–24 h. Change dialysis buffer twice, dialyzing for 12–24 h each time.

3.2.6. Determine Protein Concentration

Remove dialyzed transducin and place in microcentrifuge tubes. Determine protein content by Lowry or equivalent assay. Note that this denotes the concentration of purified protein, not the concentration of functional transducin. To determine the concentration of functional transducin, perform a transducin assay where the rhodopsin concentration is far greater than the theoretical transducin concentration. Routinely 1/3 of the protein will be functional transducin.

3.3. Transducin Assay

An assay testing for transducin activation is one of the most direct methods for testing changes in rhodopsin function. This assay can determine whether rhodopsin is constitutively active, inactive, or if the rate at which rhodopsin activates transducin has changed. Using nitrocellulose filters to bind transducin, this assay uses GTPγ^{35}S to measure the amount of transducin activated over a period of time. Here, six points are taken in the dark to measure transducin activation by dark rhodopsin, then another six points are taken in the light to determine the ability of active rhodopsin to activate transducin (Fig. 1).

3.3.1. Setup

Soak 0.45-μm nitrocellulose filters in dH_2O for at least 10 min. Prepare 150 ml of 1× G_T Buffer and 14 of 10 ml of fluid scintillation vials per assay and an additional vial for the negative control. Prepare 100 μl of GTPγS* and set up vacuum manifold with the wet filters. Make sure the vacuum is on and apparatus is sealed such that a vacuum is pulling on the filters. Begin the assay in dim red light. Prepare a reaction mix containing 150 μl total of 1× G_T Buffer, 1 mM DTT, 2 μM transducin, and 1–5 nM rhodopsin, leaving 6 μl for GTPγS* to start the reaction (see Note 5).

3.3.2. Start the Reaction

Add 6 μl of GTPγS* to the reaction and immediately start the timer. At 1 min, pipet 10 μl of reaction mix onto first filter and wash three times with 3 ml of 1× G_T Buffer. Repeat every minute for the first 6 min. At 6 min 30 s, turn on the lights and fully activate rhodopsin. Continue adding reaction to the last six filters 1/min. When finished, turn off vacuum and add filters to scintillation vials. Add 10 μl of reaction mix straight to the 13th vial and 1 μl of GTPγS* to the 14th vial and allow the vials to incubate at room temperature for at least 30 min.

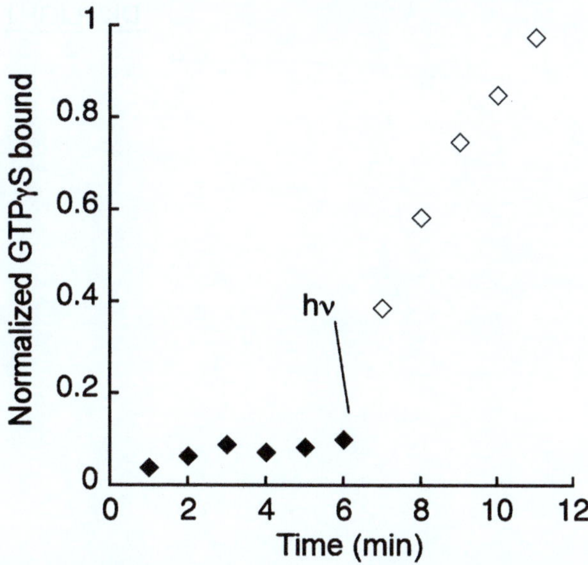

Fig. 1. Activation of transducin by rhodopsin with the transducin assay. *Filled diamonds*, rhodopsin in the *dark*; *open diamonds*, after 30 s exposure to saturating white light ($h\nu$). Activation maximum is normalized to 1.

3.3.3. Calculate Transducin Activation

The concentrations of transducin and cGMP are much greater than K_m of GDP to GTP exchange on transducin by WT rhodopsin. Thus it can be assumed that the active rhodopsin is fully saturated. Furthermore, the initial rate of GDP to GTP exchange (V_0) is the maximum rate of exchange (V_{max}) and this rate is directly proportional to rhodopsin concentration. Under these conditions, V_{max} of a reaction can be converted to K_{cat} of rhodopsin by dividing V_{max} by rhodopsin concentration. Comparison of K_{cat} sample to that of K_{cat} WT rhodopsin control will tell if the kinetics of transducin activation by rhodopsin has changed. To obtain V_{max} count scintillations with a scintillation counter, assuming that there is 30 pmol GTP in the 10 μl positive control (13th vial), and divide 30 pmol by counts per minute (CPM) reading. Multiply that number by the CPM of each time point to determine the amount of transducin bound at said time point. Under the linear range of increasing picomoles GTP bound, the slope (rise over run) of the curve is V_{max}.

3.4. Rhodopsin Kinase Assay

Phosphorylation of rhodopsin by RK is the first step in inactivation of rhodopsin. Under most circumstances, testing for rhodopsin phosphorylation is sufficient to determine changes in the kinetics of rhodopsin deactivation though it does not directly test that deactivation. Actual deactivation does not occur until arrestin binds rhodopsin and this can be determined with a PDE assay.

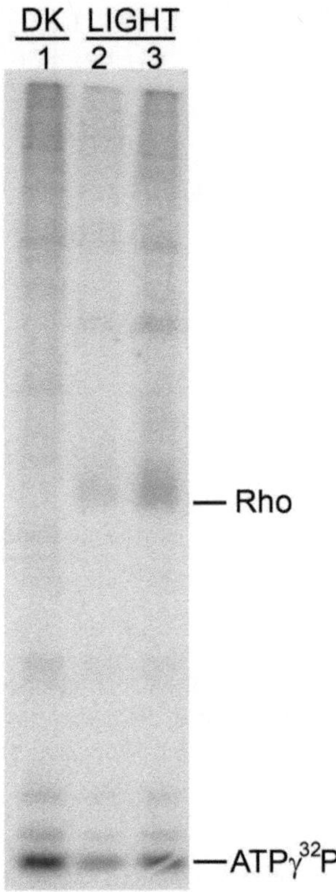

Fig. 2. Light-dependent phosphorylation of rhodopsin using the in situ rhodopsin kinase assay. Autoradiogram from wild-type mouse retina. Each *lane* represents a different retina quenched before (DK, *lane 1*) or after light activation (LIGHT, *lanes 2–3*). Note the rhodopsin bands (Rho) present in the light but not the *dark lanes*.

Two common procedures are used to determine rhodopsin phosphorylation. The first is an in situ hybridization to determine whether rhodopsin has been phosphorylated (Fig. 2) (5). The second is the isoelectric focusing (IEF) of phosphorylated rhodopsin species which allows for the separation of different phosphorylation species thereby elucidating the number of phosphorylation sites (6, 7).

3.4.1. In Situ Hybridization From Mouse Retinas

1. Incorporate ^{32}P into living retinas.

   Mice should be dark adapted for at least 12 h prior to use. Remove two retinas per sample under dim red light using forceps and a scalpel and incubate in 100 ml of phosphate-free Krebs buffer containing 1.25 mCi/ml [^{32}P] H_3PO_4 (10 mCi/ml) for at least 1 h in the dark to allow the endogenous ATP to incorporate ^{32}P.

2. Homogenize and harvest retinas.

   Wash retinas with Krebs buffer and homogenize in urea buffer either in the dark or after 5 min exposure to bright white fluorescent light. Centrifuge the homogenate at 13,000 × *g* for 20 min, wash with 20 mM Tris–HCl (pH 7.4), and resuspend in 50 ml of resuspension buffer. Incubate for 20 min with 20 U of DNaseI at room temperature.

3. Analyze results.

   Load equal amounts of each retina, between 1/8 and 1/10 usually is sufficient, onto a 12% SDS–PAGE gel. Transfer the proteins to nitrocellulose membranes in order to expose to X-ray film for detection of ^{32}P-labeled proteins and for rhodopsin quantification using Western blot analysis.

From Heterologous Expression

1. In the dark, set up a 60-μl reaction containing 1× RK buffer, 1 mM DTT, 0.01% DM, 200 ng rhodopsin, 6 μl of RK extract, and 3.2 μl of (0.1 mM) ATP*. Split the reaction into two 30-μl reactions, one dark and one light. Allow the reactions to incubate at 30°C for 30 min. Add SDS loading buffer to quench the reaction and load onto a 12% SDS–PAGE gel.

3.4.2. Isoelectric Focusing

Isoelectric focusing is a method of protein isolation using the isoelectric point of the molecule via a pH gradient; an excellent example of IEF on rhodopsin samples is shown in Shi et al. (7). Phosphorylation of a protein changes its isoelectric point allowing it to be differentiated from other species of the same protein.

1. From mouse retinas

   (a) Prepare retinas.

   Mice should be dark adapted for at least 12 h prior to use. Remove two retinas per sample under dim red light using forceps and a scalpel, expose to bright white fluorescent light for 10 min, and immediately freeze in ethanol/dry ice. As a control, freeze an additional WT retina in the dark immediately after dissection.

   (b) Homogenize and solubilize retinas.

   Homogenize two retinas per sample in 400 μl of homogenization buffer and centrifuge at 13,000 × *g* for 30 min and resuspend in 1 ml of regeneration buffer for at least 12 h at 48°C. Centrifugation at 13,000 × *g* for 30 min and solubilize for 3 h at 48°C in solubilization buffer. Centrifuge again at 13,000 × *g* for 30 min.

   (c) Separate phosphorylation species.

   Prefocus a 1-mm thick 8% polyacrylamide gel by adding six parts Pharmalyte pH 2.5–5 to four parts Pharmalyte pH 5–8 to the gel and running it for 30 min at 24 W. Solutions commonly used are 0.04 M glutamic acid for the anode solution and 1 M NaOH for the cathode solution. Load approximately 1/5 of the supernatant onto the prefocused gel for IEF. After addition of the phosphorylated rhodopsin species, run the gel in the dark at 24 W for 2 h at 4°C. A flat electrode can be used to measure the pH gradient of the gel and then the proteins will be transferred to nitrocellulose membrane for rhodopsin detection.

2. From heterologous expression
   (a) Phosphorylate rhodopsin.
   This method begins much like the in situ RK protocol, except no radioactivity is needed. In the dark, set up a 60-μl reaction containing 1× RK buffer, 1 mM DTT, 0.01% DM, 200 ng of rhodopsin, 6 μl of RK extract, and 0.1 mM ATP, and split into light and dark reactions (6, 7). The light reaction should be exposed to light for 10 min and flash frozen to prevent further reaction.
   (b) Separate phosphorylated species.
   The rhodopsin species will be separated on a 1-mm thick 8% polyacrylamide gel for IEF as previously stated and proteins transferred to a nitrocellulose membrane for analysis of different rhodopsin phosphorylation species.

4. Notes

1. When centrifuging ROS, especially for the sucrose gradient, ensure that the ultracentrifuge and rotor are both at the correct temperature. If they are not at the same temperature, centrifugation may cause condensation on the inside or outside of the rotor, which can freeze the sample leading to a significant loss of active protein. Additionally, ensure there is no condensation inside the rotor prior to centrifugation.
2. It is important to maintain an optimal environment for COS cells. Starving or over-confluent cells will lead to a reduction of protein expression, while under confluency will reduce the number of cells producing protein. Thus, monitoring cell confluency and changing media at appropriate times are important for maximum protein yields. When harvesting the cells, ensure little time is wasted while cells are in PBS because they will start dying and protein will be lost.
3. When heterologously purifying rhodopsin, ensure that enough 11-*cis* retinal is present to reconstitute all rhodopsin and enough 1D4 column is present to bind the reconstituted rhodopsin. The pH in which rhodopsin is purified is also very important. At each step during the purification process, the pH must remain between 6.4 and 7.4. Too low of a pH will denature rhodopsin inactivating it and too high of a pH will limit

the amount of rhodopsin collected. Though we recommend a pH of 7.0 to purify rhodopsin, a slightly modified pH may be more optimal in other labs.

4. When preparing transducin from bovine ROS, the DEAE column is set up such that a flow rate of 500 μl/min can be obtained without disturbing the column with the addition of protein or buffer. Disturbing the column may denature transducin giving a lower yield. After addition of Buffer K, transducin should begin eluting within the first 5–10 ml. It is important to collect small volumes of the elution to keep the concentrations high. Larger collected volumes will lower the concentration, making the sample difficult to work with.

5. Because transducin is stored in glycerol, the volume of transducin added to the transducin assay reaction should be minimized to reduce inhibition of protein–protein interactions by glycerol. Transducin concentration must be 50 times greater than rhodopsin concentration to keep transducin activation from saturating in the reactions.

References

1. Zeitz C et al (2009) Genotyping microarray for CSNB-associated genes. Invest Ophthalmol Vis Sci 50:5919–5926
2. Wright AF et al (2010) Photoreceptor degeneration: genetic and mechanistic dissection of a complex trait. Nat Rev Genet 11:273–284
3. den Hollander AI et al (2008) Leber congenital amaurosis: genes, proteins and disease mechanisms. Prog Retin Eye Res 27:391–419
4. Robinson PR (2000) Assays for detection of constitutively active opsins. Methods Enzymol 315:207–218
5. Robinson PR et al (1994) Opsins with mutations at the site of chromophore attachment constitutively activate transducin but are not phosphorylated by rhodopsin kinase. Proc Natl Acad Sci USA 91:5411–5415
6. Mendez A et al (2000) Rapid and reproducible deactivation of rhodopsin requires multiple phosphorylation sites. Neuron 28:153–164
7. Shi GW et al (2005) Light causes phosphorylation of nonactivated visual pigments in intact mouse rod photoreceptor cells. J Biol Chem 280:41184–41191
8. Papermaster DS, Dreyer WJ (1974) Rhodopsin content in the outer segment membranes of bovine and frog retinal rods. Biochemistry 13:2438–2444
9. Hong K, Hubbell WL (1972) Preparation and properties of phospholipid bilayers containing rhodopsin. Proc Natl Acad Sci USA 69:2617–2621
10. Oprian DD et al (1987) Expression of a synthetic bovine rhodopsin gene in monkey kidney cells. Proc Natl Acad Sci USA 84:8874–8878
11. Zeitz C et al (2008) Identification and functional characterization of a novel rhodopsin mutation associated with autosomal dominant CSNB. Invest Ophthalmol Vis Sci 49:4105–4114
12. Buczylko J, Palczewski K (1993) Purification of arrestin from bovine retinas. Methods Neurosci 15:226–236
13. Lorenz W et al (1991) The receptor kinase family: primary structure of rhodopsin kinase reveals similarities to the beta-adrenergic receptor kinase. Proc Natl Acad Sci USA 88:8715–8719
14. Litman BJ (1982) Purification of rhodopsin by concanavalin A affinity chromatography. Methods Enzymol 81:150–153
15. Aveldano MI (1995) Phospholipid solubilization during detergent extraction of rhodopsin from photoreceptor disk membranes. Arch Biochem Biophys 324:331–343

Chapter 12

Transfection of Primary Embryonic Chicken Retinal Cells Using Cationic Lipid

Yi-Wen Hsieh and Xian-Jie Yang

Abstract

Primary neuronal culture and transfection are useful tools in determining gene functions within specific tissue contexts and developmental stages. Chicken embryonic retinal cultures are easily obtainable and often robust as the chicken eye is relatively large compared to mouse eye at similar developmental stages. Various DNA-based constructs have been developed to overexpress or knockdown genes of interest and can be delivered into the cells using lipofectamine, a cationic lipid-based transfection system. Here, we describe a method to culture and transfect primary chicken embryonic retinal cells in order to manipulate genes involved in retinal development. This technique can simultaneously deliver multiple genes without construct-size constrains and permit the usage of tissue or cell type-specific promoters, and is thus a useful approach to explore gene functions during neural retina differentiation.

Key words: Vertebrate, Retina, Neuron, Progenitor, Primary Cell Culture, Transfection, Gene function, Immunocytochemistry

1. Introduction

Neuronal network formation in vertebrates involves complex orchestration between cell intrinsic factors and extrinsic signals to generate proper composition of various neuronal types from multipotent progenitors (1–3). Disrupting gene function has been an essential approach to elucidate molecular mechanisms that regulate neuronal fate specification and differentiation. Developmental biologists have often used transgenic technologies to express genes of interest via chosen promoters in model species such as mouse and zebrafish. In addition, homologous recombination-based gene deletion in the mouse germ line has enhanced the understanding of specific gene functions in the nervous system (4–6). Furthermore, Cre-loxP-mediated gene ablation or activation have enabled temporal and cell type-specific perturbation of gene function in the

Shu-Zhen Wang (ed.), *Retinal Development: Methods and Protocols*, Methods in Molecular Biology, vol. 884, DOI 10.1007/978-1-61779-848-1_12, © Springer Science+Business Media, LLC 2012

developing and mature nervous system (7, 8). Together, these techniques have greatly advanced our understanding of development within the in vivo context. However, the use of these techniques requires generation of transgenic animals and extensive breeding to obtain the required genotypes, which are time consuming and costly. In case of functional redundancy or complex systems that require disruption of multiple genes, the transgenic or knockout approach can be difficult or impossible to carry out. Thus, simple and efficient means to perform gene perturbation in specific cell types and temporal framework remain important.

Certain molecular mechanisms of the visual system development are evolutionarily conserved from insects to mammals (9, 10). All vertebrate retinas are similarly organized and their ontogeny follows a conserved temporal sequence. The embryonic chicken retina, compared to the rodent retina, is considerably larger at equivalent stages of development and easily accessible. Furthermore, embryological and molecular manipulation tools have been developed to study early chicken eye morphogenesis and retinogenesis (11). In contrast to the mouse retina, the chicken retina is cone-photoreceptor rich. In addition, chicken retinal cells can be easily cultured in vitro and at lower costs. In recent years, the post-hatching chicken retina has been shown to possess self-repair potentials (12). These features of the chicken retina make it a valuable vertebrate model to study neurogenesis and neural repair.

We have taken advantages of the chicken retina to address the role of Pax6 gene in early retinogenesis. Pax6 is an evolutionarily conserved paired homeodomain gene required for retinal development from the fly to humans. The dosage of Pax6 critically influences eye morphogenesis in mouse (13, 14). Conditional ablation of Pax6 in the periphery retina suggests that Pax6 plays a role to maintain the pluripotency of retinal progenitors (15). We observed that the expression level of Pax6 decreases as retinal progenitor cells enter the neurogenic phase. Furthermore, Pax6 protein levels undergo dynamic changes during the neurogenic cell cycle (16). These data indicate that Pax6 protein levels are stringently regulated during retinogenesis. To address the significance of the dynamic Pax6 protein levels in progenitor proliferation and neuronal fate specification, we perturbed Pax6 expression in primary chicken retinal cultures through transfection. Pax6 was overexpressed from a strong universal promoter, the chicken beta actin promoter (CAGp) (17) and down-regulated using U6 promoter-mediated shRNA knockdown (18). These molecular manipulations in conjunction with cell cycle analyses and confocal imaging allowed us to conclude that dynamic Pax6 expression during the neurogenic cell cycle critically influences proliferation and cell fate choices (16).

In this chapter, we describe in detail, the step-by-step procedure for delivering DNA constructs into chicken primary retinal cultures. In brief, developing chicken retinas are dissected and dissociated using enzymes. Retinal cells are cultured as a monolayer and transfected using lipofectamine PLUS with various DNA constructs along with a GFP-expressing plasmid. At different times post-transfection, cultured retinal cells are analyzed by immunocytochemistry using antibodies against GFP-positive transfected cells and other markers for proliferation, neuronal differentiation, or apoptosis to detect and quantify the effects of gene perturbation.

2. Materials

2.1. Chicken Primary Culture

1. Fertilized white leghorn chicken eggs.
2. Humidified chicken egg incubator (Lyon Electric).
3. Tissue culture incubator supplied with 5% CO_2.
4. 1× Retinal culture medium (RCM) 50 ml: in a tissue culture hood, mix 1 ml of fetal bovine serum (FBS), 0.1 ml of chicken serum (ChS), 0.5 ml of 1 M HEPES buffer pH 7.0, 0.5 ml of 100× penicillin/streptomycin, 22.9 ml of F12, and 25 ml of Dubecco Modified Eagle's Medium (DMEM) in a 50-ml sterile conical tube. Store at 4°C and warm to 37°C prior to use.

   2× RCM 25 ml: mix 1 ml of FBS, 0.1 ml of ChS, 0.5 ml of 1 M HEPES buffer, 0.5 ml of 100× penicillin/streptomycin, 10.4 ml of F12, and 12.5 ml of DMEM in a 50-ml sterile conical tube. Store at 4°C and warm to 37°C prior to use.
5. Stock solutions:
   (a) 10 and 1× Phosphate buffered saline (PBS).
   (b) 1× Dulbecco's Modified Eagle Medium (DMEM).
   (c) 1× Hank's buffered saline solution with (HBSS+) and without magnesium and calcium (HBSS−).
   (d) 100 mg/ml Trypsin type XII in HBSS−.
   (e) 1 mg/ml Soybean Trypsin Inhibitor in DMEM.
   (f) 1 g/ml DNaseI in DMEM.
   (g) 10 mg/ml poly-D-lysine in sterile tissue culture grade water.
6. Surgical tools: two #5 Inox forceps, curved scissors, spring scissor, perforated spoon, medium sterile transfer pipette.
7. Egg holder: aluminum foil shaped like a culture dish.
8. Dishes for dissection: 100, 60, and 35 mm sterile culture dishes as needed.
9. Sterile 1.5-ml Eppendorf tubes and microcentrifuge.
10. Culturing slides: 150-mm dish and 8-well Lab-Tek permnox slides (Nunc/Nalge).

2.2. Transfection Reagents

Lipofectamine PLUS (Invitrogen, Carlsbad, CA) and Opti-MEM (Invitrogen).

2.3. Immunocytochemistry

1. Fix solution: 4% paraformaldehyde in 1× PBS.

   Pre warm 300 ml of ddH_2O in a 1 L beaker to 60°C with a stirrer bar, weigh 20 g of paraformaldehyde and add to the heated water, cover with plastic wrap and stir, the solution will be cloudy with some precipitates. Using a small transfer pipette, add 10 N NaOH, a drop at a time until paraformaldehyde is completely dissolved.

   Add 50 ml of 10× PBS, mix, and measure the pH. Titrate the pH using NaOH or HCl to get a pH of 7.4. Filter the solution through funnel with a Whatman filter paper into a graduated cylinder and bring the volume up to 500 ml with ddH_2O and mix thoroughly. Aliquot into 50 ml sterile conical tubes and put on ice to use immediately or store in –20°C.
2. Wash solutions: 1× PBS; 1× PBT (1× PBS with 0.1% Tween-20).
3. Blocking solution: in 50 ml sterile conical tube, add 5 ml of FBS, 0.5 ml of Triton-X, 0.5 ml of 2% sodium azide in water, 1 ml of serum from the secondary antibody host animal and bring the volume up to 50 ml with DMEM.
4. Primary antibody: Rabbit anti-activated caspase-3 antibody (Upstate), Rabbit anti-Pax6 antibody (Chemicon), and mouse anti-GFP antibody (Molecular Probes/Invitrogen).
5. Secondary antibody: Alexa 488 anti-mouse IgG and Alexa 596 anti-rabbit IgG (Molecular Probes/Invitrogen).
6. Stock DAPI solution: 100 μg/ml 4′,6-diamidino-2-phenylindole (DAPI) in H_2O (Roche).
7. Slide mounting: Gel mount medium and glass coverslip.

3. Methods

3.1. Chicken Primary Retinal Culture (Fig. 1a)

The following procedures are performed in a dedicated surgical area and a tissue culture hood in an aseptic manner:

1. Incubate two dozen fertilized chicken eggs in the humidified chicken egg incubator at 38.5°C for 120 h (5 days). Turn off the rotator an hour before dissection (see Note 1).
2. Dissect HH stage 27 (19) chick embryo: using the curved scissors, cut away a piece of the eggshell exposing the embryo. Place the egg on an egg holder and cut through the membranes around the embryo with the curved scissors while using the forceps to hold the embryo. Scoop up the embryo using a perforated spoon and transfer the embryo into a 10-cm Petri dish containing HBSS+. Dissect away the membranes and release the embryo into the solution (see Note 2).

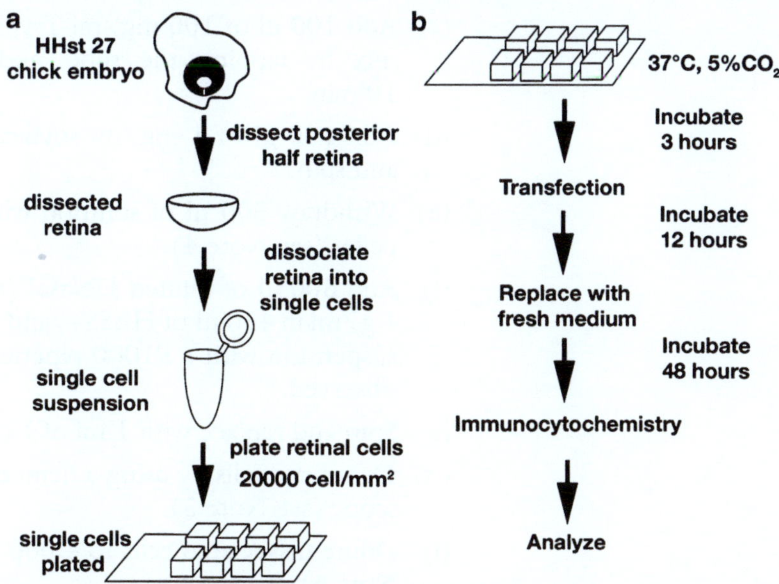

Fig. 1. Flow charts of primary chicken retinal cell transfection. (**a**) Procedures to establish dissociated primary embryonic retinal monolayer cultures. (**b**) Steps used in retinal cell transfection and immunocytochemical analysis.

3. Dissect the central retina: using the small spring scissors to cut and remove 40% of the anterior half of the eye, including the ciliary margin and the lens. Using forceps to detach the remaining 60% of the retina away from the retinal pigmented epithelium (RPE). Carefully trim away any remaining RPE still attached to the central retina. Using a sterile pipette to transfer the dissected retina to a clean 35-mm culture dish containing ~2 ml of 1× RCM prewarmed to 37°C (see Note 3).

4. Dissociate retina:

   Before starting the dissociation, make a humidified chamber for the LabTek culture slides using a 150-mm dish with a clean moist paper towel wetted with sterile water. Each 150-mm dish will hold up to 4 slides. Pre-coat the culture slide wells with 400 μl of 10 μg/ml poly-D-lysine solution in water (1:1,000 dilution of the stock solution with sterile water). Incubate at 37°C for minimum of 15 min. Aspirate right before plating.

   All retinal cell centrifugation are carried out at 327 × *g* for 2 min with a microcentrifuge.

   All solutions must be at room temperature before use.

   This process can dissociate retinas collected from up to a dozen embryos per tube.

   (a) Transfer the retinas into a 1.5-ml Eppendorf tube using a medium sterile transfer pipette.

   (b) Spin and wash with HBSS– three times with 1 ml of solution for each wash. For the last wash, take out as much of the solution as possible and add 1 ml of HBSS–.

(c) Add 100 μl of 100 mg/ml Trypsin typeXII to each tube, mix by tapping the tube, and incubate at 37°C for 10 min.

(d) Add 100 μl of 1 mg/ml soybean trypsin inhibitor, mix, and spin.

(e) Withdraw 500 μl of solution without disrupting the cell pellet (see Note 4).

(f) Add 500 μl of diluted DNaseI (50 μl of stock DNaseI at 1 g/ml in 450 μl of HBSS+) and carefully triturate the cell suspension with a P1000 pipette until no cell clumps are observed.

(g) Spin and replace with 1 ml of 1× RCM.

(h) Count the cells by using a hemocytometer under a microscope (see Note 5).

(i) Dilute dissociated cells to 4,000 cells/μl in 1× RCM (see Note 6).

(j) Plate 400 μl of cell suspension per well in the 8-well LabTek to achieve a cell plating density of 20,000 cells/mm^2. Allow the cells to settle at room temp for 30 min and then place the LabTech culture slides in the 37°C 5% CO_2 incubator for 3 h (see Note 7).

3.2. Transfection (Fig. 1b)

The preparation of DNA/PLUS reagent/Lipofectamine solution should be carried out approximately 30 min before the cells are ready. This recipe is for one well only. To accommodate multiple wells, multiply the volumes for a single well and make sure there are excess reagents, e.g. for three wells make 3.5× the amount.

1. Set up DNA and PLUS solution.

   Mix the following:

   2 μl Pax6i shRNA or Pax6 overexpression vector DNA at 500 ng/μl.

   1 μl CAGpGFP vector at 500 ng/μl.

   20 μl Opti-MEM.

   4 μl Lipofectamine PLUS reagent.

   Mix the DNA and Opti-MEM prior to the addition of the PLUS solution.

   Mix the PLUS solution prior to use.

   Add PLUS reagent directly to the DNA/Opti-MEM solution and mix.

   Incubate for 15 min at room temperature (RT).

2. In a separate tube, mix 1 μl Lipofectamine directly into 20 μl of Opti-MEM and mix.
3. Add the lipofectamine mix (step 2, 21 μl total) directly to the DNA/PLUS solution (step 1, 27 μl total) and incubate for 15 min at RT.
4. Wash cells once with Opti-MEM and add 160 μl of Opti-MEM.
5. Add all of the lipofectamine/PLUS/DNA (48 μl total volume) dropwise on the cells and incubate in 5% CO_2 incubator at 37°C for 3 h.
6. Add 200 μl 2× RCM to the cells and incubate for additional 24 h.
7. Replace with 400 μl 1× RCM and incubate for additional 48 h. At this point if the transfection was successful, GFP-positive cells can be observed under a fluorescent dissection scope.

3.3. Immunocytochemistry

All wash steps need at least 3 min of incubation (see Note 8).

1. Aspirate culture medium and wash cells with 1× PBS.
2. Fix cells with ice-cold 4% paraformaldehyde/1× PBS for 15 min at RT.
3. Wash cells with 1× PBS three times and take off the culture well away from the slide at the end of the last wash.
4. Incubate with blocking solution for 60 min at RT.
5. Incubate with primary antibody solution for 60 min at RT (e.g. to make 1 ml of primary antibody solution, add 4 μl of rabbit anti-activated Caspase-3 and 5 μl of mouse anti-GFP antibody in 1 ml of Blocking solution).
6. Wash with 1× PBT three times.
7. Incubate with secondary antibody solution for 60 min at RT (e.g. to make 1 ml of secondary antibody solution, add 2 μl of Alexa 488 conjugated anti-mouse antibody and 2 μl of Alexa 598 conjugate anti-rabbit antibody in 1 ml of blocking solution).
8. Incubate with 1× PBT containing 1 μg/ml DAPI for 10 min at RT (add 1 μl of stock 100 μl/ml DAPI solution per 100 μl of 1× PBT)
9. Wash with 1× PBT three times and remove excess solution.
10. Mount with gel mount medium and glass cover slip. Dry overnight at RT in the dark.
11. Visualize antibody labeling signals using fluorescent microscopy (Fig. 2).

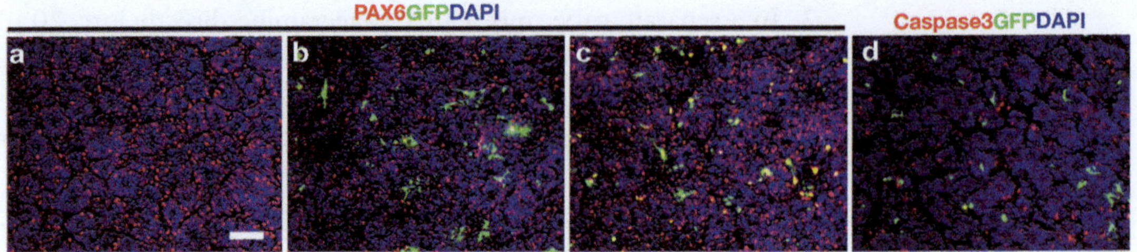

Fig. 2. Fluorescent micrograph images of primary chicken retinal cells after transfection and 72-h culture. HH stage 27 retinal cells were transfected in vitro with (**a**) no DNA, (**b**, **d**) GFP-expressing plasmid and Pax6shRNA construct, and (**c**) GFP-expressing plasmid and a Pax6-expressing construct. Cells were colabeled for Pax6, GFP, and DAPI in (**a**–**c**) and for caspase-3, GFP, and DAPI in (**d**). Scale bars = 100 m.

4. Notes

1. Chicken eggs must be rotated to prevent the embryo from sticking to the eggshell and causing deformation and abnormal development. The automatic rotator should be turned off prior to dissection so that the embryo remains at the top of the egg yolk. This will make it easier to find the embryo once the eggshell is opened.
2. If you maintain the orientation of the egg when removing it from the incubator, the embryo would be on the top center, right below the eggshell. Open a window wide enough (about 2 cm in diameter) for insertion of the perforated spoon to scoop up the embryo. Take care not to damage the embryo or disrupt the yolk by gently cracking the eggshell at an off-center position using the tip of the curved scissors, and then cut around the center using the curved scissors. Using the forceps to grab hold of the membrane close to the embryo but not the actual embryo itself. After cutting the membrane around the embryo, lift the embryo onto the perforated spoon and transfer to the dish. At this point, the embryo is surrounded by membranes. To expose the embryo, use two pairs of forceps and gently pull the membranes apart. Stage the embryos according to Hamburger and Hamilton (19) and select stage 27 embryos according to the specific external morphological features including the beak, branchial arches, and limb buds.
3. Removal of the anterior portion of the eye can facilitate the clean dissection of the retina from the RPE. There are many ways one can detach the RPE away from the retina. One effective way to remove the RPE is by inserting the forceps in between the RPE and the retina and allowing the forceps to open. However, the retina tends to be tightly adhered to the RPE at the ventral optic fissure. Further cutting along the optic fissure can help to remove the residual RPE.

4. After the trypsin incubation, the cells will be clumpy and sticky due to releasing of DNA from certain lysed cells. Therefore, DNaseI digestion, which requires Mg^{2+}, is necessary to result in single cell suspension.

5. At this point, if clumps of cells are observed, repeat dissociation from the three times HBSS– wash step.

6. The cells are diluted to 4,000 cells/µl of 1× RCM because the culturing surface area of the 8-well Lab-Tek is 0.8 cm^2 or 80 mm^2, requiring 1.6 million cells in a working volume of 400 µl to reach the plating density of 20,000 cells/mm^2.

7. Make sure to resuspend the cells before each time you take an aliquot of the cells. In order to achieve even plating of cells, avoid moving the culture dishes/slides in circular motions. At the end of the incubation time for plating, check to see if the cells have adhered to the dish by gently shaking the culture dish under an inverted microscope. Most cells that settle at the bottom should not move.

8. We put enough solution to cover the cells and aspirate the solutions from a fixed position (i.e. top left corner of each well) each time to minimally disturb the cells.

References

1. Agathocleous M, Harris WA (2009) From progenitors to differentiated cells in the vertebrate retina. Annu Rev Cell Dev Biol 25:45–69
2. Livesey FJ, Cepko CL (2001) Vertebrate neural cell-fate determination: lessons from the retina. Nat Rev Neurosci 2:109–118
3. Yang XJ (2004) Roles of cell-extrinsic growth factors in vertebrate eye pattern formation and retinogenesis. Semin Cell Dev Biol 15:91–103
4. Koller BH, Hagemann LJ, Doetschman T, Hagaman JR, Huang S, Williams PJ, First NL, Maeda N, Smithies O (1989) Germ-line transmission of a planned alteration made in a hypoxanthine phosphoribosyltransferase gene by homologous recombination in embryonic stem cells. Proc Natl Acad Sci USA 86:8927–8931
5. Koller BH, Smithies O (1989) Inactivating the beta 2-microglobulin locus in mouse embryonic stem cells by homologous recombination. Proc Natl Acad Sci USA 86:8932–8935
6. Mansour SL, Thomas KR, Capecchi MR (1988) Disruption of the proto-oncogene int-2 in mouse embryo-derived stem cells: a general strategy for targeting mutations to non-selectable genes. Nature 336:348–352
7. Sauer B, Henderson N (1988) Site-specific DNA recombination in mammalian cells by the Cre recombinase of bacteriophage P1. Proc Natl Acad Sci USA 85:5166–5170
8. Wang X (2009) Cre transgenic mouse lines. Methods Mol Biol 561:265–273
9. Gehring WJ (2002) The genetic control of eye development and its implications for the evolution of the various eye-types. Int J Dev Biol 46:65–73
10. Sanes JR, Zipursky SL (2010) Design principles of insect and vertebrate visual systems. Neuron 66:15–36
11. Yang XJ (2002) Retrovirus-mediated gene expression during chick visual system development. Methods 28:396–401
12. Fischer AJ, Reh TA (2001) Muller glia are a potential source of neural regeneration in the postnatal chicken retina. Nat Neurosci 4:247–252
13. Hill RE, Favor J, Hogan BL, Ton CC, Saunders GF, Hanson IM, Prosser J, Jordan T, Hastie ND, van Heyningen V (1991) Mouse small eye results from mutations in a paired-like homeobox-containing gene. Nature 354:522–525
14. Schedl A, Ross A, Lee M, Engelkamp D, Rashbass P, van Heyningen V, Hastie ND (1996) Influence of PAX6 gene dosage on development: overexpression causes severe eye abnormalities. Cell 86:71–82
15. Marquardt T, Ashery-Padan R, Andrejewski N, Scardigli R, Guillemot F, Gruss P (2001) Pax6 is required for the multipotent state of retinal progenitor cells. Cell 105:43–55

16. Hsieh YW, Yang XJ (2009) Dynamic Pax6 expression during the neurogenic cell cycle influences proliferation and cell fate choices of retinal progenitors. Neural Dev 4:32
17. Niwa H, Yamamura K, Miyazaki J (1991) Efficient selection for high-expression transfectants with a novel eukaryotic vector. Gene 108:193–199
18. Sui G, Soohoo C, Affar el B, Gay F, Shi Y, Forrester WC (2002) A DNA vector-based RNAi technology to suppress gene expression in mammalian cells. Proc Natl Acad Sci USA 99:5515–5520
19. Hamburger V, Hamilton HL (1992) A series of normal stages in the development of the chick embryo. 1951. Dev Dyn 195:231–272

Chapter 13

Production of High-Titer RCAS Retrovirus

Run-Tao Yan and Shu-Zhen Wang

Abstract

RCAS (B/P) is a replication-competent avian retrovirus engineered by Hughes et al. (J Virol 61:3004–3012, 1987) and is referred to in this chapter as RCAS for simplicity. The RCAS retrovirus has been widely used as a vehicle for stable transduction of a gene into cells both in the developing chick embryo and tissue/cell culture. It can be used for both gain- and loss-function experiments. The ability of this virus to spread among proliferating cells makes it possible to achieve widespread gene transduction in the developing retina. The transduction efficiency of RCAS is highly depending on the titer of the viral stock, particularly for experiments involving solid tissues such as the developing retina. Here, we describe the procedure that we have used for 15 years to generate RCAS viral stocks with a titer of $1–5 \times 10^8$ pfu/ml.

Key words: Avian retrovirus, Replication competent, High titer, Retroviral transduction, Chick embryos, CEF cells

1. Introduction

RCAS is a replication-competent avian retrovirus engineered from ALV LTR with splice acceptor retrovirus (1). The natural ability of this virus to spread in a population of dividing cells makes it possible to achieve widespread viral infection and, hence, gene transduction in chick embryos (2). A widespread gene transduction in turn renders it possible for phenotypic changes, provided that the gene is capable of inducing such change. Since the infectious viral particles can be administered into the developing eye after it has assumed its basic structure, RCAS retrovirus can be particularly useful in elucidating genetic regulation of retinal development. RCAS viral infection has no obvious effect on retinal development; we have observed no abnormalities in the developing chick retina either at the gross level or at the microscopic level from hundreds

Shu-Zhen Wang (ed.), *Retinal Development: Methods and Protocols*, Methods in Molecular Biology, vol. 884, DOI 10.1007/978-1-61779-848-1_13, © Springer Science+Business Media, LLC 2012

of embryos infected with RCAS or RCAS expressing the green fluorescent protein (RCAS-GFP).

The RCAS system offers a powerful tool for gain-of-function analysis. In gain-of-function studies using embryonic chick eyes, RCAS transduction of various regulatory genes has produced overt, readily detectable phenotypes, including a thickened outer nuclear layer (3), microphthalmia (4), or corneal extrusion (5). Notably, a widespread transduction of certain genes may detrimentally affect the embryos. For instance, infection of chick embryos with RCAS expressing *neurognin3* (*ngn3*) (6), a gene well known for its proendocrine role by determining which precursor cells in the developing pancreas will become insulin-producing cells of the islets of Langerhans, is embryonic lethal (7). Embryonic lethality also occurs with infection of RCAS expressing *cNSCL1* (8) or *ash1* (5). Compared to the transgenic mouse approach, the RCAS system is more economical and less time consuming.

When engineered to express a dominant negative construct, the RCAS retrovirus can be an effective alternative to other loss-of-function approaches. In our experience, it is particularly potent when coupled with Drosophila Engrailed-mediated active repression if the gene of interest is a transcription activator. For example, infection of the developing retina with RCAS expressing the construct of Drosophila Engrailed (En) fused with NeuroD lacking the C-terminal region (En-NeuroDΔC) results in photoreceptor deficits (9), and widespread infection with RCAS-En-Cbx(HD) is embryonic lethal while that of RCAS-Cbx is tolerated (10).

Like any other experimental system, the RCAS retroviral system has inherent limitations. For example, it creates ectopic expression of a gene (or its dominant negative construct) outside the temporal window and the spatial locale of its normal expression. In addition, the level of expression is not readily alterable or regulated, and the time of onset of gene expression in a specific cell may not be simply calculated from the time of the initial viral administration due to viral spreading through secondary and tertiary infections. Therefore, while RCAS offers opportunities for functional studies on a number of fronts, its limitations should be kept in mind in experimental design and data interpretation.

2. Materials

Prepare all solutions using ultrapure water and analytical grade reagents. Use sterilized reagents in cell culture and preparation of viral stocks.

2.1. Chick Embryos

1. Pathogne-free, fertilized chick eggs.
2. Chick egg incubator, set at 38°C with a water reservoir for humidity.

3. Sterile (autoclaved) forceps/tweezers with curve tips (Dumont #7 tweezers).
4. Sterile (autoclaved) razor blades.

2.2. Cell Culture Medium

1. Medium 199.
2. Medium 199 plus 10% fetal bovine serum (Medium 199 + 10% FBS).
3. FBS.
4. Phosphate-buffered saline (PBS).
5. 0.25% Trypsin/EDTA.
6. Cell-Strainer (Falcon 352235).
7. 15-ml plastic tube.
8. Cell culture flasks, 25 and 75 cm^2.
9. Plastic transfer pipettes.
10. 37°C Water bath.

2.3. Transfection

1. Purified RCAS DNA prepared using a commercial plasmid DNA preparation kit.
2. Fugene 6 (Roche Biochemicals).

2.4. Centrifugation and Ultracentrifugation

1. High-speed centrifuge J2-HS (Beckman Instruments).
2. Ulultracentrifuge L-80 (Beckman Instruments).
3. Ultra-Clear centrifuge tubes (Cat# 344058 for SW32Ti rotor, Beckman Instruments).

3. Methods

There are four main steps in producing high-titer RCAS viral stocks: (a) set up primary culture of chick embryonic fibroblast (CEF) cells, (b) transfect the cells with RCAS DNA, (c) harvest cell culture medium containing viral particles, and (d) concentrate the viral particles to obtain a high-titer viral stock. All steps need to be carried out with care to avoid microbial contaminations.

3.1. Establishing CEF Cell Culture

1. Take one day-8 chick embryo (see Note 1) out of egg incubator and clean the center top of its shell with 70% ethanol.
2. Gently knock at the cleaned center top with a pair of sterile forceps to introduce small cracks in the area.
3. Remove the shell pieces.
4. Scoop out the embryo with the curved tips of a pair of sterile forceps and place it in a sterile plastic dish.

5. Decapitate the embryo with a sterile razor blade, remove the gust with tweezers, and transfer the remaining tissue (body trunk with limbs) into another dish.
6. In a cell culture hood, mince the tissue with a razor blade into fine pieces, and transfer them into a 15-ml tube.
7. Wash the tissue twice with PBS.
8. Add 2.5 ml of 0.25% trypsin/EDTA.
9. Place the tube in a 37°C water bath for 5 min.
10. Bring the tube back in cell culture hood and triturate the tissue ten times with a plastic transfer pipette.
11. Add 10 ml of Medium 199 + 10% FBS. Mix by pipetting.
12. Pass portion (~3 ml) of the cell/tissue suspension through a Cell-Strainer.
13. Transfer 1 ml of the pass through (containing CEF cells) into a 25-cm^2 flask with 4 ml of Medium 199 + 10% FBS.
14. Place the cell culture flask in a 37°C incubator with 5% CO_2.
15. When the CEF cell culture reaches 70% confluence, begin the DNA transfection step as described next.

3.2. Transfecting CEF Cells with Virus DNA

Fugene 6 (see Note 2) is used for transfection of the CEF cells with RCAS DNA. Detailed protocol provided by Roche Biochemicals is followed and is not repeated here.

3.3. Harvesting RCAS Retrovirus

1. When the transfected cell culture becomes confluent (about 2–3 days after DNA transfection), split the 25-cm^2 flask of CEF cells into two 75-cm^2 flasks.
2. Culture the cells in 75-cm^2 flasks with 10 ml of Medium 199 + 10% FBS. Change medium every other day.
3. When the culture is confluent, split the culture 1–5.
4. Change medium every other day with 10 ml of Medium 199 + 10% FBS.
5. When the culture reaches 90% confluence, change medium with 5 ml of Medium 199 + 10% FBS.
6. After 24 h, collect the culture medium from each flask and replenish each flask with 5 ml of Medium 199 + 10% FBS.
7. Centrifuge the collected medium at 8,000 × g (5,500 rpm if using a JA-17 rotor) for 10 min at 4°C to remove cells/cell debris (see Note 3). Combine and store the harvest at –80°C (see Note 4).
8. Continue the harvesting step daily for another 5 days (see Notes 4 and 5).

3.4. Concentrating RCAS Retrovirus (See Note 6)

1. Thaw the collected culture medium (with virus) in a water-filled container at room temperature.
2. Centrifuge the collected medium at 67,000–81,000×*g* (20,000–22,000 rpm for sw41 Ti or sw32 Ti rotor) at 4°C for 20–30 min (see Note 7).
3. Discard the supernatant, followed by one fling of the centrifuge tube.
4. Use the small amount of the medium left in the tube to resuspend the viral particles by repeated pipetting for a few times with a 1-ml plastic pipette. The final volume should be 1% of the original (see Note 8).
5. Store the concentrated virus at −80°C in aliquots of 50 μl (see Notes 9 and 10).

4. Notes

1. We find that CEF cell culture from day-8 chick embryo works the best. Cells in culture established with younger embryos peel off culture flasks earlier, thus shortening the time during which virus is harvested. On the other hand, cells in a culture with older embryos often grow slower, lengthening the time required to generate virus.
2. $CaCl_2$ precipitation is commonly used to transfect cultured cells for the production of retroviruses. We have experimented with different transfection methods and found that results using Fugene 6 are comparable to those using CaCl2 precipitation. The advantages of Fugene 6 over CaCl2 precipitation are (a) it needs less DNA, (b) it is easier to perform, and (c) the results are more reproducible. Thus, we recommend Fugene 6 as the transfection reagent in this protocol.
3. Some published protocols call for the use of ultracentrifugation at 67,000×*g* for removing cells/cell debris. However, we found this to be detrimental to the yield of RCAS retrovirus, as it perhaps removes 90% of the virus along with cells/cell debris.
4. We routinely combine all harvests into a 500-ml bottle.
5. At this point, cells start to peel off the culture flask. Otherwise, one may continue the incubation and harvesting for another couple of days until the cells start to peel off.
6. It is important to have concentrated viral stocks free from microbial contamination for late in vivo and in vitro experiments. Therefore, cautions are to be exercised to avoid microbial contamination during ultracentrifugation steps. Use

autoclaved centrifuge tubes/bottles. When possible, operate in a cell culture hood.

7. Avoid longer than 30 min of ultracentrifugation at 67,000–81,000 × *g*, as longer ultracentrifugation induces clumping of the viral particles and inadvertently reduces the yield. Clumping of viral particles is evident by difficulty in resuspending. If clumping occurs with centrifugation for 30 min, reduce the time to 25 or 20 min. In our experiences, a 20-min centrifugation yields ~90% recovery of the viral particles, whereas a 25-min one gives >95% recovery.
8. In our experience, RCAS retrovirus stock with a high titer of $1–5 \times 10^8$ pfu/ml is required for a widespread viral infection of retinal cells by delivering the virus through microinjection into the developing eye. By resuspending into 1% of the original volume, the titer of the viral stock can be as high as 5×10^8 pfu/ml, since the concentration of the retrovirus in the harvested medium is $\sim 5 \times 10^6$ pfu/ml.
9. Our working experience tells us that the RCAS retrovirus is more stable than MSCV retrovirus. The procedure described here is developed over the past 15 years. It is simple and reliable.
10. Overall, three key points in achieving 98–100% recovery during the viral concentration step are those specified in Notes 3, 7, and 8.

Acknowledgments

This work is supported by NIH/NEI EY011640, EyeSight Foundation of Alabama FY2011-12-276, and an unrestricted grant to UAB Department of Ophthalmology from Research to Prevent Blindness.

References

1. Hughes SH, Greenhouse JJ, Petropoulos CJ, Sutrave P (1987) Adaptor plasmids simplify the insertion of foreign DNA into helper-independent retroviral vectors. J Virol 61:3004–3012
2. Fekete DM, Cepko CL (1993) Replication-competent retroviral vectors encoding alkaline phosphatase reveal spatial restriction of viral gene expression/transduction in the chick embryo. Mol Cell Biol 13:2604–2613
3. Yan R-T, Wang S-Z (1998) neuroD induces photoreceptor cell overproduction in vivo and de novo generation in vitro. J Neurobiol 36:485–496
4. Li C-M, Yan R-T, Wang S-Z (1999) Misexpression of cNSCL1 disrupts retinal development. Mol Cell Neurosci 14:17–27
5. Mao W, Yan R-T, Wang S-Z (2009) Proneural gene ash1 promotes amacrine cell production in the chick retina. Dev Neurobiol 69:88–104
6. Sommer L, Ma Q, Anderson DJ (1996) neurogenins, a novel family of atonal-related bHLH transcription factors, are putative mammalian

neuronal determination genes that reveal progenitor cell heterogeneity in the developing CNS and PNS. Mol Cell Neurosci 8:221–241

7. Ma W, Yan R-T, Mao W, Wang S-Z (2009) Neurogenin3 promotes early retinal neurogenesis. Mol Cell Neurosci 40:187–198
8. Yan R-T, Wang S-Z (2001) Embryonic abnormalities from misexpression of *cNSCL1*. Biochem Biophys Res Cummun 287: 949–955
9. Yan R-T, Wang S-Z (2004) Requirement of NeuroD for photoreceptor formation in the chick retina. Invest Ophthalmol Vis Sci 45:48–58
10. Li C, Yan R-T, Wang S-Z (2002) Chick homeobox gene cbx and its role in the development of retinal bipolar cells. Mech Dev 116:85–94

Chapter 14

Chick Retinal Pigment Epithelium Transdifferentiation Assay for Proneural Activities

Shu-Zhen Wang and Run-Tao Yan

Abstract

We describe a cell culture system for assaying proneural activities of genes hypothesized to play instrumental roles in neuronal fate specification during vertebrate retinal development. The retinal pigment epithelium (RPE) is collected from embryonic day 6 (E6) chick to establish a primary RPE cell culture. The culture is then infected with a replication competent retrovirus RCAS expressing the gene of interest. The presence of retinal neurons in the otherwise nonneural, RPE cell culture is examined between 4 and 10 days after the administration of the virus. Taking advantage of the plasticity and the relative simplicity of RPE cells, this method offers an informative assay for proneural activities prior to planning for large-scale in vivo experiments.

Key words: Proneural genes, Transcription factor, Transdifferentiation, Photoreceptor, RPE, Neuron, Retina

1. Introduction

Deciphering the role of a transcription factor in retinal cell fate specification during vertebrate retinal development constitutes an important area of study in the broad field of neural development. One of the main hurdles encountered in this area of research is a lack of an effective assay for the many factors implicated to play inductive roles during retinal neurogenesis. The neural retina, due to its natural expression of many neural genes, including proneural genes, and its composition of varied cell types, can be ill-suited for such a study, as it is often difficult to dissect the role of each individual gene, much less a combination of genes and their hierarchies. To address this issue, we have been exploring the possibility of using the retinal pigment epithelium (RPE) as an alternative medium to illustrate proneural activities of factors implicated in retinal neurogenesis.

Shu-Zhen Wang (ed.), *Retinal Development: Methods and Protocols*, Methods in Molecular Biology, vol. 884,
DOI 10.1007/978-1-61779-848-1_14, © Springer Science+Business Media, LLC 2012

The RPE consists of darkly pigmented cells organized as a single-layered, transporting epithelium with important roles in retinal physiology. Anatomically, the RPE lies immediately adjacent to neural retina and forms the outer blood–retinal barrier. Developmentally, the nonneural RPE and the neural retina originate from the same structure—the optic vesicle. The identities of RPE vs. neural retina is established during the transformation of the optic vesicle into the double-layered optic cup, with the outer layer forming the RPE and the inner layer developing into the neural retina.

Three well-known properties of the RPE render it a suitable medium for assaying proneural activities. First, the nonneural, single-layered epithelium lacks the expression of many proneural genes. Second, unlike retinal neurons, RPE cells can reenter the cell cycle upon stimulation. Third, progeny cells of RPE may, under appropriate conditions, differentiate into cell types other than RPE (1–7). Classic experiments have revealed tissue transdifferentiation into a neural retina from embryonic chick RPE (8–13), embryonic rodent RPE (3, 14), and amphibian RPE (2, 15).

In this chapter, we describe an RPE transdifferentiation assay that uses dissociated chick RPE cell culture as a medium to reveal proneural activity through RCAS (16) retroviral transduction of a gene of interest. Over the years, we have observed the following regarding this assay. (a) The type of neural cells resulted from gene-directed RPE transdifferentiation appears to reflect, to some extent, the function of the gene in retinal neurogenesis (see Note 1). (b) Not all genes important for retinal development are effective in inducing RPE transdifferentiation (see Note 2). (c) RPE transdifferentiation can be induced by extrinsic factors (see Note 3). (d) Co-treatment with two or more factors can be applied to the assay (see Note 4). (e) The efficiency and the extent of neural differentiation show gene dependency (see Note 5).

It should be kept in mind that RPE cells are not retinal neuroblasts. In addition, while the simplicity of RPE makes it an attractive system for experimental manipulations, this very simplicity may become a drawback. Caution should be exercised when attempting to apply results from RPE experiments to the developing retina. To this end, RPE transdifferentiation assays should be complemented with and supported by experiments using the neural retina and retinal cells. Nonetheless, the RPE transdifferentiation assay is a useful tool in revealing the potential roles of genes and factors in retinal neurogenesis.

2. Materials

2.1. Chick Embryo Incubation

1. Chick egg incubator.
2. Pathogen free, fertilized chick eggs (see Note 6).
3. Gooseneck fiber light source for "candling" chick embryo in egg.

2.2. Tissue Dissection

1. Dissecting microscope with gooseneck fiber light source.
2. Sterile (autoclaved) dissecting tools: Dumont #7 (curve), #3, and #5 tweezers (see Note 7).
3. Sterile 60-mm dishes.
4. 15-ml Sterile, disposable centrifuge tubes.
5. Sterile, plastic transfer pipette.

2.3. Cell Culture

1. Equipment:

   Cell culture hood (i.e. biological safety cabinet).

   37°C cell culture incubator with 5% CO_2.

   Countertop centrifuge.

   Inverted microscope.
2. Cell culture dishes (35-mm diameter) or six-well culture plates.
3. Cell culture medium (see Note 8):

   Medium 199.

   Medium 199 supplemented with 10% fetal calf serum (199 + 10% FCS).

   Knock-out D-MEM supplemented with 20% serum replacement (KO/SR).

   0.25% Trypsin/EDTA.

   Hank's balanced salt solution (HBSS).

   Ca^{2+}, Mg^{2+}-free HBSS (CMF).
4. Concentrated RCAS (16) retrovirus expressing a gene of interest and RCAS virus expressing a control gene, such as GFP, with a titer of $1–5 \times 10^8$ pfu/ml (see Note 9).

2.4. Neural Detection

1. Fixation solution: ice-cold, 4% paraformaldehyde in phosphate buffered saline (PBS), pH 7.4.
2. Antibodies or antisense RNA probes specific for retinal neurons, and buffers and reagents for routine immunochemistry or in situ hybridization.

3. Methods (See Note 10)

1. In a cell culture hood (biological safety cabinet), add 1 ml of KO/SR to each 35-mm culture dish (see Note 11). Place the dishes in a 37°C cell incubator (for seeding cells later of the day).

2. Warm the following solutions in a 37°C water bath for 10–20 min:

   Medium 199 (see Note 12).

   Medium 199 + 10% FCS.

   CMF HBSS.

   0.25% Trypsin/EDTA.

3. Wipe the outside of the bottles with 70% ethanol and Kimwipes, flame the mouth of the bottles, and place them in a cell culture hood.

4. Add 10 ml of Medium 199 to each of three 60-mm dishes (for dissection), and add 3 ml of Medium 199 + 10% FCS to a 35-mm dish (for collecting isolated RPE tissue).

5. Transfer the dishes to a dissecting bench surface-cleaned with 70% ethanol.

6. Candle chick eggs at E6 (see Note 13) to select for those with viable embryos and mark the position of the embryo with a permanent marker. Lay the chosen ones horizontally on an egg carton.

7. Clean egg top with 70% ethanol. Gently knock at the center-top of shell with a pair of tweezers to produce cracks in shell, and remove shell pieces with a pair of sterilized Dumont #7 tweezers. With another pair, break the vitelline membrane and scoop out the embryo. Place the embryo in a 60-mm dish.

8. Decapitate the embryo, enucleate the eyes (see Note 14), and place the eyes in the first of the three 60-mm dish with 10 ml of Medium 199.

9. Under a dissecting microscope, remove the sclera using Dumont #3 tweezers. Transfer the RPE-retina-vitreous-lens into the second 60-mm dish with 10 ml of Medium 199 (see Note 14).

10. Using Dumont #5 tweezers, make an incision in the RPE + retina along the ora serrata to rid of the ciliary epithelium, the associated periphery retina, and the lens. Place the RPE-retina-vitreous into the third 60-mm dish with 10 ml of Medium 199.

11. Separate RPE from the retina and vitreous. Make sure that no retinal tissue is attached to the RPE, particularly at the periphery region. Place the isolated RPE in the 35-mm dish with 199 + 10% FCS (see Note 15).

12. After enough RPE tissues have been collected (see Note 16), take the 35-mm dish with RPE to the cell culture hood. Transfer the RPE into a sterile 15-ml tube using a sterile, plastic transfer pipette.

13. Let the tissue sink down, and take out the residual solution. Rinse the RPE twice with 5–10 ml of CMF.

14. Add trypsin/EDTA, triturate the tissue about 20 times with a plastic transfer pipette, and watch carefully not to over-trypsinize the tissue (see Note 17).
15. As soon as all big pieces of RPE tissue become small and barely visible (about 5 min in the hood), add 10 ml of 199 + 10% FCS to stop the trypsin digestion. Mix by pipetting a couple times.
16. Centrifuge at 650 × *g* for 5 min at room temperature in a countertop centrifuge.
17. In the culture hood, flame around the cap area of the tube, and gently reverse the tube to discard the supernatant. Use the transfer pipette to take away any extra solution at the mouth of the tube.
18. Resuspend the cell pellet with 199 + 10% FCS (1.5 ml for each E6 RPE) by pipetting with a transfer pipette.
19. Seed 0.5 ml cells into each of the 35-mm dishes placed in the 37°C cell incubator and containing 1 ml of KO/SR.
20. Culture the cells and change medium every other day with 1.5 ml KO/SR (see Note 18).
21. Examine the culture with an inverted microscope. At ~50% confluency (see Note 19), add 10–20 μl RCAS virus. Swirl the dish gently 50 revolutions. Repeat the swirling four times during the day (see Note 20).
22. On the following day, repeat the swirling (four times during the day, each with 50 revolutions).
23. Change medium every other day with 1.5 ml of KO/SR.
24. Between 8 and 10 days after the administration of the virus (see Note 21), fix the cells with ice-cold 4% paraformaldehyde in PBS for 30 min, and proceed with immnocytochemistry, in situ hybridization, or physiological (e.g. Ca^{2+} imaging; see Note 22) analysis for the presence of retinal neurons in the culture (see Note 23).

4. Notes

1. When ectopically expressing *neuroD*, the RPE cells begin to express a photoreceptor phenotype, an observation consistent with *neuroD* expression in young photoreceptor cells and their precursors and its selective promotion of photoreceptor production in the retina (17–20). On the other hand, ectopic expression of *neurogenin2* (*ngn2*) in RPE cell culture induces de novo appearance of molecularly and morphologically different types of cells, including photoreceptor cells, retinal ganglion cells (RGCs), and in a smaller number, amacrine cells (21). This is consistent with *ngn2*'s expression in proliferating,

multipotent progenitors as shown by a fate mapping study using the Cre-ER™—LacZ system (22). Yet, another proneural bHLH gene, *ash1*, induces RPE cells to transdifferentiate into cells resembling neither photoreceptor cells nor RGCs, but rather amacrine cells (23). This agrees with the observation that *ash1* mis- or overexpression in the developing chick retina increases amacrine cell production while causing photoreceptor deficiency (24).

2. We have observed that not all bHLH genes that are expressed in the developing retina and are homologous to Drosophila proneural genes are capable of initiating detectable RPE transdifferentiation toward retinal neurons. These "ineffectual" bHLH genes include *NSCL1* and *NSCL2*. In addition, several homeodomain genes well known for their roles in eye/retina development, such as *Rax*, *RaxL*, and *six3*, also appear "ineffectual." The negative outcome, however, does not undermine the importance of these genes/factors in retinal development. Rather, it shows their ineffectiveness in inducing neurogenesis in the context of RPE cells.
3. Under the induction of bFGF, which is believed to potentiate RGC fate (12), RPE cells begin to transdifferentiate in the direction of becoming RGCs (25), although the extent of transdifferentiation is very limited.
4. The extent of bFGF-induced transdifferentiation towards RGCs can be enhanced by RCAS transduction of *ath5* and *NSCL1* (26), bHLH genes expressed in developing RGCs, and to a greater extent by their cotransduction (27).
5. Under the induction of *ngn1* or *ngn3*, over 80% of the cells present in a dish may display a noticeable neural trait (28). The neural differentiation initiated by *ngn1* can proceed to advanced stages examined at the molecular, morphological, and physiological levels (28). On the other hand, transdifferentiation initiated by *sox2* seems to stall at primitive stages (29).
6. Pathogen-free chick embryos are important for experiments using retrovirus RCAS as a vehicle for gene transduction.
7. Tweezers with fine tips are vital for isolating the single-layered RPE.
8. These solutions must be free of viral contamination.
9. In our hand, a viral stock with a high titer, $1–5 \times 10^8$ pfu/ml, plays an important role in the success of the transdifferentiation assay. For method on producing high titer RCAS virus, see the chapter by Yan and Wang in this volume.
10. Because the delicacy of the tissue being handled, investigators may refrain from caffeine intake before and during dissection.
11. Each investigator will decide on number of dishes needed for one particular experiment. Depending on the specifics of each

experiment, 6-well plates may work better than individual 35-mm dishes. Cell culture vessels with smaller culture area may be used, but they often produce larger variations in cell densities. It is important to minimize variations in cell densities among the dishes in a given experiment to avoid undue complication in data interpretation.

12. Although we routinely use Medium 199 for chick cells, we have found that other cell culture medium, such as D-MEM:F12 or HBSS, can also be used.

13. We prefer to use E6 RPE for easy handling of the embryos and for effective transdifferentiation. While RPE from younger embryo may be more receptive to transdifferentiation, it, nevertheless, contains fewer cells and is technically more difficult to handle during the dissection process.

14. It is important to keep the eye ball intact during this step. Any puncture will result in collapsing of the eyecup during the following steps, and that will make it very difficult to isolate the PRE.

15. We found that the presence of 10% FCS improves the viability of RPE cells.

16. It is very important to have high quality RPE cell culture for the RPE transdifferentiation assay to be successful. To have high-quality RPE cell culture, the first step is to isolate sufficient amount of RPE tissues free from retinal contamination in a relatively short period of time. We recommend practicing RPE isolation a few times before carrying out assay experiment.

17. Over-trypsiniziation of the tissue is evident when the tissue/cell suspension becomes slimy after a gentle swirling.

18. Maintaining healthy cells in the culture improves the experimental outcome. We recommend two simple measures: (a) keep medium as fresh as possible (take out and warm up only the amount you need each time, and never use medium older than a month) and (b) minimize the time of cells being outside the incubator.

19. It usually takes 3 days for the culture to reach ~50% confluency. A culture that reaches ~50% confluency too early (e.g. in 2 days) or too late (e.g. in more than 4 days) tends not to work well.

20. This is to increase the chance of RCAS to attach to cells for subsequent infection.

21. As short as 4 days after the administration of the RCAS viruses expressing a proneural gene, the culture may display visible signs of neural transdifferentiation. One such sign is lower cell density in the transdifferentiating dish than the control dish infected with RCAS-GFP. Another sign is, under an inverted microscope, transdifferentiating dish contains many clusters of cells with compact cell body and displaying long processes,

reminiscent of neuronal clusters. The presence of these two foresees a highly effective neural transdifferentiation. However, the absence of these two does not necessarily indicate a lack of neural transdifferentiation. Instead, it may be a sign of relatively low efficiency due to the gene or low viral transduction.

22. Details on Ca^{2+} imaging analysis of the transdifferentiated culture can be found in refs. 28, 30.

23. This chapter focuses on steps of establishing and transducing RPE cell culture for assaying proneural activities, with a premise that neural detection is a routine procedure in a laboratory interested in the RPE transdifferentiation assay. Therefore, details in neural detection are not provided.

Acknowledgments

This work is supported by NIH/NEI EY011640, EyeSight Foundation of Alabama FY2011-12-276, and an unrestricted grant to UAB Department of Ophthalmology from Research to Prevent Blindness.

References

1. Eguchi G (1986) Instability in cell commitment of vertebrate pigmented epithelial cells and their transdifferentiation into lens cells. In: Moscona AA, Monroy A (eds) Current topics in developmental biology. Academic Press, New York, pp 21–37
2. Reh TA, Nagy T, Gretton H (1987) Retinal pigmented epithelial cells induced to transdifferentiate to neurons by laminin. Nature 330:68–71
3. Dutt K, Scott M, Sternberg PP, Linser PJ, Srinivasan A (1993) Transdifferentiation of adult human pigment epithelium into retinal cells by transfection with an activated H-ras proto-oncogene. DNA Cell Biol 12:667–673
4. Zhao S, Thornquist SC, Barnstable CJ (1995) In vitro transdifferentiation of embryonic rat pigment epithelium to neural retina. Brain Res 677:300–310
5. Crisant S, Guidry C (1995) Transdifferentiation of retinal pigment epithelial cells from epithelial to mesenchymal phenotype. Invest Ophthalmol Vis Sci 36:391–405
6. Zhao S, Rizzolo LJ, Barnstable CJ (1997) Differentiation and transdifferentiation of the retina pigment epithelium. Int Rev Cytol 171:225–265
7. Araki M, Yamao M, Tsudzuki M (1998) Early embryonic interaction of retinal pigment epithelium and mesenchymal tissue induces conversion of pigment epithelium to neural retinal fate in the silver mutation of the Japanese quail. Dev Growth Differ 40:167–176
8. Orts-Llorca F, Genis-Galvez JM (1960) Experimental production of retinal septa in the chick embryo. Differentiation of pigment epithelium into neural retina. Acta Anat 42: 31–70
9. Coulombre JL, Coulombre AJ (1965) Regeneration of neural retina from the pigmented epithelium in the chick embryo. Dev Biol 12:79–92
10. Park CM, Hollenberg MJ (1989) Basic fibroblast growth factor induces retinal regeneration in vivo. Dev Biol 134:201–205
11. Pittack C, Jones M, Reh TA (1991) Basic fibroblast growth factor induces retinal pigment epithelium to generate neural retina in vitro. Development 113:577–588
12. Guillemot F, Cepko CL (1992) Retinal fate and ganglion cell differentiation are potentiated by acidic FGF in an in vitro assay of early retinal development. Development 114: 743–754

13. Opas M, Dziak E (1994) bFGF-induced transdifferentiation of RPE to neuronal progenitors is regulated by the mechanical properties of the substratum. Dev Biol 161:440–454
14. Sakami S, Etter P, Reh TA (2008) Activin signaling limits the competence for retinal regeneration from the pigmented epithelium. Mech Dev 125:106–116
15. Vergara MN, Del Rio-Tsonis K (2009) Retinal regeneration in the *Xenopus laevis* tadpole: a new model system. Mol Vis 15:1000–1013
16. Hughes SH, Greenhouse JJ, Petropoulos CJ, Sutrave P (1987) Adaptor plasmids simplify the insertion of foreign DNA into helper-independent retroviral vectors. J Virol 61:3004–3012
17. Yan R-T, Wang S-Z (1998) DeuroD induces photoreceptor cell overproduction in vivo and de novo generation in vitro. J Neurobiol 36:485–496
18. Yan R-T, Wang S-Z (2000) Expression of an array of photoreceptor genes in chick embryonic RPE cell cultures under the induction of neuroD. Neurosci Lett 280:83–86
19. Pennesi ME et al (2003) BETA2/NeuroD1 null mice: a new model for transcription factor-dependent photoreceptor degeneration. J Neurosci 23:453–461
20. Yan R-T, Wang S-Z (2004) Requirement of NeuroD for photoreceptor formation in the chick retina. Invest Ophthalmol Vis Sci 45: 48–58
21. Yan R-T, Ma W, Wang S-Z (2001) Neurogenin2 elicits the genesis of retinal neurons from cultures of non-neural cells. Proc Natl Acad Sci USA 98:15014–15019
22. Ma W, Wang S-Z (2006) The final fates of neurogenin2-expressing cells include all major neuron types in the mouse retina. Mol Cell Neurosci 31:463–469
23. Mao W, Yan R-T, Wang S-Z (2008) Reprogramming chick RPE progeny cells to differentiate towards retinal neurons by ash1. Mol Vis 14:2309–2320
24. Mao W, Yan R-T, Wang S-Z (2009) Proneural gene ash1 promotes amacrine cell production in the chick retina. Dev Neurobiol 69:88–104
25. Yan R-T, Wang S-Z (2000) Differential induction of gene expression by basic fibroblast growth factor and neuroD in cultured retinal pigment epithelial cells. Visual Neurosci 17: 157–164
26. Ma W, Yan R-T, Xie W, Wang S-Z (2004) bHLH genes *cath5* and *cNSCL1* promote bFGF-stimulated RPE cells to transdifferentiate towards retinal ganglion cells. Dev Biol 265:320–328
27. Xie W, Yan R-T, Ma W, Wang S-Z (2004) Enhanced retinal ganglion cell differentiation by ath5 and NSCL1 coexpression. Invest Ophthalmol Vis Sci 45:2922–2928
28. Yan R-T et al (2010) Neurogenin1 effectively reprograms cultured chick RPE cells to differentiate towards photoreceptors. J Comp Neurol 518:526–546
29. Ma W, Yan R-T, Li X, Wang S-Z (2009) Reprogramming RPE cell differentiation in vivo and in vitro with Sox2. Stem Cells 27:1376–1387
30. Liang L, Yan R-T, Lim X, Chimento M, Wang S-Z (2008) Reprogramming progeny cells of embryonic RPE to produce photoreceptors: development of advanced photoreceptor traits under the induction of neuroD. Invest Ophthalmol Vis Sci 49:4145–4153

Part IV

Regeneration/Stem Cells

Chapter 15

Studying the Generation of Regenerated Retinal Neuron from Müller Glia in the Mouse Eye

Mike O. Karl and Thomas A. Reh

Abstract

Retinal regeneration has been studied for decades in nonmammalian species. From these studies, we learned that retinal Müller glia are a potential source of neuronal regeneration by de novo neurogenesis. Although spontaneous regeneration in mammals is absent after retinal damage, we discovered that certain manipulations stimulate a limited regenerative program in adult mice. This allows the study of cellular and molecular barriers that limit regeneration in mice and man as well as to compare it to regenerative species with more complete repair—like fish and birds. Thereby, it may offer novel means to the better understanding of retinal regeneration and develop successful therapies of retinal diseases. In this chapter, we describe the methods to study Müller glia-derived regeneration in adult mice in vivo and discuss potential caveats.

Key words: Regeneration, Glia, Stem cell, Neuron, Retina, Mouse

1. Introduction

Like most parts of the central nervous system, the mammalian neuronal retina is devoid of physiological or regenerative adult neurogenesis. Multiple conditions due to genetic, metabolic, or age-related changes can lead to neurodegeneration of the retina, which often leads to visual impairment. While gene therapy has made substantial progress towards treatment of inherited diseases, alternative therapies will be necessary as treatment options. Specifically, patients with already severe cell loss, and therefore loss of vision, might need to receive cell replacement. Exogenous cell transplantation as well as endogenous cell replacement by regeneration are both possible future approaches to retinal repair. Studying retinal regeneration by de novo neurogenesis from an

Shu-Zhen Wang (ed.), *Retinal Development: Methods and Protocols*, Methods in Molecular Biology, vol. 884, DOI 10.1007/978-1-61779-848-1_15, © Springer Science+Business Media, LLC 2012

already existing endogenous cell might not only reveal novel strategies for these two approaches, but also open our eyes to new ways to protect and slow down neuronal cell loss. Cell transplantation to date is limited by very low cell integration efficiency (1–4). Although various cell sources for transplantation have been proposed and investigated, studies of the limited endogenous regenerative response to injury in mammals might provide additional insights into methods to improve the production and integration of cell transplantation approaches. Ultimately, studies on regeneration might allow us to glean multiple new strategies for regenerative medicine, including therapies to prevent the onset and progression of neurodegenerations in the first place efficiently.

In fish, a damaged retina can heal itself, since it is able to generate new neurons that structurally and functionally integrate into the existing circuitry and restore vision (5–8). All types of neurons can be replaced depending on the types of neurons lost. More extensive damage induces a greater regenerative response. Studies on regeneration revealed that Müller glia can function as an adult stem cell source in fish and chick. Unfortunately, in the mammals that have been analyzed, this endogenous regeneration does not automatically occur. Future comparative studies may reveal the underlying mechanisms that specifically enable retinal repair by integration of new neurons into an existing circuitry in fish and those that limit it in mammals.

Although there is no spontaneous regeneration in mammalian retina, interestingly, many studies in various species including humans have shown that Müller glia are highly responsive to all kinds of retinal diseases and stresses (9–11). More recent research has challenged the barriers that limit neuronal repair in rats and mice by Müller glia-derived adult regenerative neurogenesis (12, 13). Based on the experience from studies in fish and chick, we developed a strategy following along those lines that enabled us to stimulate neuronal regeneration in adult mice in vivo (Figs. 1 and 2) (14, 15).

The purposes of our regeneration experiments are to selectively and acutely damage specific retinal cell types and then try to stimulate retinal regeneration with various factors. These factors are part of pathways that are known to influence retinal development in mice and/or regeneration in other species. We hypothesize that a defined program regulates neuronal regeneration of the retina on multiple cellular and molecular levels. Although we expect that regenerative programs shows strong similarities to embryonic retinogenesis, an obvious difference is the fact that, if only rod photoreceptors are lost, Müller glia progeny can (re)generate one cell type at very high efficiency specifically. This capacity offers the opportunity to learn how to generate specific retinal neurons in high numbers. Moreover, although cell loss is part of embryonic retinogenesis, it does not initiate it in the first place, which obviously

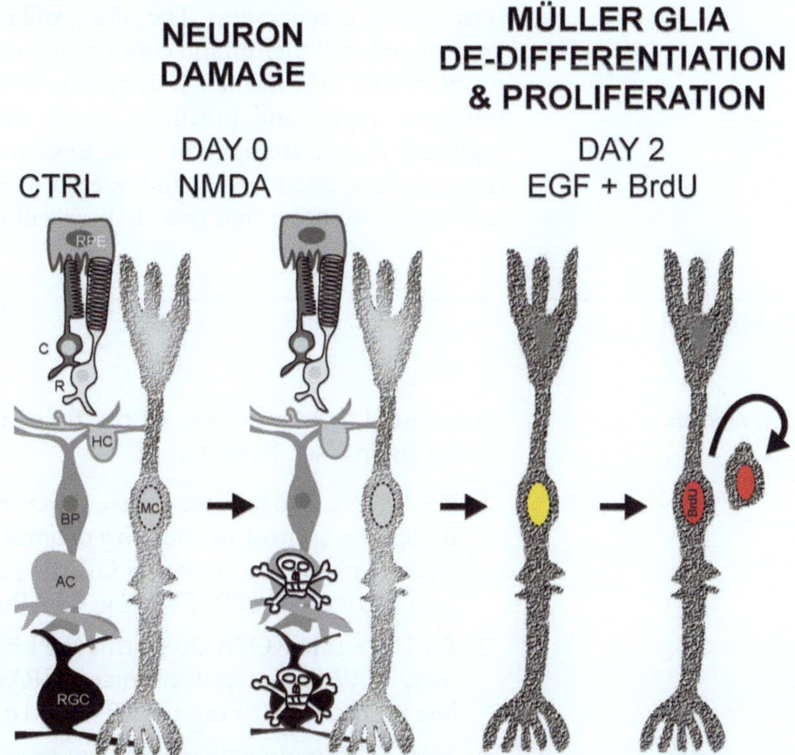

Fig. 1. Overview of retinal regeneration. In the adult mouse retina, Müller glia are quiescent cells. Upon neurotoxic damage (NMDA) and mitogen (EGF) stimulation, Müller glia dedifferentiate into a progenitor cell-like state, reenter the cell cycle, and divide.

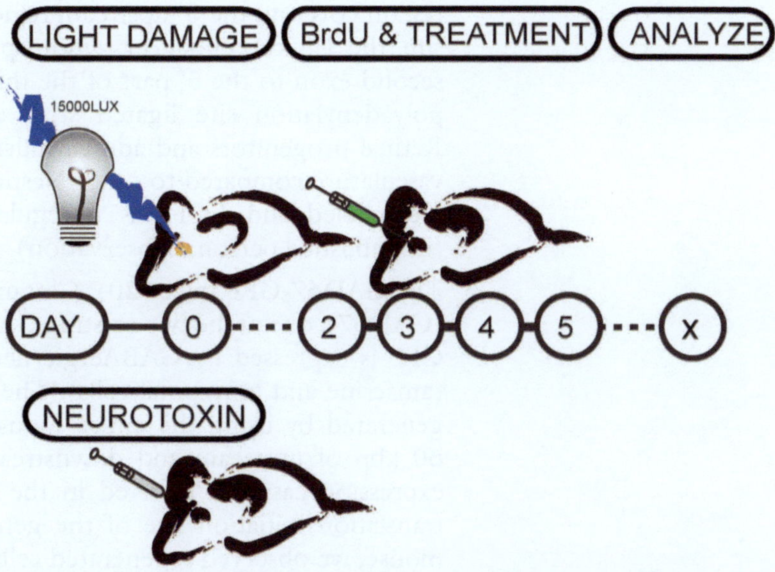

Fig. 2. Protocol for stimulation of retinal regeneration in vivo. Neurotoxic (NMDA) damage or light damage is induced in adult (P30) mice, and 48 h later, BrdU and mitogenic factors are injected intraocularly. BrdU is injected intraperitoneally. Treatment is applied up to four subsequent days.

is the case in regeneration. Therefore, still unknown extrinsic soluble and/or cell–cell membrane contact signals initiate the regenerative program and inform the Müller glia and its progeny about the cell numbers, types, and positions in the circuitry that need to be replaced. Future studies still need to identify the specific regenerative program, and identify and overcome the remaining barriers in mammals. We hope that this chapter will contribute to this quest.

2. Material

2.1. Animals

Our study involves the use of the following previously published mouse strains (see Note 1).

1. GFAP-Cre (16) mice: These mice carry a cre-recombinase under the control of the *gfap* promoter, a 2.2 kbp 5′ flanking region from the human GFAP (hGFAP) gene (Jackson Laboratory, #04600, FVB-Tg(GFAP-cre)25Mes/J).
2. B6.129×1Gt(ROSA)26Sortm1(EYFP)Cos/J (Jackson Laboratory, #006148) reporter mice (GFAP-Cre::RosaEYFP). This line is crossed with the GFAP-Cre (16) line (see Note 2).
3. Hes5-GFP transgenic mice: They were generated using a 3-kbp portion of the *Hes5* gene, including 1.6 kbp of the 5′ flanking region, with eGFP cloned into the translational start site (17, 18). In them, retinal progenitors and adult Müller glia are labeled.
4. Nestin-EGFP mice (19): They carry the rat *Nestin* genomic region covering the 5′ upstream region of the translation start site, the 1.8-kbp enhancer region spanning the 3′ part of the second exon to the 5′ part of the third intron, and the rabbit polyadenylation site ligated with enhanced GFP (EGFP). Retinal progenitors and adult Müller glia (not astrocytes and vasculature compared to other Nestin reporter mice available) are labeled and EGFP is upregulated in damaged retinas (unpublished personal observation).
5. The GAD67-GFP mice (20): Glutamic acid decarboxylase 67 (GAD67), one of the two existing GAD isoforms, and GAD67-GFP is expressed by GABAergic neurons in the adult retina (amacrine and horizontal cells). The GAD67-GFP mice were generated by using the entire mouse *Gad67* gene including 60 kbp of upstream and downstream regions and the GFP expression cassette, inserted in the first coding exon at the translation initiation site of the gene of interest. Using this mouse, we observed regenerated cells (15).
6. The transgenic mouse line Tg(Grm6-EGFP)5Var (21, 22), in which ON bipolar cells express EGFP at maturity, carries EGFP under the control of the promoter of the *Grm6* gene encoding mGluR6.

7. Nrl-L-EGFP (23) mice: Mice of this line carry a 2.5-kbp upstream segment of the mouse Nrl gene cloned into the pEGFP1 vector, which is expressed in the retina in newly born and mature rod photoreceptors.
8. Thy1-CFP-4 (24) transgenic mice: The animals contain 6.5 kbp of the murine *Thy1.2* gene ranging from the promoter to the intron following exon 4, but lacking exon 3 and its flanking introns driving ECFP. In adult retina, Thy1-CFP is expressed in retinal ganglion cells.

2.2. Treatment of Animals

1. Ethicon suture: PERMA-HAND, silk-suture, taper point needle, BB, 17 mm, 3/8 circle, black braided, 60 cm, 5-0, K580H.
2. Animal scale.
3. Common cotton swabs.
4. Micropipettes, 1–5 μl capillary, Wiretrol II, WHITE (# 5-000-2005, Drummond).
5. Pipette puller (Sutter).
6. Insulin syringes, 26.5-G needle for intraperitoneal (i.p.) injections, and 31.5-G needle for conjunctiva and sclera perforation.
7. Cold light source, stereomicroscope ×25 with long working distance objective.
8. Mice sterotactic device with manipulator and mouse adaptor (51600 Series Lab Standard, Stolting).
9. Four standard rat animal cages, including wire lid (floor area 530 cm^2).
10. Light damage: 15k lux are generated by eight parallel (5 cm apart) regular fluorescent daylight white color light bulbs (Philips, USA: F32/T865 Plus/ALTO, 32 W, 6500K, CRI86, 48″ length, UV-impermeable diffuser) mounted above rat animal cages with their insides lined by a reflective interior (mylar foil, BoPET).
11. Neurotoxins: *N*-methyl-D-aspartic acid (NMDA, Sigma Aldrich, M3262), Kainic acid (KA, Sigma Aldrich, K0250), stock solutions (store at –20°C): 100 mM in sterile 1× PBS.
12. Phosphate-buffered saline (PBS, Invitrogen).
13. BrdU (Sigma), stock solutions (store at –20°C): 10 mg/ml, use 50 mg BrdU/kg animal weight if injected i.p. and 1 μl stock per eye if intraocular injected. Store aliquots at –20°C for up to 6 months.
14. Growth factors: rhEGF, FGF1 (R&D), insulin (Sigma): stock solutions (store at –20°C): 1 μg/μl in sterile PBS with 0.2% BSA as control or growth factor carrier. Store aliquots at –20°C for up to 6 months.

15. General mouse anesthesia: 0.13 mg ketamine/g body weight (BW) plus 8.8 μg xylazine/g BW applied by i.p. injection.
16. Paralube Ointment (petrolatum grease, Fougera; or Ad-Lube).
17. Local anesthesia: Proparacaine HCl 0.5% (Bausch & Lomb), eye drops.
18. Local antiobiotics: Bacitracin (500 U/g) (Fougera), Gentamytrex 0.3%.
19. Pupil dilation: Phenylephrin 2.5%, Tropicamid 0.5%.
20. Visidic Gel eye ointment (prevents dry eye during surgery).

2.3. Retinal Analysis

1. PBS including 0.02% NaN.
2. PBS with 0.3% Triton-X (PBST) and 5% animal serum (e.g., bovine).
3. 50% Glycerol in PBS (refractive index of 1.408, own measurements).
4. Paraformaldehyde (32%, #15714, EMS).

Table 1
Antibodies used for the study of retinal regeneration, including sources and working dilutions

Antibody	Usage	Source
Rat anti-BrdU	1:100	Covance
Rabbit anti-Calretinin	1:1,000	SWANT
Rabbit anti-Cralb	1:1,000	Dr. Jack Saari (USA)
Chicken anti-GFP	1:400	Abcam
Mouse anti-Hu C/D	1:200	Invitrogen
Mouse anti-Islet1	1:10	DHSB
Mouse anti-NeuN	1:500	Chemicon
Rabbit anti-Neurofilament	1:500	Chemicon
Mouse anti-Pax6	1:250	DHSB
Rabbit anti-Pax6	1:400	Covance
Rabbit anti-phospho-histone H3	1:400	Novus
Rabbit anti-Prox1	1:200	Chemicon
Rabbit anti-Sox9	1:500	Chemicon
Goat anti-Sox2	1:200	Santa Cruz
Secondary antibodies	1:500	Alexa Invitrogen
Secondary antibodies	1:500	Jackson Immuno

5. Slow rocking platform or shaker at 4°C.
6. 22×22-mm² glass coverslips, thickness #1 (0.15 mm).
7. High vacuum grease (Dow Corning).
8. Secure-seal spacer 13 mm, 0.12 mm deep (Invitrogen).
9. Cryomold (TissueTek).
10. Small chemical spatula (width 5 mm, length >5 mm).
11. Iridectomy scissor.
12. Tweezer No. 3 (one dull and one sharp).
13. Antibodies (Table 1).

3. Methods

3.1. Retinal Damage

We found that an acute damage is important for the study and that NMDA, kainate, and light damage are suitable.

3.1.1. Neurotoxic Damage

Retinal neurotoxic damage can be induced by various neurotransmitters. Commonly used are kainate and NMDA (see Note 3). Aseptic precautions must include the method of instrument sterilization prior to initial use and between animals, if applicable. We use fresh sterile material for each mouse.

1. Sterilize micropipettes for intraocular injection with UV light overnight (see Note 4).
2. Thaw neurotoxin solutions: 0.1 M NMDA and 0.1 M kainite in PBS.
3. Anesthetize adult mice (postnatal day 30) using 0.13 mg ketamine/g BW plus 8.8 μg xylazine/g BW by an intraperitoneal injection (see Note 5).
4. Apply Proparacaine topical anesthetic to the eyes prior to injection.
5. Apply Visidisc eye ointment throughout to prevent dry eye.
6. Place the animal under a warm red light in separate boxes until fully anesthetized, which can be tested by gently pinching the front toe.
7. Place the mouse on a stereotactic device with micromanipulator fitted with a mouse adaptor, which allows accurate positioning of the micropipette for injections. The mouse is placed with the teeth passing through the gap in tooth bar and nose clamp to secure the head in position. The ear bars are positioned up to the bony part of the ear to stabilize the head.
8. Place the stereotactic device under a stereoscope (with an objective of at least 60-mm working distance).

9. Using two tweezers, suture the eyelid with a single loop and gently restrain by fastening the suture at the stereotactic device so that the upper eye half is well exposed. This usually is sufficient to slightly pop the eye out of its socket. Otherwise, carefully use a cotton swab or tweezers to apply gentle pressure to the periorbital area.
10. Using a 30-G needle, carefully make a very small incision at the upper temporal (about 02:00 hours) ora serrata/pars plana to gain access to the back part of the eye. It is most important not to damage the lens and also to make sure that the incision is not only through the sclera, but also gains access to the vitreal chamber. The aperture of the hole should allow almost a seal with the injection pipette diameter.
11. Mount micropipette with neurotoxin solution on micromanipulator that comes with the stereotactic device, as an angle of about 45–60°.
12. Insert the micropipette in the scleral hole and vitreal chamber.
13. Inject the neurotoxin solution into the eyes with high precision using graded glass micropipettes with a fine tip diameter (see Note 4).
14. Examine the eye carefully through the stereoscope to assess whether the injection is successful (see Note 6).
15. Apply Bacitracin antibiotic ointment to the eye after injection.
16. After intraocular injections, remove the clamps and ear bars before the animal wakes up.
17. Keep the animals under a warm red light until they regain consciousness (see Note 7).
18. After damages, wait for 48 h before any further intraocular treatments (Fig. 2).
19. Continue monitoring the animals for several days following the injections (see Note 8).

3.1.2. Acute Light Damage

Light damage has the advantage that no intraocular injection is necessary and it is less dependent on the experimenter's experience to generate comparable levels of damage between animals. Air can fully circulate and temperature of the environment and animals is monitored throughout the experiments (experiments are conducted in a temperature-regulated room).

1. One day before the experiment, dark-adapt the animal overnight (>12 h) (see Note 9).
2. Under dim red background light, dilate the pupils with a mydriaticum, 15 min before light exposure.
3. Under dim red background light, place the animals into rat cages (one animal each) with a regular grid cover. The interiors

of the cages are covered by reflective mylar foil and the corners are altered to prevent animals from hiding their eyes from the light.

4. Expose animals to light (see Note 10). The distance of the light bulbs is adjusted to generate 15k lux in the center of all cages. Food and water are placed directly in separate cage corners ad libitum. Up to four cages with one animal each are exposed.
5. Monitor the animals until they regain consciousness (see Notes 8 and 11).

3.2. Induce and Monitor Müller Glia Proliferation and De-differentiation

Following retinal damage, we typically inject mitogens (see Note 12) to enhance Müller glia proliferation and de-differentiation (Fig. 2). To monitor cell proliferation, we inject BrdU together with the first mitogen application (Fig. 2). BrdU incorporation allows permanent labeling of a Müller glia subpopulation in adult mice in vivo (Fig. 3c) (see Note 13).

1. Between 2 and 5 days after the induction of retinal damage, inject mitogens plus BrdU (1 μg/μl) intraocularly, following the dictions detailed in Subheading 3.1.1.
2. Using a 26.5-G needle and an insulin syringe, carry out intraperitoneal injection of BrdU (1 μg/μl) so that the animal receives 50 μg/g BW BrdU i.p.
3. If desired, repeat the injection daily up to 4 days.

3.3. Analysis of Neuronal Regeneration

3.3.1. Preparation of Retinal Flatmount

1. Euthanatize mice by methods approved by your institutional animal use and care committee.
2. Enucleate the eyes and place them in PBS.
3. Under a dissection microscope, puncture the cornea with a 30-G ½″ syringe needle. Cut the cornea towards its margin using blunt tweezers or a pair of iridectomy scissors. Peel the sclera and the RPE off the retina. Remove the lens and the ciliary body, including the zonular fibers. Never directly poke or squeeze the retina at any time. This should leave a perfect shaped retinal cup in one piece.
4. Transfer the retinal tissue into PBS using a transfer pipette or spatula.
5. Make four radial cuts half way towards the optic nerve head, so that the retina looks like a flower, with iridectomy scissors.
6. Transfer the retina with its outer nuclear layer down onto a glass slide. Straighten the retina out carefully with a blunt forceps. Use tissue paper to remove the remaining PBS media. Be careful not to contact the retina with the tissue paper. Flatten the retina with forceps if the edges overlap.

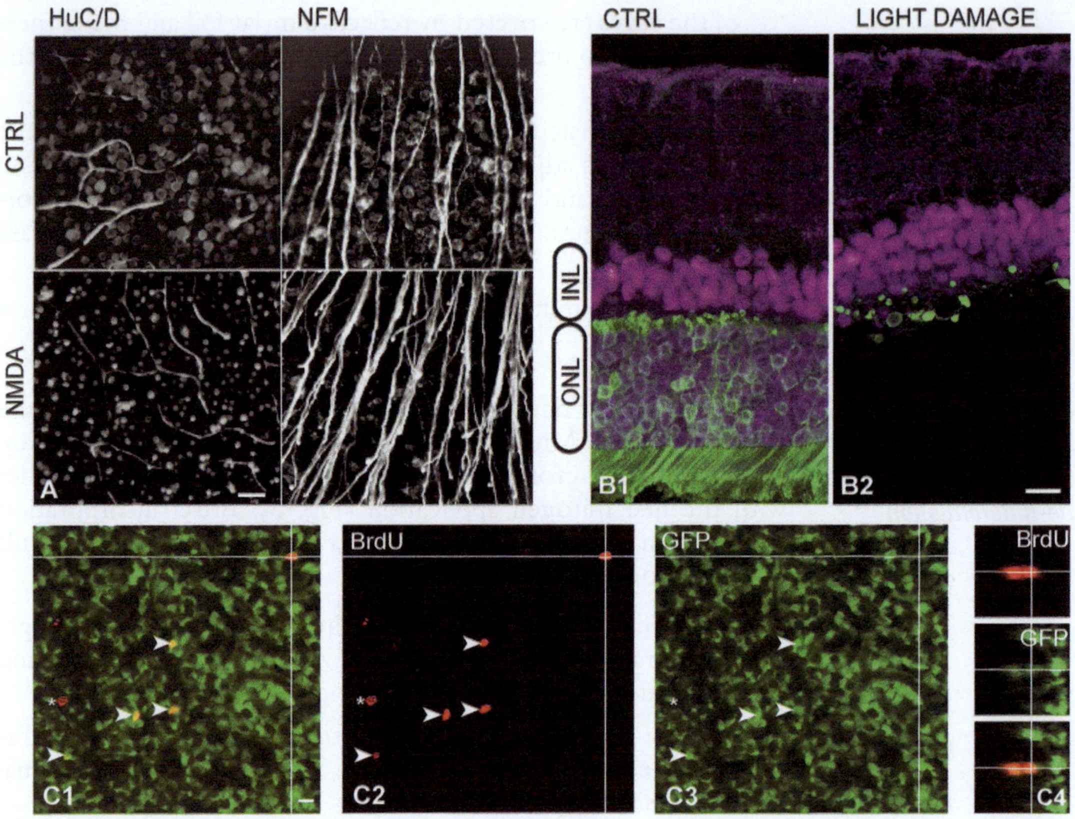

Fig. 3. Confocal imaging in retinal regeneration. (**a**) NMDA-induced neuronal cell damage. Neurofilament M (NFM) and HuC/D immunostaining (both labeling many ganglion and amacrine cells) shows cell loss in damaged versus control (CTRL) retina. (**b**) Retinal light damage. High-intensity light damage leads to strong degeneration of photoreceptors (Nrl-GFP labels all rod photoreceptors in the outer nuclear layer, ONL, *green*) in central retina compared to control. OTX2 labels bipolar cell nuclei in the inner nuclear layer (INL, *magenta*) and rod photoreceptor nuclei. (**c**) Müller glia and potential progeny: proliferation analysis using BrdU and lineage-trace (GFAP-Cre::ROSA-YFP) colabeling. Shown is a 3 × 1-μm confocal image merged *z*-stack at the Müller glia cell layer in the INL [C1: *crosshair* and *arrowheads* show BrdU (C2) colabeled GFAP-Cre::ROSA-YFP Müller glia (C3); *asterisk* represents some YFP-negative BrdU cell]. The cell in the *crosshair* is shown in the *yz*-view to provide evidence of colabeling. Scale bars in (**a**–**c**) are 20 μm.

7. Slowly add some 4% paraformaldehyde in PBS next to the retina so that it will float gently and stretch out on top of the solution surface. Fix the retina for 30 min at room temperature (RT).
8. Transfer to PBS and store in PBS with 0.02% sodium azide at 4°C.

3.4. Flatmount Immunohistochemistry

1. Place each retina in its own well of a 24-well plate.
2. Permeabilize the retina with a 1% SDS solution in PBS for not longer than 3 min.
3. Wash three times thoroughly for 15 min in PBS.
4. Block nonspecific binding sites with 500 μl of 5% serum, 0.3% Triton X-100, and 0.02% sodium azide in PBS at RT for 1 h.

5. Immunostaining for certain transcription factors and BrdU requires additional treatment with 1,000 U/ml DNase, which can be added to the blocking solution and incubated at 37°C for 1 h.
6. Add 200 μl per well of primary antibody in PBS with 0.3% Triton X-100 and 5% serum.
7. Incubate on nutator for 1–2 days.
8. Wash 3 × 30 min in PBS in a cold room.
9. Add 200 μl per well secondary antibody. We use Alexa or Jackson Immuno fluorescent secondary antibodies.
10. Incubate at least overnight in the cold room.
11. Wash 3 × 30 min in PBS.
12. For imaging, transfer the retinal flatmout with a drop of PBS to a glass coverslip that has a spacer or a ring of grease on it. It acts as a spacer so that each retina is sandwiched between two glass coverslips.
13. Before coverslipping, remove PBS and replace it by a small amount of 50% glycerol in PBS (this improves retinal transparency for imaging). Sandwiched this way, the retina can be imaged from both sides, if necessary.

3.5. Retinal Flatmount Confocal Imaging

For this step, we offer the following points.

1. Retinas are best imaged with the ganglion layer closest to the laser beam light path because the photoreceptor outer segments reduce imaging depth.
2. With adult retinas being about 100-μm thick after fixation, confocal microscopy can be achieved across its entire depth.
3. Each adult mouse retina has an area of about 16 mm^2 so that 3D confocal imaging of an entire retina at ×40 magnification on standard microscope setups generates about 70 image stacks.
4. Depending on the experimenter's goal, the resolution setting, imaging speed, number of fluorescent channels recorded, and optical slide thickness, each image stack may take between 5 min and up to several hours (which is limited due to fluorescent bleaching).
5. For general screens, for example, the total number of proliferating cells (BrdU) and the number of regenerated neurons (BrdU and GAD56-GFP double-positive cells), we use ×40, with the maximum scan speed, 1,024 × 1,024 pixel, and set the pinhole to 2-μm optical slice thickness to image the entire retina (about 7 min/stack).

6. All potential double-labeled cells are confirmed at higher resolution and magnification. For the higher resolution images, we use a ×63 objective, increase the image averaging, and decrease the pinhole to 0.25–1-μm optical slice thickness to achieve proper 3D reconstruction of cell nuclei and morphology.

4. Notes

1. The use of animals requires an approved protocol and is in accordance with the guidelines established by institutional IACUC and the National Institute of Health. For experiments described in this chapter, we use mice at least 30 days old. Mice are maintained under specific pathogen-free (SPF) conditions and handled under institutional protocols according to international guidelines and under a 12-h light/dark cycle with access to food and water ad libitum (including time during the light damage experiments).
2. In various parts of the brain, astrocytes are YEFP labeled (16) and in the healthy retina occasionally some Müller glia cells, but no astrocytes, are labeled. Upon retinal damage, most, if not all, Müller glia and its progeny are EYFP labeled (13).
3. NMDA induces loss of amacrine, ganglion, and horizontal cells; kainate, which is much more potent, in addition leads to bipolar cell death and possibly subsequent rod photoreceptor loss. Various neurotoxic retinal damages have been described in the literature and their effects depend on species and age. For example, NMDA damage can be induced by a single injection of 2 μl of 0.1 M NMDA, which induces the maximal cell loss by NMDA in adult mice (Fig. 3a). Within 24 h, cell death is apparent throughout the retina and 5 days later the number of cells present is significantly decreased (15). Kainate may produce a much higher loss of neurons within 2 weeks, leaving behind mostly Müller glia after an injection of 2 μl of 0.1 M kainate (15).
4. The micropipettes are similar to those used for patch-clamp electrophysiology with a slightly larger tip aperture. The micropipettes have a stamp that allows precise injection of 0.5–5-μl volume.
5. In 25–30-g animal, 100 μl anesthesia solution last for 2–3 h, sometimes up to 6 h; monitor animals at all times. The deep anesthesia needs to last for at least 20 min, although with practice the injection only takes 3–5 min. Any other reliable method of anesthesia can be substituted. Stress during the time after

the injection of the anesthesia may reduce its full effect, which can be tested by gently pinching the front toe.

6. A successful injection will lead to perturbation in the anterior chamber (turbid during the injection for a few seconds). Further, a slight transient swelling of the entire eye may occur. Instant backflush might indicate a failed injection into a scleral pocket (e.g., between sclera and choroid) instead of injecting into vitreal space. Usually, we inject a defined higher volume than the eye actually holds to wash out the existing aqueous humor so that a delayed backflush is expected.
7. Postsurgical monitoring depends on anesthesia depth, but animals should be awake after 10–30 min.
8. Monitoring protocol: Animals should be monitored after each procedure for several hours for general signs of distress, such as hunching, lethargy, piloerection, ungroomed coat, scratching of injection site, dehydration, feeding behavior, and agitated movements. Steps should be taken to alleviate any suffering (topical anesthesia or, if required, euthanasia). For several days following the injections, animals should also be monitored for eye-specific inflammation and infection.
9. It is necessary to reduce stress for the animals (stress reduces the amount of retina damage) and to keep the animals in total darkness (to standardize and reduce variability in pre-bleaching of the animals' retina rhodopsin) until acute exposure to the bright light.
10. Light exposure time varies between mouse strains (25), and the animals are sacrificed at various times and their retinas are analyzed. Depending on the genetic background of each strain, retinal damage is observable after 1 (e.g., albino mice) to 8 h (e.g., C57/B6). Various gene sequence variants have been described and most likely further still unknown modify the light damage response (25). Retinal light damage may lead to loss of the entire outer nuclear layer in the central retina (Fig. 3b).
11. The animal will not experience any pain from retina degeneration because the retina is devoid of pain sensitivity (all retinal diseases are pain free, no nociceptive nerves are residing in the retina). The animals will not suffer from photokeratitis, which is a burn of the cornea (the clear front surface of the eye) and is also called radiation keratitis or snow blindness. This is a painful condition caused by exposure of unprotected eyes to the ultraviolet (UV) rays in bright sunlight, which is not related to retina degeneration. Photokeratitis is not induced because the UV light is absorbed by the bulb's fluorescent coating and, along with the bulb's glass, prevents the harmful UV light from escaping.

12. Mitogenic factors that successfully induce Müller glia proliferation include at least EGF, insulin, FGF1, Wnt3a, Shh, but not TGFbeta, or various control solutions (PBS, 1% BSA, and BrdU). An additional vehicle control, using the solvent in which the factor was dissolved, is also included in our experimental design. We observed the highest number of proliferating Müller glia using EGF, fewer with the combination of FGF1 and insulin, and even fewer with either Wnt3a or Shh. To maximize and prolong cell proliferation as well as to induce neuronal differentiation, animals received up to five doses injected intraocularly once daily, between 2 and 5 days after the induction of retinal damage (Fig. 2).
13. It has been shown that BrdU given during light damage may lead to incorporation into dying photoreceptors (26). We typically inject BrdU together with the first mitogen application and never before or during induction of cell damage. Therefore, we have not seen any BrdU labeling of neurons before day 6 after NMDA or light damage, although most neurons have already died and disappeared during this time frame. In our experiments, about 10% of all Müller glia are BrdU labeled the third day after damage and about 1% can still be found 5 days later (15).

Acknowledgments

We would like to thank the past and present members of the Reh lab for help on paving the way to study retinal regeneration in mice in vivo. This work was supported by grants to MOK (DFG KA 2794/1-1, CRTD, and ProRetina Germany) and TAR (NIH R01 EY 021482).

References

1. Reh TA, Lamba D, Gust J (2010) Directing human embryonic stem cells to a retinal fate. In: Ding S (ed) Cellular programming and reprogramming methods and protocols. Humana Press, New York
2. Bartsch U et al (2008) Retinal cells integrate into the outer nuclear layer and differentiate into mature photoreceptors after subretinal transplantation into adult mice. Exp Eye Res 86:691–700
3. Lamba DA, Gust J, Reh TA (2009) Transplantation of human embryonic stem cell-derived photoreceptors restores some visual function in Crx-deficient mice. Cell Stem Cell 4:73–79
4. MacLaren RE et al (2006) Retinal repair by transplantation of photoreceptor precursors. Nature 444:203–207
5. Fischer AJ (2005) Neural regeneration in the chick retina. Prog Retin Eye Res 24:161–182
6. Hitchcock P et al (2004) Persistent and injury-induced neurogenesis in the vertebrate retina. Prog Retin Eye Res 23:183–194
7. Raymond PA et al (2006) Molecular characterization of retinal stem cells and their niches in adult zebrafish. BMC Dev Biol 6:36
8. Senut MC, Fausett B, Veldman M, Goldman D (2007) Gene regulation during axon and tissue regeneration in the retina of zebrafish. In: Becker CG, Becker T (eds) Model organisms in spinal cord regeneration. Wiley-VCH. Hoboken, NJ
9. Bringmann A et al (2006) Müller cells in the healthy and diseased retina. Prog Retin Eye Res 25:397–424

10. Garcia M, Vecino E (2003) Role of Müller glia in neuroprotection and regeneration in the retina. Histol Histopathol 18:1205–1218
11. Pekny M, Nilsson M (2005) Astrocyte activation and reactive gliosis. Glia 50:427–434
12. Lamba D, Karl M, Reh TA (2008) Neural regeneration and cell replacement: a view from the eye. Cell Stem Cell 2:538–549
13. Karl MO, Reh TA (2010) Regenerative medicine for retinal diseases: activating endogenous repair mechanisms. Trends Mol Med 16:193–202
14. Lamba DA et al (2008) Baf60c is a component of the neural progenitor-specific BAF complex in developing retina. Dev Dyn 237:3016–3023
15. Karl MO et al (2008) Stimulation of neural regeneration in the mouse retina. Proc Natl Acad Sci USA 105:19508–19513
16. Zhuo L et al (1997) Live astrocytes visualized by green fluorescent protein in transgenic mice. Dev Biol 187:36–42
17. Basak O, Taylor V (2007) Identification of self-replicating multipotent progenitors in the embryonic nervous system by high Notch activity and Hes5 expression. Eur J Neurosci 25:1006–1022
18. Nelson BR et al (2011) Genome-wide analysis of Müller glial differentiation reveals a requirement for Notch signaling in postmitotic cells to maintain the glial fate. PLoS One 6:e22817
19. Yamaguchi M et al (2000) Visualization of neurogenesis in the central nervous system using nestin promoter-GFP transgenic mice. Neuroreport 11:1991–1996
20. Chattopadhyaya B et al (2004) Experience and activity-dependent maturation of perisomatic GABAergic innervation in primary visual cortex during a postnatal critical period. J Neurosci 24:9598–9611
21. Dhingra A et al (2008) Probing neurochemical structure and function of retinal ON bipolar cells with a transgenic mouse. J Comp Neurol 510:484–496
22. Morgan JL et al (2006) Axons and dendrites originate from neuroepithelial-like processes of retinal bipolar cells. Nat Neurosci 9:85–92
23. Akimoto M et al (2006) Targeting of GFP to newborn rods by Nrl promoter and temporal expression profiling of flow-sorted photoreceptors. Proc Natl Acad Sci USA 103:3890–3895
24. Feng G et al (2000) Imaging neuronal subsets in transgenic mice expressing multiple spectral variants of GFP. Neuron 28:41–51
25. Wenzel A et al (2005) Molecular mechanisms of light-induced photoreceptor apoptosis and neuroprotection for retinal degeneration. Prog Retin Eye Res 24:275–306
26. Joly S et al (2011) Pax6-positive Müller glia cells express cell cycle markers but do not proliferate after photoreceptor injury in the mouse retina. Glia 59:1033–1046

Chapter 16

Production and Transplantation of Retinal Cells from Human and Mouse Embryonic Stem Cells

Anna La Torre, Deepak A. Lamba, Anu Jayabalu, and Thomas A. Reh

Abstract

Over the last few years, numerous studies have introduced strategies for the generation of neuronal populations from embryonic stem cells. These techniques are valuable both in the study of early neurogenesis and in the generation of an unlimited source of donor cells for replacement therapies. We have developed a protocol to direct mouse and human embryonic stem cells to retinal fates by using the current model of eye specification. Our method is a multistep protocol in which the cultures are treated with IGF1 and a combination of BMP and Wnt inhibitors to promote the expression of key retinal progenitor genes, as assayed by RT-PCR and immunofluorescence microscopy. The retinal progenitor population spontaneously undergoes differentiation towards various types of retinal neurons, including photoreceptors.

Key words: Eye field, Retinal progenitor, Photoreceptor, mES Cells, hES Cells

1. Introduction

The retina is subject to a number of age-related and hereditary diseases that lead to irreversible cell loss and subsequent visual impairment. These devastating conditions affect millions of people worldwide. Age-related macular degeneration, the leading cause of blindness in the elderly, causes cone photoreceptor loss, while many inherited conditions result in rod photoreceptor degeneration. Significant advances have been made to understand these visual disorders at a molecular level. Some of the inherited ocular pathologies are linked to specific mutations in retinal-specific genes; for example, mutations in the rhodopsin gene underlie many forms of Retinitis Pigmentosa. For these individuals, many groups are exploring gene therapy and other medical treatments in an effort to prevent or slow the neuronal degeneration; however, once the photoreceptor cells have degenerated, any therapeutic strategy to

Shu-Zhen Wang (ed.), *Retinal Development: Methods and Protocols*, Methods in Molecular Biology, vol. 884, DOI 10.1007/978-1-61779-848-1_16, © Springer Science+Business Media, LLC 2012

restore vision would require prosthetics or cell replacement by transplantation (1).

Stem cells have the ability of indefinite self-renewal and the potential to differentiate into any cell type of the body providing an ideal source for cellular replacement therapies. The knowledge of the developmental biology provides mechanistic information for designing methods to direct embryonic stem cells towards neuronal cell fates. Protocols for deriving various types of central nervous system neurons from embryonic stem cells have now been published, including motor neurons, dopaminergic midbrain neurons, and cerebral cortical neurons (2–4). All of these protocols rely on manipulation of developmental signaling pathways that are normally critical for regionalization of the CNS.

Eye development begins at late gastrula stages with the specification of the eye field, a single domain of retinal precursors positioned in the anterior neural plate. Although the first morphological evidence of eye development is a bilateral evagination from the early forebrain to form the optic vesicles, the prospective eye tissue exists prior to the formation of the optic vesicles. The eye field is located in a medial position and is surrounded rostrally and laterally by telencephalic progenitors and caudally by diencephalic cells. Two critical developmental pathways regulate the formation of the anterior–posterior and lateral–medial axis in the nervous system: BMP and Wnt (5). Inhibition of the BMP pathway promotes neural plate development medially, while higher levels of BMP laterally promote epidermal development. The overall level of canonical Wnt signaling is higher in posterior regions of the embryo, promoting hindbrain and spinal cord development. Endogenous inhibitors of both BMP and Wnt signaling are, therefore, important patterning morphogens; targeted deletion of the genes coding for Dkk1, a potent Wnt inhibitor, along with Noggin, a BMP antagonist, leads to almost the complete absence of the head in mice (6). These results, along with many others, have led to the current model of neural specification, in which anterior neural structures develop when both BMP and Wnt signaling are blocked. Additionally, insulin-like growth factor (IGF) signaling has been shown to play a role in head and eye formation (7).

Classic transplantation experiments performed over a century ago demonstrated that the eye field region of the neural plate of amphibian embryos can be transplanted to the flank of another embryo and give rise to an ectopic eye. However, only recently the transcription factors that are both necessary and in some cases sufficient for eye development have been identified (8); these are collectively named the eye field transcription factors (EFTFs). During embryonic development, the expression of the EFTFs is defined by a specific set of signals (9, 10), which are particularly effective at inhibiting the canonical Wnt signaling pathway. At least three mechanisms inhibit Wnt signaling in the developing eye field: (1) the Wnt inhibitor, Sfrp1, is expressed within the eye field; (2) a signal from Wnt11 or Wnt4 activates the non-cannonical Wnt pathway via Frizzled 5 or Frizzled 3, which also inhibits the canonical Wnt signal; and (3) IGF activates the Akt pathway and this also blocks canonical Wnt signaling.

Based on these previous studies, we developed a protocol to promote eye field development from embryonic stem cells (11, 12). We treat the undifferentiated embryonic stem cells with inhibitors of BMP and Wnt signaling along with IGF1 and then assay for the expression of EFTFs in the resulting cells. The methods described in this chapter outline the basic protocol for generating retinal progenitors, confirming commitment to this state with RT-PCR and immunofluorescence, and additional information for further directing and analyzing their subsequent differentiation into retinal neurons using immunofluorescence and intraocular transplantation.

2. Materials

2.1. Culture of Undifferentiated Mouse Embryonic Stem Cells

1. Cells: R1 and G4 mESC lines from Andras Nagy.
2. The mouse ES cell media contains 80 ml of Dulbecco's Modified Eagle's Medium (DMEM) from Invitrogen, 20 ml of fetal bovine serum (ES qualified, Invitrogen), 1 ml of nonessential amino acids (Invitrogen), 1 ml of sodium pyruvate (Invitrogen) (see Note 1), and 100 μl of β-mercaptoethanol (0.1 M, Sigma) (see Note 2).
3. Media is supplemented with 100 μl of leukemia inhibitory factor (LIF, ESGRO Millipore ESG1106, 10 million units/ml), 3 μM of GSK3β inhibitor Stemolecule CHIR99021 (Stemgent), and 0.4 μM of Stemolecule PD0325901 (Stemgent) (see Note 3).
4. Penicillin and streptomycin (Invitrogen) may be added to all media solutions to prevent contamination.
5. All growth factors are dissolved in DMEM containing 0.1% of bovine serum albumin (BSA, Sigma) and maintained as stocks at a concentration of 100 ng/μl in −80°C freezer.

2.2. Generation of Retinal Cells from Undifferentiated Mouse ES Cells

1. Trypsin (Invitrogen).
2. Retinal induction (RI) media contains DMEM with F12 (Invitrogen), 10% of fetal bovine serum, N2 supplement, B27 supplement, 1× sodium pyruvate, 1× nonessential amino acids, 1 ng/ml mouse Noggin (R&D Systems), 1 ng/ml mouse recombinant Dkk-1 (R&D Systems), and 1 ng/ml of recombinant IGF-1 (R&D Systems) (see Note 4).
3. Embryoid bodies (EBs) are generated and cultured in Ultra low-attachment plates (Costar, VWR).

4. Retinal differentiation (RD) media contains DMEM/F12, N2 supplement, B27 supplement, 1× sodium pyruvate, 1× nonessential amino acids, 10 ng/ml mouse Noggin, 10 ng/ml mouse recombinant Dkk-1, and 10 ng/ml of recombinant IGF-1 (see Note 4).

2.3. Culture of Undifferentiated Human Embryonic Stem Cells

1. Cells: H-1 (WA-01) human embryonic stem cell line from WiCell.
2. Human ES cell media: TESR2 Media (Stem Cell Technologies).

2.4. Generation of Retinal Cells from Undifferentiated Human ES Cells

1. Collagenase/Dispase mix contains 10 mg/ml of Dispase (Invitrogen) and 10 mg/ml of Collagenase type IV (Invitrogen).
2. RI media contains DMEM/F12, 10% knockout serum replacer, N2 supplement, B27 supplement, 1× sodium pyruvate, 1× nonessential amino acids, 1 ng/ml mouse noggin, 1 ng/ml human recombinant Dkk-1 (R&D Systems), and 1 ng/ml human recombinant IGF-1 (R&D Systems) (see Note 4).
3. RD media contains DMEM/F12, N2 supplement, B27 supplement, 1× sodium pyruvate, 1× nonessential amino acids, 10 ng/ml mouse noggin, 10 ng/ml human recombinant Dkk-1, and 10 ng/ml human recombinant IGF-1 (see Note 4).
4. Penicillin and streptomycin may be added to all media solutions to prevent contamination.
5. All growth factors are dissolved in DMEM containing 0.1% of BSA and maintained as stocks at a concentration of 100 ng/ml in –80°C freezer.

2.5. Coating for Adherent Culture of Cells

1. Poly-D-lysine hydrobromide MW 30–70 kDa, lyophilized powder, cell culture tested (Sigma). Poly-D-lysine is dissolved in sterile water at a concentration of 0.5 mg/ml and a 1-ml aliquot is stored in 15-ml conical tubes at –20°C.
2. Coverslips (12-mm circular) sterilized by autoclaving.
3. Tissue culture plates.
4. Growth-Factor Reduced Matrigel (Collaborative Research): Matrigel is supplied by the manufacturer as a frozen solution. Thaw the bottle slowly on ice (for several hours) to prevent gel formation. Make small (200 μl) aliquots using precooled tubes (15 ml) on ice and a prechilled pipette. If the Matrigel warms during the aliquotting, it will gel and not be effective for the cell cultures. Store aliquots at –20°C for up to 6 months.
5. DMEM media.

3. Methods

3.1. Substrate for Adherent Culture of Cells

Adherent cultures of undifferentiated ES or retinal progenitors can be carried out either on glass coverslips coated with poly-D-lysine and Matrigel or tissue culture plates coated with Matrigel.

1. To coat the coverslips, remove one aliquot of Matrigel from the −20°C freezer and place on ice for 30 min to thaw (200 μl is used for one 24-well plate).
2. Add 10 ml of ice-cold HBSS+ to the 15-ml tube containing 200 μl of thawed Matrigel. Mix gently.
3. Immediately, put 1 ml of the dilute Matrigel solution into each well of a 6-well plate and place the plate in the incubator for 30 min at 37°C.
4. Remove the plate from the incubator and, under the sterile hood, remove all of the liquid from the wells.
5. Plate cells onto the Matrigel.

3.2. Culture of Undifferentiated Mouse Embryonic Stem Cells

R1 and G4 cells are cultured in mouse ES cell media supplemented with LIF and the two inhibitors (2i, GSK3 inhibitor, and MAPK inhibitor) in a 37°C incubator in 5% CO_2.

1. Thaw a cryogenic straw containing undifferentiated mouse ES cells by fast transferring from liquid nitrogen tank to room-temperature water.
2. After a few seconds, rinse the straw with 70% ethanol, cut the edges of the straw, and transfer the cells into a 15-ml tube containing 5 ml of mouse ES media (no supplements) (see Note 5).
3. Centrifuge the tube at 1,200 rpm for 3 min.
4. Resuspend the cells in fresh mouse ES media (with LIF and 2i) and transfer the cells to Matrigel-coated plates (see Note 6). Put in an incubator at 37°C and 5% CO_2.
5. Change the media the next day and after that every 2–3 days, until the cells are 70–80% confluent.

3.3. Passage of Undifferentiated Mouse Embryonic Stem Cells

1. Remove media from a semi-confluent plate of mouse ES cells (70–80% confluence), wash twice with PBS, and treat the plate with trypsin (2–3 ml of trypsin per 10-cm plate or 1 ml per 35-mm plate).
2. Place the plate back to incubator and wait for 4–8 min or until the cells start to lift. Meanwhile, prepare a 15-ml tube with 5 ml of mouse ES media (no supplements).
3. Disrupt the cells from the plate using a long glass pipette, rubber bulb, and by tapping the sides of the plate. Observe that the cells are removed from the plate, and gently triturate to achieve a single cell suspension.

4. Pipette the removed cells into the tube containing the 5 ml of mouse ES media. Add an additional 3 ml/10 cm or 1 ml/35 mm of media to rinse the plate and pipette the remaining wash media into the 15-ml tube.
5. Balance the centrifuge and spin at 1,200 rpm for 3 min.
6. Aspirate supernatant being careful not to disturb the pellet.
7. Carefully disperse the pellet using a 5-ml pipette with 5 ml of mouse ES media supplemented with LIF and the two inhibitors (GSK3β and MAPK inhibitors) (see Note 7).
8. Passage at a ratio of 1:20–1:50.
9. Change the media the following day.

3.4. Generation of Retinal Cells from Undifferentiated Mouse ES Cells

1. Remove the media from a semi-confluent plate of mouse ES cells (70–80% confluence), wash twice with PBS, and treat the plate with trypsin (2–3 ml of trypsin per 10-cm plate or 1 ml/35-mm plate).
2. Place the plate back into the incubator and wait for 4–8 min or until the cells start to lift. Meanwhile, prepare a 15-ml tube with 5 ml of mouse ES media (no supplements).
3. Disrupt the cells from the plate using a long glass pipette, rubber bulb, and by tapping the sides of the plate. Observe that the cells are removed from the plate, and gently triturate to achieve a single cell suspension.
4. Pipette the removed cells into the tube containing the 5 ml of mouse ES media. Add an additional 3 ml/10 cm or 1 ml/35 mm of media to rinse the plate and pipette the remaining wash media into the 15-ml tube.
5. Balance centrifuge and spin at 1,200 rpm for 3 min.
6. Aspirate supernatant being careful not to disturb the pellet.
7. Carefully disperse the pellet using a 5-ml pipette with 5 ml of RI media. Plate the cell suspension in low-attachment plates (typically one million cells/well of a 6-well plate). Floating mouse ES cells spontaneously form 3D aggregates (EBs) within 12 h under these conditions (see Note 8).
8. Twenty-four hours later, change media by placing the plate at an angle so as to allow the cells to settle down at the edge of the plate and carefully remove most of the media.
9. Seventy-two hours later, transfer the EBs to a plate previously coated with Matrigel. Evenly distribute the cells by moving the plate front to back and side to side a few times.
10. After overnight incubation, all the EBs should stick down and start spreading out on the Matrigel. Change the media with fresh RD media (see Note 9).

11. Change the media every 2–3 days for up to 12–15 days. As the cells grow, they form rosette-like structures.
12. After 15 days, the cells can be cultured in media without any growth factor.

3.5. Passaging of mESC-Derived Retinal Cells

Do not passage the cells for at least 10 days as this interferes with cell-to-cell interactions and in turn retinal progenitor proliferation.

1. Rinse the plate once with PBS.
2. Add 1 ml of 0.25% trypsin to the plate and incubate at 37°C for 2 min.
3. Carefully remove the trypsin from the wells and add DMEM + 10% FBS.
4. Scrape the cells off the plate gently using the tip of a 5-ml pipette and transfer the contents to a 15-ml tube.
5. Centrifuge at 1,200 rpm for 3–5 min.
6. Remove supernatant and add fresh media to the tube.
7. Gently mix the cells and passage at a ratio of 1:3–1:4.
8. Change media the following day.

3.6. Culture of Undifferentiated Human Embryonic Stem Cells

H-1 cells are cultured in the TESR2 media in a 37°C incubator in 5% CO_2.

1. Thaw a cryogenic straw containing undifferentiated human ES cells by quickly transferring from liquid nitrogen tank to room-temperature water.
2. After 30 s, rinse the straw with 70% ethanol, cut the edges of the straw, and transfer the cells to a 15-ml tube. Add human ES media dropwise slowly (see Note 5).
3. Centrifuge the tube at 1,200 rpm for 3 min.
4. Resuspend the cells in fresh human ES media and transfer the cells to Matrigel-coated plates. Put the cells in incubator at 37°C and 5% CO_2.
5. Change the media next day and thereafter every other day until they are 60–80% confluent.

3.7. Generation of Retinal Cells from Undifferentiated Human ES Cells

1. Remove the media from a confluent plate of human ES cells and treat the plate with a combination of collagenase IV and dispase solution.
2. Replace the plate back to incubator and wait for 3–5 min until the edges of the human ES cell colonies just start to lift.
3. Remove the collagenase/dispase solution and gently and thoroughly rinse the cells twice with PBS.

4. Add TESR2 media to the plate and gently scrape the cells off the plate using the tip of a 5-ml pipette.
5. Triturate the colonies such that the final size of the colonies is approximately 200–250 cells (see Note 10).
6. Plate ~50,000 cells/well of a Matrigel-coated 6-well plate (see Note 11). Evenly distribute the cells by moving the plate front to back and side to side a few times.
7. Twenty-four hours later, change the media to RI media.
8. On the fourth day, change the media to RD media.
9. Change three-fourth of the media every 2–3 days for up to 3 weeks. As the cells grow out, they form rosette-like structure throughout the plate starting around 14–16 days of induction (see Note 12).
10. After 3 weeks, the cells can be cultured in media without any growth factors.

3.8. Passaging of hESC-Derived Retinal Cells

Do not passage the cells for at least 2 weeks as this interferes with cell-to-cell interactions and in turn retinal progenitor proliferation.

1. Rinse the plate once with PBS.
2. Add 1 ml of 0.25% trypsin to the plate and incubate at 37°C for 2–3 min.
3. Carefully remove the trypsin from the wells and add DMEM + 10% FBS.
4. Scrape the cells off the plate gently using the tip of a 5-ml pipette and transfer the contents to a 15-ml tube.
5. Centrifuge at 1,200 rpm for 3–5 min.
6. Remove supernatant and add fresh media to the tube.
7. Gently mix the cells and passage at a ratio of 1:3–1:4 (see Note 13).
8. Change the media the following day.

3.9. Analysis of Retinal Determination Using Quantitative Real-Time PCR

1. Harvest the ES-derived retinal cells from the plate using the Collagenase/Dispase mix and collect the cells as a pellet by centrifugation at 1,200 rpm for 5 min.
2. Resuspend the pellet into 500 μl of TRIzol in a 1.5-ml RNase-free tube.
3. Homogenize tissue thoroughly with a Pellet Pestle Motor and fresh, RNase-free Pellet Pestles; samples can be frozen (−80°C) at this point indefinitely.
4. Add 200 μl RNase-free chloroform, vortex and centrifuge to separate layers, and transfer the top aqueous layer to a fresh tube.

5. Re-extract a second time with chloroform to clear any remaining impurities.
6. Add an equal volume of 100% isopropanol, mix, and centrifuge at 15,000 × *g* for 10 min.
7. Wash the pellet with 70% ethanol (RNase free), decant, and air-dry (see Note 14). Resuspend in 40 μl of RNase-free H_2O. Samples can be stored at −80°C at this point.
8. Digest genomic DNA by adding RiboLock RNase-inhibitor, 10× RQ1 DNAse buffer, and RQ1 RNase-free DNase.
9. Incubate at 37°C for 30–60 min.
10. Remove genomic DNA by using the RNAeasy-cleanup procedure, part of the RNAeasy mini kit, according to manufacturer's instructions.
11. Elute in 20 μl of RNase-free water.
12. Synthesize cDNA using standard oligo-dT-primed cDNA synthesis reaction with SuperScript II Reverse Transcriptase. Standard positive RT reaction mix:

    1 μg of total RNA from above dissolved in 10 μl RNAse-free water

    1 μl oligo-dT primer (0.5 mg/ml)

    1 μl dNTPs (10 mM)

    Denature at 65°C, 5 min, place on ice, and then add the following:

    4 μl 5× SSII First Strand Buffer

    2 μl DTT (100 mM)

    1 μl RiboLock RNase-inhibitor

    1 μl SuperScript II RT
13. Incubate for 50–75 min at 42°C, and heat inactivate the RT at 70°C for 15 min (always include a no-RT control). We typically use half the amount for the no-RT reaction. Dilute reactions 1:3 or 1:4 with H_2O to prepare them for normalization via qPCR (store at −20°C).
14. The reaction mix for qPCR reaction is as follows:

    1 μl cDNA

    1 μl Forward primer (20 mM)

    1 μl Reverse primer (20 mM)

    7 μl H_2O

    10 μl SYBR Green PCR Master Mix

    20-μl Total volume

Table 1
Primers for quantitative PCR of ES-derived mouse retinal cells

Gene	Primer pairs (mouse)
β-Actin	F-ctaaggccaaccgtgaaaag R-accagaggcatacagggaca
Pax-6	F-ctggagaaagagtttgagagg R-tgataggaatgtgactaggag
Lhx-2	F-ctgttccagagtctgtcggg R-cagcaggtagtagcggtcag
Six-3	F-ggtttaagaaggccgctgac R-taccgagaggatcgaagtgc
Rx	F-ttcgagaagtcccactaccc R-ttcatggacgacacttccag
Crx	F-tctgtgtgttacagacatgaccactaa R-catcaagcttcttttgcattttgt
Rhodopsin	F-tcaagcctgaggtcaacaagc R-acttccttctctgccttctgagtg
S-Opsin	F-cagcatccgcttcaactccaa R-gcagatgagggaaagaggaatga
Recoverin	F-gcagcttcgatgccaacag R- tcatgtgcagagcaatcacgta
Engrailed-1	F-atgggacattggacacttcttc R-cccacagaccaaataggagcta
Otx-2	F-ccgccttacgcagtcaatg R-gagggatgcagcaagtccata
Emx-1	F-gaagaagaagggttcccaccat R-ccgttggcctgcttcgt

15. In order to compare samples, normalize by assaying levels of a control gene, i.e., β-actin. Adjust the sample concentrations according to the ratio of the cycle numbers at which the control transcript(s) exhibits log scale increases in amplification as measured by fluorescence. Set the threshold at the level that the fluorescence increase has reached a maximal slope. The cycle number difference in transcript levels measured between the experimental and control cDNA samples is used to calculate the fold difference. This fold difference is used in conjunction with the original sample volumes to dilute the more concentrated sample to that of the less concentrated sample. After sample concentrations are adjusted, retest β-actin levels to determine how well they were normalized.
16. Each positive control sample should be run in duplicate or triplicate and one negative RT reaction to check genomic DNA contamination.
17. Further qPCR analysis with additional primer pairs for retinal candidate genes, including Pax6, Six3, Rx, and Lhx2 (see Tables 1 and 2): qPCR primer sets should be designed to amplify 50–200-bp amplicons, and should always be checked for specificity. Additionally, always include β-actin in each run to allow for more precise normalization of sample concentrations and accurate values in test primer sets.

Table 2
Primers for quantitative PCR of ES-derived human retinal cells

Gene	Primer pairs (human)
β-Actin	F-actcttccagccttccttc R-atctccttctgcatcctgtc
Pax-6	F-tctaatcgaagggccaaatg R-tgtgagggctgtgtctgttc
Lhx-2	F-tagcatctactgcaaggaagac R-gtgataaaccaagtcccgag
Six-3	F-ggaatgtgatgtatgatagcc R-tgatttcggtttgttctgg
Rx	F-gaatctcgaaatctcagccc R-cttcactaatttgctcaggac
Crx	F-atgatggcgtatatgaaccc R-tcttgaaccaaacctgaacc
Rhodopsin	F-tcatcatggtcatcgctttc R-catgaagatgggaccgaagt
S-Opsin	F-gatgaatccgacacatgcag R-ctgttgcaaacaggccaata
Recoverin	F-ccagagcatctacgccaagt R-cacgtcgtagagggagaagg
Engrailed-1	F-ccgcaccaccaactttttcat R-tggacagggtctctacctgc
Otx-2	F-gcagaggtcctatcccatga R-ctgggtggaaagagaagctg
Emx-1	F-aggtgaaggtgtggttccag R-agtcattggaggtgacatcg

18. Upon comparing expression levels of the EFTFs, Pax6, Rx, Lhx2, and Six3, there should be an 80- to 160-fold up-regulation of all of these genes after differentiation (Figs. 1 and 2). We also found that IGF-1, Dkk1, and Noggin were each required for this retinal determination of the undifferentiated ES cells.

3.10. Analysis of Retinal Determination Using Fluorescent Immunohistochemistry

1. Fix the cells with 4% paraformaldehyde at 4°C for 30–60 min.
2. Rinse cells with 1× PBS twice with PBS.
3. Incubate the cells in block solution (PBS with 5% serum and 0.5% Triton X) for 30 min at room temperature.
4. Add primary antibody solution. The antibody is diluted to the suggested working concentration (Table 3) in the block solution.
5. Incubate overnight at 4°C.
6. Wash three times, 5 min each in PBS.
7. Add the secondary antibody solution (usually 1:500 Alexa-fluor conjugated at 1:500 diluted in block solution) and incubate in a humidified chamber for about 1 h in the dark at room temperature.
8. Wash three times, 5 min each in PBS.
9. Mount in Fluoromount-G.

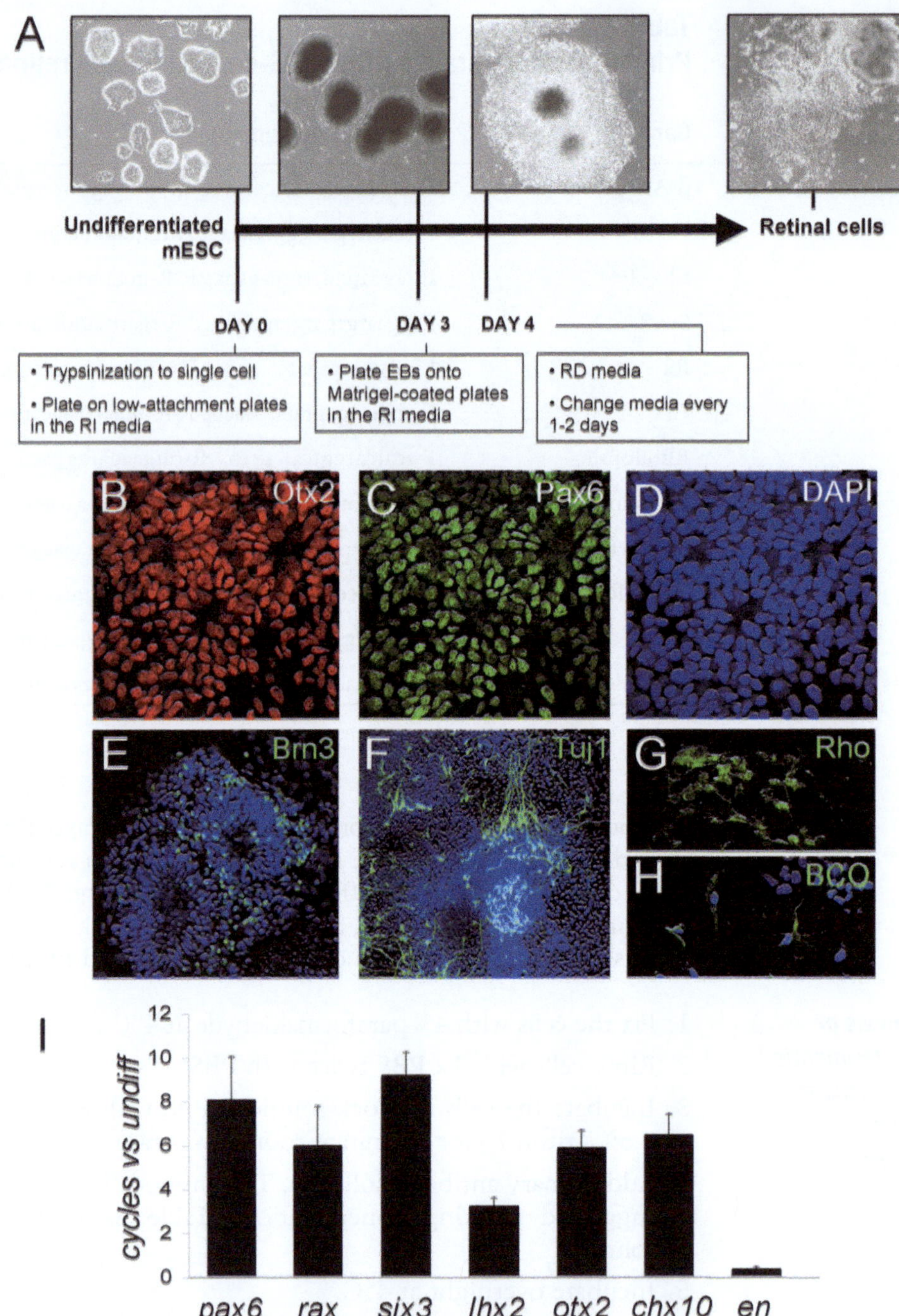

Fig. 1. Generation of retinal cells from mouse ES cells. (**a**) Schematic showing the steps involved in differentiating retinal cells from mouse ES cells. After 7 days of differentiation, the majority of the cells express retinal progenitor markers (**b–d**). The cells express Otx2 (**b**) as well as Pax6 (**c**). In 12 days, the cells express the differentiation markers Brn3 (**e**), Tuj1 (**f**), Rhodopsin (**g**), and Blue Cone Opsin (**h**). (**I**) Quantitative PCR data showing expression of EFTFs; the mid-hindbrain gene *engrailed* was not up-regulated.

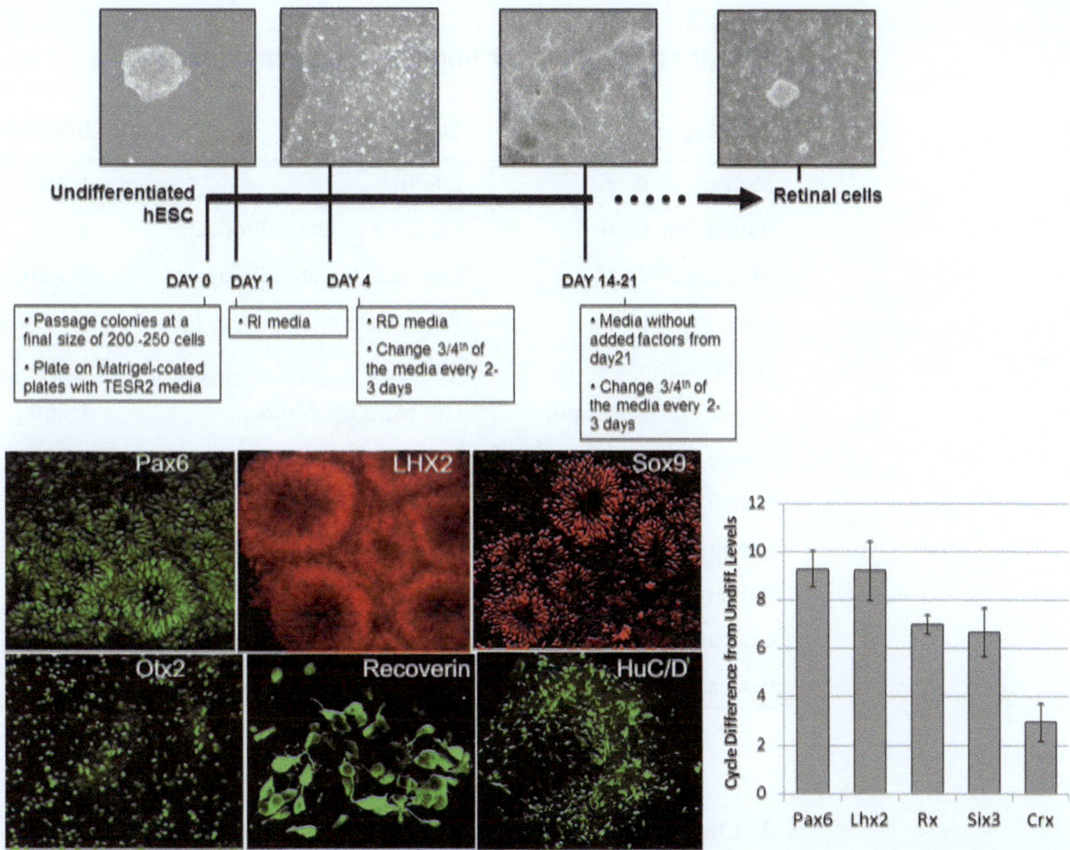

Fig. 2. Generation of retinal cells from human ES cells. (**a**) Schematic showing the steps involved in differentiating retinal cells from human ES cells. The retinal cells express progenitor markers Pax6 (**b**), Lhx2 (**c**), and Sox9 (**d**), as well as differentiation markers Otx2 (**e**), recoverin (**f**), and HuC/D (**g**). (**h**) Quantitative PCR data showing expression of eye field transcription factors and Crx.

10. After 7 days of the retinal determination protocol (mouse cells) or 3 weeks (human ES cells), the majority of the cells should show co-expression of Otx2, Pax6, Lhx2, and Sox9 [~70–80% cells (Figs. 1 and 2)]. Also, many of the cells should be labeled with neuronal markers like Hu C/D, Neurofilament, and Tuj1, as well as photoreceptor markers like Crx and Nrl.

3.11. Subretinal Injection of Dissociated Cells into Adult Mouse Recipients

One day prior to transplantation, the mice are started on Cyclosporine A injections (10 mg/kg/day) daily until euthanasia.

1. Prepare the dissociated cell suspension by rinsing the cells with PBS followed by incubation in 0.2% trypsin for 5 min. As the cells lift off the plate, inhibit the trypsin activity with 10% FBS in DMEM. Centrifuge the cells at 1,200 rpm for 5 min and resuspend in media at a concentration of 80,000–100,000 cells/μl.

Table 3
Primary antibodies for immunocytochemistry

Antibody	Company	Dilution
Mouse anti-Pax6	DHSB	1:250
Rabbit anti-Pax6	Covance Research Products	1:400
Mouse anti-Tuj-1	Covance Research Products	1:1,000
Goat anti-Otx2	R&D Systems	1:500
Rabbit anti-Sox9	Millipore	1:200
Mouse anti-Rhodopsin	Gift from Dr. Molday	1:750
Mouse anti-Hu C/D	Invitrogen	1:200
Rabbit anti-Nrl	Gift from Dr. A. Swaroop	1:500
Rabbit anti-Neurofilament	Chemicon	1:500
Rabbit anti-Crx	Gift from Dr. C. Gregory-Evans	1:2,000
Goat anti-Lhx2	Santa Cruz Biotechnologies	1:100
Goat anti-Opn1sw	Santa Cruz Biotechnologies	1:100

2. One eye per animal is injected with cells, while the other eye serves as a control. We always inject the left eye to avoid confusion.
3. Prepare pipettes by pulling them to a long (10 μm), gentle taper. The tip is then broken with forceps to an ~30–50-μm opening, which will be small enough to minimize injury, but large enough to let cells pass through easily. Just before starting the injections, mount the pipette into the holder, insert the plunger, and draw up 2–3 μl of cell suspension.
4. Anesthetize the transplant recipient mouse with isoflurane gas anesthesia. Any other reliable method of anesthesia can be substituted. Use a toe pinch to assure that the animal is fully anesthetized. Apply petrolatum ophthalmic ointment to the control eye to keep it moist, and apply a topical anesthetic (0.5% proparacaine hydrochloride ophthalmic solution) to the eye to be injected.
5. Position the stereotaxic apparatus under a dissection microscope. Mount the animal's head in the head holder. First, insert the upper teeth into the notch of the tooth bar, and tighten the nose clamp just enough to keep the head level. Then, insert the ear bars into the bony ear canal on each side, and tighten gently. Make sure that the skull is firmly in place before proceeding.
6. From here on, work while observing through the microscope with good bright lighting. Gently lift the upper eyelid using

forceps, and then pass the suture through the upper eyelid. Fasten both ends of the suture material to the screws of the stereotaxic to pull the eyelid open. Put gentle pressure on the periorbital area with the blunt end of your forceps to pop the eye out of the socket.

7. Using a sterile #11 scalpel blade very carefully and gently cut away the sclera on the dorsal surface of the eye. The blade should be repeatedly passed across the sclera to gradually thin the tissue in a small area approximately 0.5 μm in diameter (see Note 15). If using a pigmented mouse, the dark coloration of the pigment epithelium should be increasingly visible. Make sure that the surface of the eye does not dry out (see Note 16). Continue cutting until a tiny area of retina (approximately the diameter of the pipette tip) is exposed at the center of the thinned area. Due to the high intraocular pressure, the retina will bleb out of the opening in the sclera, so it is very important to keep the cut as small as possible. Use a Kimwipe to wick up any blood.
8. Advance the pipette tip towards the opening in the sclera. Keep it at a shallow angle (10–20°) relative to the surface to the eye to avoid puncturing the retina. Position the tip directly over the opening, and then slip the tip just under the sclera with a tiny down-and-forward motion. Now, apply very gentle outward traction on the eye by pulling the pipette back up. This helps to open up a space under the retina. Advance the pipette tip about 250 μm into the subretinal space (see Note 17).
9. Pressing on the plunger very slowly and lightly, inject 1 ml of cells into the subretinal space (Fig. 3). Expect some of the injected cells to squirt back out around the needle tip because the subretinal space is a potential space and does not admit much volume. Let the pipette rest in place for about 30 s to allow the injected volume to disperse in the subretinal space, and then gently withdraw the pipette.
10. Remove the suture, and gently push the eyeball back into the socket. Remove the mouse from the holder, apply bacitracin ophthalmic ointment to the injected eye, and let the animal recover from the surgery. Using this method, we have not observed any intraocular or periocular infections. The injection site will heal within a few days, and the mice tolerate the procedure very well.
11. After survival periods of 1–4 weeks, the animals are sacrificed by a protocol that has been previously approved by the institutional IACUC (or similar animal care committee) and the eyes removed for analysis of transplanted cells by fixation (2 h to overnight 2–4% paraformaldehyde in PBS at 4°C), standard cryostat sectioning, and immunofluorescent labeling as described above.

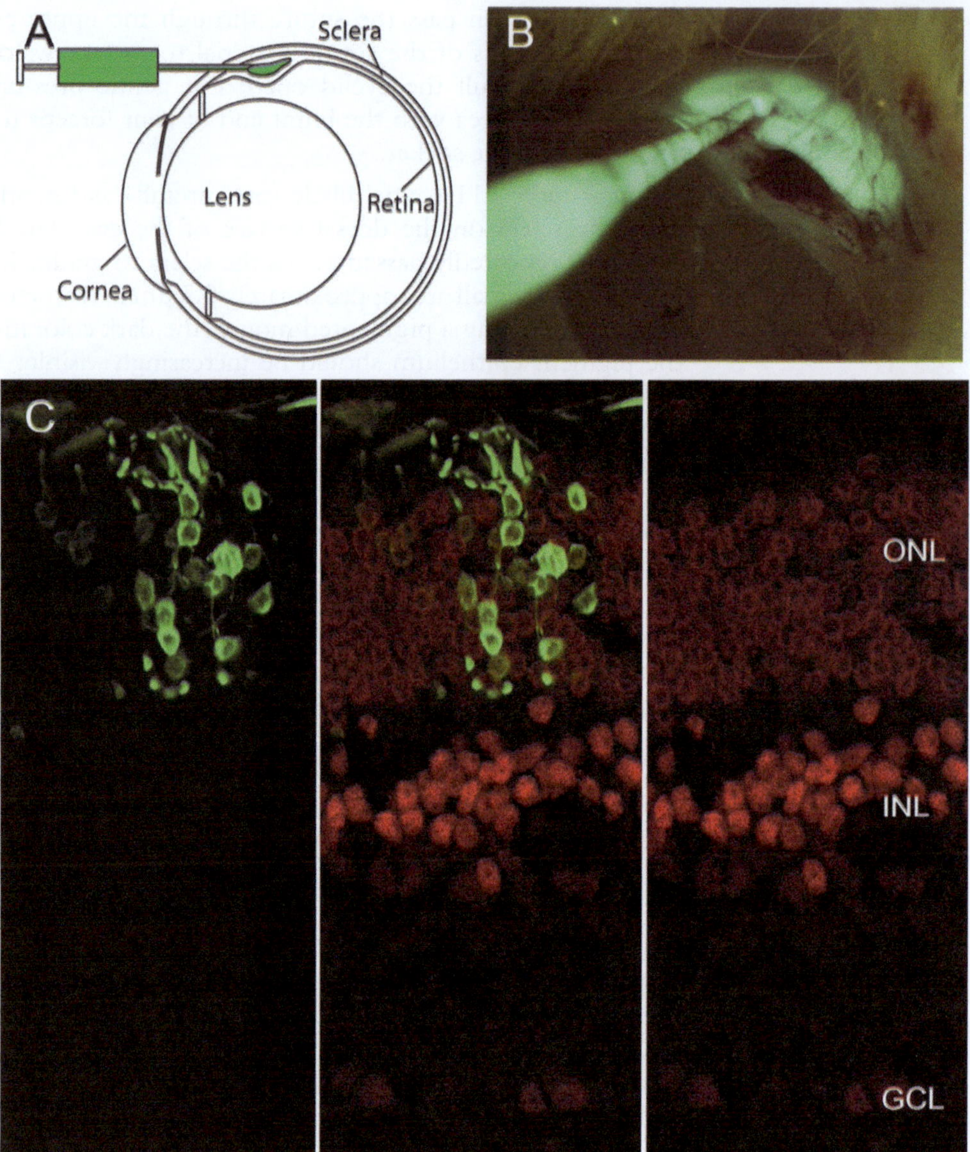

Fig. 3. Transplantation of human ESC-derived retinal cells. (**a**) Schematic of sub-retinal route of transplantation of GFP-expressing human retinal cells. (**b**) Image of transplantation procedure showing the glass pipette filled with GFP-expressing cells along with cells in the sub-retinal space of an albino mouse. (**c**) A section of mouse retina containing transplanted GFP-expressing human ESC-derived photoreceptors (*green*; *left panel*) that have integrated into the outer nuclear layer and are also expressing Otx2 (*red*; *right panel*); the *middle panel* is a merged image.

4. Notes

1. We find it is best to aliquot all the components of the media to avoid contamination.
2. The beta-mercaptoethanol stock solution (0.1 M) can be stored at room temperature for no more than 1 month.

3. Freeze-thawing the LIF can affect its efficiency; for long-term storage, it is recommended to store concentrated LIF at 4°C. The GSK3β inhibitor, CHIR99021, and the MEK inhibitor, PD0325901, are reconstituted in DMSO as 1,000× stocks. Following reconstitution, the aliquots can be stored at −20°C protected from light. We find it important to gently vortex these solutions because some precipitation can occur.
4. We find that it is best to prepare this media fresh on a weekly basis.
5. Cut first the bottom edge of the straw and place it right on top of a falcon tube. Cut the top edge and the liquid will flow out by gravity. Gently tapping the straw with your fingers will help with the last few drops.
6. Carefully disperse the pellet using a 5-ml pipette. Do it thoroughly, ensuring that no cell clumps are left, but avoid air bubbles.
7. Differentiation of mouse ES cells can be prevented by supplementing the media with LIF. However, if the undifferentiated mouse ES cells are passaged as small aggregates instead of single cells, some colonies can undergo spontaneous differentiation. To avoid these undesirable colonies, it is important to trypsinize the cells thoroughly and separate them with a long glass pipette. Additionally, differentiated colonies showing flattened morphology can be manually removed with a glass pipette.
8. In general, free-floating mouse ES cells will aggregate to form EBs within 12 h. However, if there are too many cells in one well, after 24 h, the individual EBs will further fuse forming fishnet-like aggregates. These 3D structures are too big and some cell death occurs in the inner part of the aggregate. The fishnet-like aggregates can be divided by gently pipetting with a 5-ml pipette tip and a 1:3 passage of the EBs will be necessary.
9. Some cell death may occur. To remove all the cell debris, rinse twice with PBS before adding the RD media.
10. If the colonies are too small, they may not survive passage. Alternatively, if they are too large, the center of colony may not differentiate as well or undergo cell death.
11. If >50,000 cells are plated per well, they may turn superconfluent too early, requiring passaging earlier than 2 weeks. This usually results in inefficient RI.
12. By the end of the second week, the cells may need daily media change.
13. Do not over-titurate as this will make the colonies too small which can result in both increased cell death as well as inefficient RI.
14. Do not overdry the RNA pellet, or it will be difficult to redissolve. If the pellet is difficult to redissolve, the tube can be incubated at 55–65°C for 10 min.

15. Do not apply too much pressure as you may puncture the eye.
16. Do not let the surface dry.
17. Move the tip very gently, or the tip will puncture the retina resulting in intravitreal transplantation.

Acknowledgments

The authors would like to thank the members of the Reh and Bermingham-McDonogh lab for their input and advice during the derivation of the methods described in this chapter. This work was funded by grants from the NIH (PO1 GM081619) and the Foundation Fighting Blindness (Wynn/Gund Translational Award TA-CBT-0608-0464-UWA-WG) to T.A.R.

References

1. Lamba D, Karl M, Reh T (2008) Neural regeneration and cell replacement: a view from the eye. Cell Stem Cell 2:538–549
2. Cho MS, Hwang DY, Kim DW (2008) Efficient derivation of functional dopaminergic neurons from human embryonic stem cells on a large scale. Nat Protoc 3:1888–1894
3. Dimos JT, Rodolfa KT, Niakan KK, Weisenthal LM, Mitsumoto H, Chung W, Croft GF, Saphier G, Leibel R, Goland R et al (2008) Induced pluripotent stem cells generated from patients with ALS can be differentiated into motor neurons. Science 321:1218–1221
4. Park CH, Minn YK, Lee JY, Choi DH, Chang MY, Shim JW, Ko JY, Koh HC, Kang MJ, Kang JS et al (2005) In vitro and in vivo analyses of human embryonic stem cell-derived dopamine neurons. J Neurochem 92: 1265–1276
5. Niehrs C (2001) Developmental biology. Solving a sticky problem. Nature 413: 787–788
6. del Barco Barrantes I, Davidson G, Grone HJ, Westphal H, Niehrs C (2003) Dkk1 and noggin cooperate in mammalian head induction. Genes Dev 17:2239–2244
7. Pera EM, Wessely O, Li SY, De Robertis EM (2001) Neural and head induction by insulin-like growth factor signals. Dev Cell 1:655–665
8. Zuber ME, Gestri G, Viczian AS, Barsacchi G, Harris WA (2003) Specification of the vertebrate eye by a network of eye field transcription factors. Development 130:5155–5167
9. Cavodeassi F, Carreira-Barbosa F, Young RM, Concha ML, Allende ML, Houart C, Tada M, Wilson SW (2005) Early stages of zebrafish eye formation require the coordinated activity of Wnt11, Fz5, and the Wnt/beta-catenin pathway. Neuron 47:43–56
10. Esteve P, Bovolenta P (2006) Secreted inducers in vertebrate eye development: more functions for old morphogens. Curr Opin Neurobiol 16:13–19
11. Lamba DA, Gust J, Reh TA (2009) Transplantation of human embryonic stem cell-derived photoreceptors restores some visual function in Crx-deficient mice. Cell Stem Cell 4:1–7
12. Lamba DA, Karl MO, Ware CB, Reh TA (2006) Efficient generation of retinal progenitor cells from human embryonic stem cells. Proc Natl Acad Sci USA 103:12769–12774

Chapter 17

Light-Induced Photoreceptor Degeneration in the Retina of the Zebrafish

Scott Taylor, Jing Chen, Jing Luo, and Peter Hitchcock

Abstract

Exposure of the zebrafish retina to intense light is a noninvasive method to elicit the selective degeneration of photoreceptors. In zebrafish, photoreceptor degeneration is followed by robust photoreceptor regeneration from stem cells that are intrinsic to the teleost retina. Two recent light-lesioning methods have been developed, each with specific advantages. The first involves a prolonged period of dark adaptation followed by exposure to high-intensity halogen lighting at ~3,000–20,000 lux for 3–4 days. This causes widespread degeneration of rod and cone cells in the dorsal and central regions of the retina. The second method uses ultrahigh-intensity lighting at intensities greater than 120,000 lux, with an exposure time of 30 min. This causes degeneration of rod and cone cells within a relatively narrow naso-temporal band in the central retina. The advantages of the first (lower light intensity) technique are the widespread destruction of photoreceptors and the lower cost of equipment. The advantages of the second (higher light intensity) technique are the elimination of the prolonged dark adaptation and short duration of the exposure, thereby allowing experiments to be conducted more rapidly.

Key words: Light lesion, Photoreceptor degeneration, Photoreceptor regeneration, Stem cells

1. Introduction

Determining the molecular mechanisms that govern neuronal regeneration in the retina of the zebrafish requires techniques that provide control of the nature and the degree of neuronal death. Many investigators studying this phenomenon employ photolytic lesions to induce the selective death of photoreceptors while leaving other cell types relatively uninjured. Photolytic lesions model some aspects of photoreceptor degenerations in humans, and the absence of death among other cell types allows changes in gene expression to be attributed largely to the process of photoreceptor

Shu-Zhen Wang (ed.), *Retinal Development: Methods and Protocols*, Methods in Molecular Biology, vol. 884, DOI 10.1007/978-1-61779-848-1_17, © Springer Science+Business Media, LLC 2012

death and regeneration (e.g., ref. 1). Further, photolytic lesions have the advantages of being noninvasive, and the necessary equipment relatively inexpensive.

Other techniques have been employed to induce retinal damage, and all have advantages for certain types of research. Surgical (2, 3) and chemical (4) means of inducing retinal damage are rapid and effective, but they are not selective for photoreceptors and are therefore more useful for studying whole retinal regeneration. Laser ablation (5) and thermal lesions (6) can damage specific target areas in the retina, sometimes only involving photoreceptors and the RPE, but lesions are very localized and damage to other cell types cannot be excluded. A transgenic line has been created that results in the rapid degeneration of rod cells (7) and induces a continuous cycle of rod cell degeneration and regeneration throughout life. This allows the study of factors associated with rod-specific degeneration and regeneration; but without a specific triggering event, both processes are occurring simultaneously and cannot be studied separately. Therefore, exposure to high-intensity light is presently the only method of inducing degeneration and subsequent regeneration in photoreceptors on demand, without damage to other retinal layers. This allows the study of molecular mechanisms that are specific to the degeneration and regeneration of photoreceptor cells.

Two basic light-lesioning paradigms have been described in the literature, and several labs have used variations of these techniques for studying photoreceptor regeneration. Vihtelic and Hyde (8) and Vihtelic et al. (9) established a technique that uses high-intensity (20–21,000 lux) halogen lamps over a relatively long duration (3–7 days) to cause damage to photoreceptors in the retinas of zebrafish carrying the *albino* mutation. *Albino* zebrafish lack melanin in the retinal pigmented epithelium, and therefore photoreceptors are not protected by this light-absorbing pigment and are subjected to constant light. This paradigm has been shown to cause widespread death of rod and cone photoreceptors in the dorsal and central regions of the retina while causing very little photoreceptor damage ventrally. The regional differences in photoreceptor degeneration might be attributable to the presence of a reflective tapetum in the dorsal retina, to longer photoreceptor outer segments in the dorsal retina, to the distribution of other pigment types in the retina, or to other factors not yet understood (9).

The other widely used light-lesioning technique was developed by Bernardos et al. (10), and involves exposure to ultrahigh-intensity (>120,000 lux) halogen lights over a much shorter time period (30 min). This procedure results in the destruction of rod and cone photoreceptors within a narrow band along the naso-temporal axis of the retina (1).

A problem that must be overcome with either technique is heating of the water and maintenance of appropriate water quality

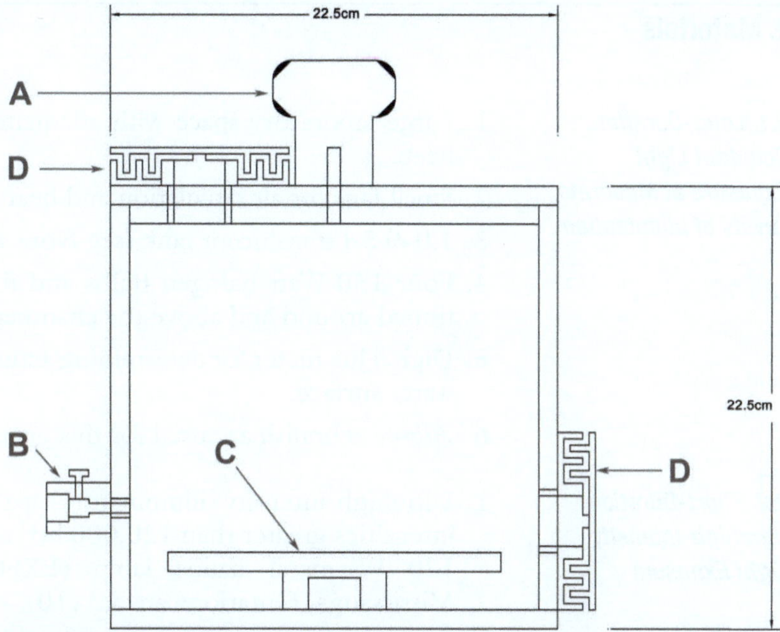

Fig. 1. Light lesion box for short-duration ultrahigh-intensity light exposure. *A* lid, *B* port for the fiber-optic line, *C* adjustable platform, *D* Baffles to release heat.

parameters during the light treatment. Zebrafish can tolerate a wide temperature range (6.7–41.7°C), but the recommended standard for husbandry based on survival and growth rates is 28.5°C (11). As light exposure time increases, heating of the water can become problematic. To maintain water temperature within a reasonable range, the light exposure procedure should be conducted in a well-ventilated room so that heat is quickly dissipated. Fans, heat-conductive materials in contact with the tanks, air conditioning, and other methods should be considered to help with heat dissipation. For ultrahigh-intensity but short-duration exposures, heat can still be a problem due to the intense output of the lighting mechanism and the smaller volume of water. To reduce the heating effect, Bernardos et al. (10) focused the ultrahigh-intensity light on fish contained in a 100-ml beaker, and placed the beaker into a 15-cm glass dish filled with water to conduct heat away from the experimental chamber.

Our laboratory utilizes a custom-built chamber (Fig. 1) and ultrahigh-intensity light (see Subheadings 2.2 and 3.2 below) to produce photolytic lesions. However, because two procedures are commonly in use to create photolytic lesions, both approaches and their advantages and limitations are detailed below.

2. Materials

2.1. Long-Duration, Constant Light Exposure at Moderate Levels of Illumination

1. Large laboratory space with adequate ventilation to dissipate heat.
2. Small fans for air circulation and heat dissipation.
3. 1.0–9.5-l translucent tank (see Note 1).
4. Four 150-Watt halogen bulbs and fixtures that can be positioned around and above the chamber.
5. Digital lux meter for determining actual light intensity at glass/water surface.
6. *Albino* zebrafish are used for this procedure (see Note 2).

2.2. Short-Duration Ultrahigh-Intensity Light Exposure

1. Ultrahigh-intensity illuminator capable of generating light intensities greater than 120,000 lux, such as the EXFO X-Cite 120 W metal halide lamp (EXFO Photonic Solutions, Mississauga, Ontario, Canada) (10).
2. 3-mm-diameter fiber-optic liquid light line connected to the illuminator (Fig. 1).
3. 100-ml glass beaker containing 80 ml of aquarium water.
4. 15-cm-diameter (or similar) shallow glass dish containing tap water.
5. Digital lux meter for determining actual light intensity at glass/water surface.

3. Methods

3.1. Long-Duration, Constant Light Exposure at Moderate Levels of Illumination

1. Fish are held in constant darkness for 7–14 days prior to light treatment in small aquaria (1.0–9.5-l containers have been used).
2. Aquaria are removed from the dark environment and immediately placed in the photolytic environment.
3. Throughout the procedure, appropriate water quality conditions must be maintained with aeration, filtration, and/or water changes as appropriate.
4. Fluorescent or metal halide lights of moderate intensities are placed facing two or four sides of the aquarium, 50–80 cm from the tank. This ensures equal light intensity reaching the fish from all directions. Standard 150-Watt halogen bulbs and fixtures are used to generate these light intensities (see Note 3).

5. Photoreceptor cell death is evident by 24 h after the onset of light exposure (8). Severe damage both to rod and cone photoreceptors is evident in the dorsal and central retinas by 3 days after the light onset (9). Maximum loss of photoreceptor-specific markers occurs 4 days after the light onset. Photoreceptors are largely regenerated by 7–11 days of the light onset (12).
6. Specific advantages of this procedure are described in Note 4, and specific disadvantages are described in Note 5.

3.2. Short-Duration Ultrahigh-Intensity Light Exposure

1. A few fish (~4–6) are placed in a 100-ml glass beaker, backed with reflective aluminium foil, containing 80 ml of aquarium water.
2. The end of the fiber-optic line is placed 5 cm from the beaker and the illuminator is turned on.
3. Fish are exposed for 30 min.
4. Fish are returned to the normal aquarium water for recovery.
5. By 24 h post light exposure, rod and cone cells are degenerated in a relatively narrow band across a naso-temporal band of central retina. Proliferation of photoreceptor cells at 72 h after a short-duration ultrahigh-intensity light exposure is shown by PCNA staining in Fig. 2.
6. Specific advantages of this procedure are described in Note 6, and specific disadvantages are described in Note 7.

4. Notes

1. Various sizes of tanks have been used for this procedure to hold fish during light treatments. The size of the tank depends partially on the number of fish to be held in the tank and treated at once. Larger sizes hold more fish and facilitate maintenance of temperature and other water quality parameters, but light intensity on the fish is decreased with increasing volume of water.
2. Albino zebrafish have typically been used for this procedure because photoreceptors are not protected by melanin, but Craig et al. (12) used light exposure of 27,000 lux on fully pigmented fish and achieved similar results.
3. In earlier studies (1, 8, 9, 13, 14), light intensity at the glass/water surface was maintained within the range of 18,000–30,000 lux. However, in some of the more recent studies, lower intensities (2,800–8,000 lux) have been used with comparable results (12, 15–18).

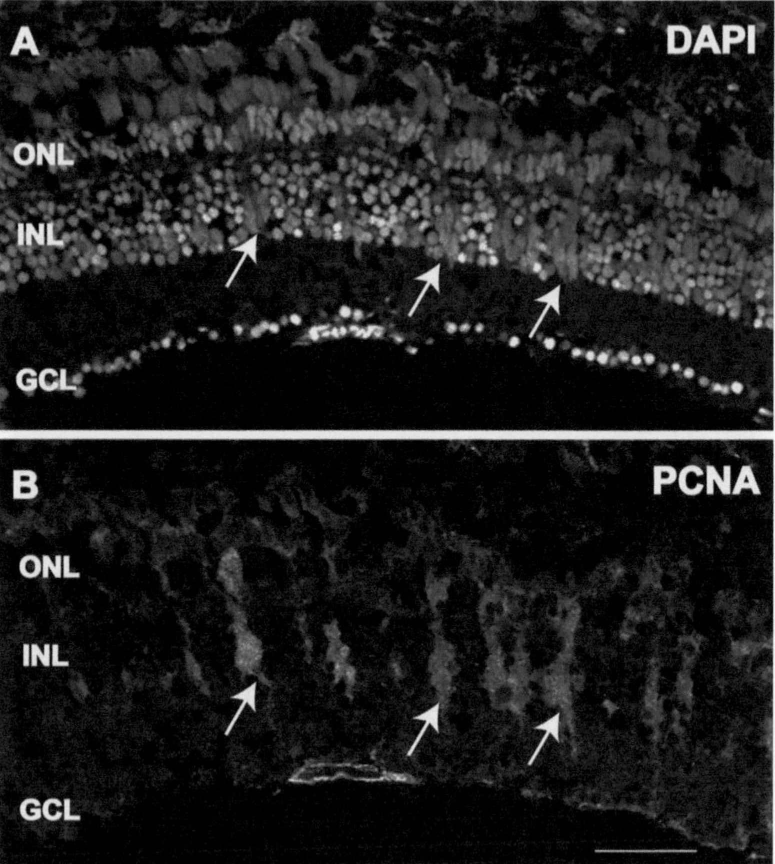

Fig. 2. Fluorescence photomicrograph of a section through the retina of an adult zebrafish 72 h after a short-duration ultrahigh-intensity photolytic lesion. *Panel* (**a**) Nuclei stained with DAPI. *Arrows* identify radial clusters of proliferating cells. (**b**) Same section as in *panel* (**a**), with a different fluorescence filter set, illustrating cells immunostained with antibodies against proliferating cell nuclear antigen (PCNA; *arrows*). *ONL* outer nuclear layer, *INL* inner nuclear layer, *GCL* ganglion cell layer. Scale bar equals 50 μm.

4. Specific advantages of using the lower intensity, longer duration procedure described in Subheading 3.1 include the following.
   (a) The lighting apparatus is relatively inexpensive to construct compared with the higher light intensity technique. The 150-Watt halogen bulbs and fixtures can be purchased from a local hardware store.
   (b) Photoreceptor degeneration occurs over a wide area of retina (the entire dorsal half of the retina). Experiments can be conducted with the knowledge that photoreceptors will be destroyed over a broad area, and this precludes the need to search the retina for individual cells or isolated areas of damage.
   (c) A large number of fish can be placed in the treatment tank due to the relatively large size of the tank and wide area of exposure of light.

5. Specific disadvantages of using the lower intensity, longer duration procedure described in Subheading 3.1 relate to the longer duration of treatment. Fish must be held in complete darkness (dark adaptation) for 7–14 days prior to light treatment, and then the light treatment must be maintained for 3–4 days to cause substantial photoreceptor degeneration. This results in the following difficulties.
   (a) Maintenance of appropriate water temperatures (around 28.5°C) is difficult due to the high-intensity lighting and long duration of exposure. Therefore, the light treatment must be conducted in a large, well-ventilated laboratory, and other cooling mechanisms (such as fans and/or frequent water changes) must be employed.
   (b) Maintenance of other water quality parameters (dissolved oxygen, pH, low-nitrogenous waste) is difficult with several fish in a relatively small aquarium. Aeration, filtration, and water changes must be used to maintain these parameters appropriately.
6. Specific advantages of using the ultrahigh-intensity, shorter duration method described in Subheading 3.2 include the following.
   (a) Short light-exposure times and no need for a prolonged dark-adaptation period prior to light treatment make this procedure rapid and convenient.
   (b) The short exposure time makes temperature control and other husbandry issues less of a concern. The ultrahigh-intensity light source heats the water faster than the lower intensity procedure, but there are several simple ways to maintain the temperature within an acceptable range for 30 min. Heating of the entire room is not a problem with this procedure and, therefore, it can be conducted in a laboratory setting.
   (c) Fully pigmented fish can be lesioned, which alleviates the need to maintain a separate line of animals for regeneration studies.
7. Specific disadvantages of using the ultrahigh-intensity, shorter duration method described in Subheading 3.2 include the following.
   (a) Photoreceptor damage occurs over a smaller area (only in the central retina in a naso-temporal band) than that with the previous procedure. This can make it challenging to find sufficient numbers of damaged/regenerating cells for study.
   (b) Equipment is more expensive—the ultrahigh-intensity light source costs several thousand dollars verses a few hundred dollars for the previous procedure; however, this is a standard lamp for modern fluorescence microscopes and will be available in many laboratories.

(c) Only a few fish can be treated at a time due to the small volume of water.

(d) Caution must be used to insure that laboratory personnel are not exposed to the ultrahigh light source.

Note Added in Proof:

Thomas et al. (19) recently demonstrated that dark adaptation combined with 30-minute exposure to ~100,000 lux light followed by 4-day exposure to ~8000 lux light resulted in more complete photoreceptor ablation than either light treatment alone.

References

1. Craig SEL, Calinescu A-A, Hitchcock P (2008) Identification of the molecular signatures integral to regenerating photoreceptors in the retina of the zebra fish. J Ocul Biol Dis Inform 1:73–84
2. Liu Q (2002) Up-regulation of cadherin-2 and cadherin-4 in regenerating visual structures of adult zebrafish. Exp Neurol 177:396–406
3. Senut MC, Gulati-Leekha A, Goldman D (2004) An element in the alpha1-tubulin promoter is necessary for retinal expression during optic nerve regeneration but not after eye injury in the adult zebrafish. J Neurosci Off J Soc Neurosci 24:7663–7673
4. Fimbel SM et al (2007) Regeneration of inner retinal neurons after intravitreal injection of ouabain in zebrafish. J Neurosci Off J Soc Neurosci 27:1712–1724
5. Wu DM et al (2001) Cones regenerate from retinal stem cells sequestered in the inner nuclear layer of adult goldfish retina. Invest Ophthalmol Vis Sci 42:2115–2124
6. Raymond P et al (2006) Molecular characterization of retinal stem cells and their niches in adult zebrafish. BMC Dev Biol 6:36
7. Morris A et al (2005) Cone survival despite rod degeneration in XOPS-mCFP transgenic zebrafish. Invest Ophthalmol Vis Sci 46: 4762–4771
8. Vihtelic T, Hyde D (2000) Light-induced rod and cone cell death and regeneration in the adult albino zebrafish (*Danio rerio*) retina. J Neurobiol 44:289–307
9. Vihtelic T et al (2006) Retinal regional differences in photoreceptor cell death and regeneration in light-lesioned albino zebrafish. Exp Eye Res 82:558–575
10. Bernardos R et al (2007) Late-stage neuronal progenitors in the retina are radial Müller glia that function as retinal stem cells. J Neurosci Off J Soc Neurosci 27:7028–7040
11. Lawrence C (2007) The husbandry of zebrafish (*Danio rerio*): a review. Aquaculture 269:1–20
12. Craig SE et al (2010) The zebrafish galectin Drgal1-l2 is expressed by proliferating Müller glia and photoreceptor progenitors and regulates the regeneration of rod photoreceptors. Invest Ophthalmol Vis Sci 51:3244–3252
13. Kassen SC et al (2008) The Tg(ccnb1:EGFP) transgenic zebrafish line labels proliferating cells during retinal development and regeneration. Mol Vis 14:951–963
14. Calinescu A et al (2009) Cellular expression of Midkine-a and Midkine-b during retinal development and photoreceptor regeneration in zebrafish. J Comp Neurol 514:1–10
15. Kassen S et al (2007) Time course analysis of gene expression during light-induced photoreceptor cell death and regeneration in albino zebrafish. Dev Neurobiol 67:1009–1031
16. Thummel R et al (2008) Inhibition of Müller glial cell division blocks regeneration of the light-damaged zebrafish retina. Dev Neurobiol 68:392–408
17. Thummel R et al (2008) Characterization of Müller glia and neuronal progenitors during adult zebrafish retinal regeneration. Exp Eye Res 87:433–444
18. Thummel R et al (2010) Pax6a and Pax6b are required at different points in neuronal progenitor cell proliferation during zebrafish photoreceptor regeneration. Exp Eye Res 90: 572–582
19. Thomas J et al (2012) Characterization of multiple light damage paradigms reveals regional differences in photoreceptor loss. Exp Eye Res 97:105–116

Chapter 18

Microarray-Based Gene Profiling Analysis of Müller Glia-Derived Retinal Stem Cells in Light-Damaged Retinas from Adult Zebrafish

Zhao Qin and Pamela A. Raymond

Abstract

Microarray-based gene profiling has become an important technique to measure changes in gene expression on a genome-wide scale. Recently, cell-specific microarrays have been reported to study changes in gene expression of a particular cell type in several model organisms. Here, we describe a protocol to prepare RNA samples for microarray analysis of isolated Müller glia-derived retinal stem cells from light-damaged adult zebrafish expressing a fluorescent marker in Müller cells using enzymatic retinal dissociation followed by fluorescence-activated cell sorting (FACS).

Key words: Zebrafish retina, Müller glia, Retinal lesion, Retinal dissociation, Microarray

1. Introduction

Microarray-based gene profiling has become a widely used technique to measure and compare gene expression levels to define "transcriptomes." In this analysis, RNA samples from cells or tissues under different experimental conditions are prepared and reverse transcribed to generate fluorescently tagged cDNA samples that are hybridized with DNA oligonucleotide probes spotted on microarrays. Expression profiles are compared between different conditions based on hybridization signal intensities on the arrays. Recently, cell-specific microarrays have been reported in several model organisms to study changes in gene expression of a particular cell type. Although newer and more powerful methods of transcriptional gene profiling are now available with the development of high-throughput sequencing technologies that allow whole transcriptome shotgun sequencing, known as RNA-seq, these methods are more costly and not as widely available.

Shu-Zhen Wang (ed.), *Retinal Development: Methods and Protocols*, Methods in Molecular Biology, vol. 884, DOI 10.1007/978-1-61779-848-1_18, © Springer Science+Business Media, LLC 2012

Three different approaches have been used to isolate or enrich RNA samples from a particular cell type in preparation for gene profiling. The first is to generate transgenic reporter lines in which a cell-specific promoter is used to express a fluorescent marker (such as green fluorescent protein, GFP), then dissociate the tissue and purify the labeled cells with fluorescence-activated cell sorting, FACS (1, 2). A second method is to enrich cells based on their anatomical location using laser-capture microdissection on tissue sections (3). The third method is to isolate cell-specific transcripts by an mRNA-tagging technique. To do this, an epitope-tagged mRNA-binding protein (e.g., FLAG-PAB) is expressed in the cells of interest, then FLAG-PAB-bound transcripts are immunoprecipitated and used for microarray experiments (4).

Retinal cell-specific microarray gene profiles using the first two approaches have been reported (1–3). Here, we describe in detail a protocol used to prepare high-quality RNA samples for microarray analysis of isolated Müller glia-derived retinal stem cells from light-damaged adult *Tg(gfap:GFP)mi2002* zebrafish expressing a fluorescent reporter in Müller cells using enzymatic retinal dissociation and FACS.

2. Materials

2.1. Retinal Lesion

- 100-ml Glass beaker (half of outer surface covered with foil and glued in the center of a 15-cm-diameter glass dish).
- Fiber optic liquid light line (3-mm diameter) connected to an EXFO X-Cite 120W metal halide lamp (EXFO Photonic Solutions, Mississauga, Ontario, Canada) (5).

2.2. Retinal Dissociation

Prepare the Following Solutions Using Autoclaved Glass-Distilled Water, Aliquot, and Store at –20°C

- 10× Papain (Worthington): 160 U/ml.
- 10× l-Cysteine (Sigma-Aldrich): 55 mM.
- 100× Dispase (Worthington): 19 U/ml.
- 10× Papain inhibitor: 10 mg/ml papain inhibitor (Worthington), 10 mg/ml bovine serum albumin, BSA (Sigma-Aldrich).
- 100× DNase I (Sigma-Aldrich): 10 mg/ml.

Prepare the Following Solution Using Autoclaved, Glass-Distilled Water, and Store at 4°C

- 25× $MgCl_2$: 50 mM.

Prepare the Following Solutions Using Milli-Q Water and Store at 4°C

- 10× Phosphate buffered saline (PBS): 0.02 M NaH_2PO_4, 0.08 M Na_2HPO_4, 1.5 M NaCl, and 0.025 M KCl.
- 1× PBS, pH 6.5: dilute one part 10× PBS stock with nine parts water, pH to 6.5.
- 1× PBS, pH 7.4: dilute one part 10× PBS stock with nine parts water, pH to 7.4.

Additional Materials Include

- Dissection tools: microscalpel, microscissors, forceps with curved fine tips.
- 6-cm Petri dish.
- 1.5-ml Siliconized tube.
- 30 cc Syringe (with 16 G needle).
- Single-edge razor blade.
- Microscope slide.
- Fire-polished 5 3/4″ glass Pasteur pipette (see Note 1).
- Nutating mixer placed in a 28.5°C incubator.
- Standard tabletop centrifuge.

2.3. Fluorescence-Activated Cell Sorting

- Vantage SE cell sorter (BD Biosciences).

2.4. RNA Extraction

- RNAqueous-4PCR kit (Ambion).
- 2100 BioAnalyzer (Agilent Technologies).

2.5. Microarray

- Ovation Biotin Labeling System (NuGEN).
- GeneChip Zebrafish Genome Array (Affymetrix).

3. Methods

3.1. Retinal Lesion (Fig. 1)

1. Add 50 ml of aquarium system water into the inner beaker.
2. Place 5–6 adult zebrafish (3-month to 1-year old) in the beaker (see Note 2).
3. Fill the outer dish with aquarium system water to the same level as the inner beaker (see Note 3).
4. Position the tip of the fiber optic liquid light line outside the dish at the midpoint of the water level. Adjust the orientation of the dish so that the foil on the inner beaker is on the side opposite the light (see Note 4).
5. Illuminate the fish for 30 min (see Note 5).
6. Return the fish to aquarium system to recover.

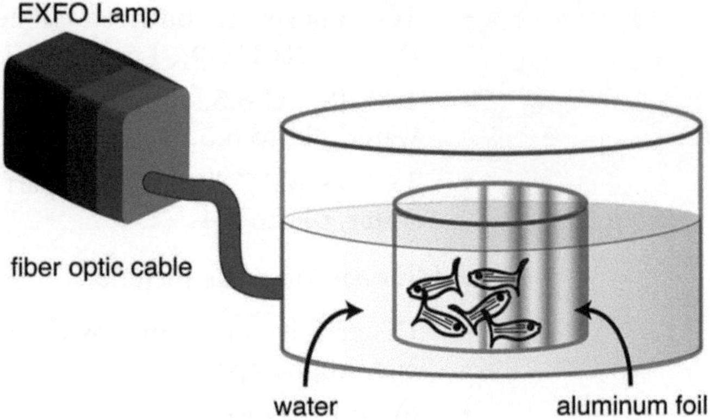

Fig. 1. Apparatus for delivering high-intensity light to freely swimming adult zebrafish. See text for details.

3.2. Retinal Dissociation

1. Dark adapt fish 2 h to overnight (see Note 6).
2. For each sample (3–4 fish, 6–8 retinas), prepare 500 μl of papain/dispase solution in a 1.5-ml siliconized tube: add 50 μl of 10× papain, 50 μl of 10× l-cysteine, 5 μl of 100× dispase into 400 μl of 1× PBS, pH 6.5, mix and incubate at 28.5°C (ref. 6; see Note 7).
3. Dissect retinas:
   - Anesthetize the fish by submerging in ice water until respiration ceases followed by cervical dislocation.
   - Poke a hole approximately 1 mm in length in the cornea over the pupil with a microscalpel.
   - Insert the microscissors into the hole, and cut through the cornea and into the iris up to the limbal junction between cornea and sclera.
   - Cut circumferentially along the limbus to the site of the initial incision and remove the anterior portion of the eye (cornea, iris, and lens).
   - With a 30 cc syringe direct a gentle stream of 1× PBS, pH 7.4 into the subretinal space between the neural retina and the retinal pigmented epithelium.
   - Cut the optic nerve to release the neural retina and flush it out of the eye cup.
   - Place isolated retinas in a 6-cm Petri dish with 1× PBS, pH 7.4 on ice.
4. Transfer the retinas onto a microscope slide using a pair of forceps with curved fine tips. Mince the retinas with a razor blade.
5. Transfer the pieces of tissue into the pre-activated papain/dispase solution with a glass Pasteur pipette. Incubate at 28.5°C on a nutating mixer for 30 min.

6. For each sample, prepare another 500 μl of papain/dispase solution as in step 2. Prepare 500 μl of DNase solution: add 50 μl of 10× papain inhibitor, 20 μl of 25× $MgCl_2$, 5 μl of 100× DNase I into 425 μl of 1× PBS, pH 7.4, mix and put on ice (see Note 8).
7. Triturate the tissue three times using a glass Pasteur pipette. Sit the tube for 2 min at room temperature.
8. Transfer ~400 μl of supernatant to a fresh tube. Pellet cells at 3,500 × *g* for 3 min at room temperature.
9. Add 500 μl of freshly activated papain/dispase solution to the remaining tissue (~100 μl). Incubate at 28.5°C on a nutating mixer for 30 min.
10. Take out the tube from the centrifuge (step 8), remove supernatant, and resuspend cells in 100 μl of DNase solution (see Note 9). Incubate at room temperature for 10 min. Triturate once using a glass Pasteur pipette and put on ice (see Note 10).
11. Take out the tube with remaining tissue from the incubator. Triturate three times using a glass Pasteur pipette and incubate for 2 min at room temperature.
12. Transfer ~500 μl of supernatant to a fresh tube and leave ~100 μl of papain/dispase solution with the remaining tissue. Pellet cells from supernatant at 6,000 rpm for 3 min at room temperature. Remove supernatant and resuspend cells in 100 μl of DNase solution. Incubate at room temperature for 10 min. Triturate once using a glass Pasteur pipette and put on ice.
13. Let the tube with the remaining tissue stand at room temperature for 10 min. Tap the tube until the tissue looks very fluffy. Add 200 μl of DNase solution. Tap the tube to mix. Incubate at room temperature for 10 min. Tap the tube again until the tissue is almost "dissolved." Triturate once using a glass Pasteur pipette and put on ice (see Note 11).
14. Combine resuspended cells from steps 10 and 12 with "dissolved" tissue in step 13 (total volume ~500 μl) and put on ice (see Note 12).

3.3. Fluorescence-Activated Cell Sorting

1. Set gating parameters (cell size and fluorescence intensity per manufacturer's instructions) by reference to a control sample of dissociated, unlabeled retinal cells from adult wild-type zebrafish.
2. Sort GFP[+] cells from the sample of dissociated retinal cells from *Tg(gfap:GFP)mi2002* zebrafish (Fig. 2, see Note 13).

3.4. RNA Extraction

1. Extract and purify total RNA from freshly sorted GFP[+] cells using the RNAqueous-4PCR kit per manufacturer's instructions (see Notes 14 and 15).

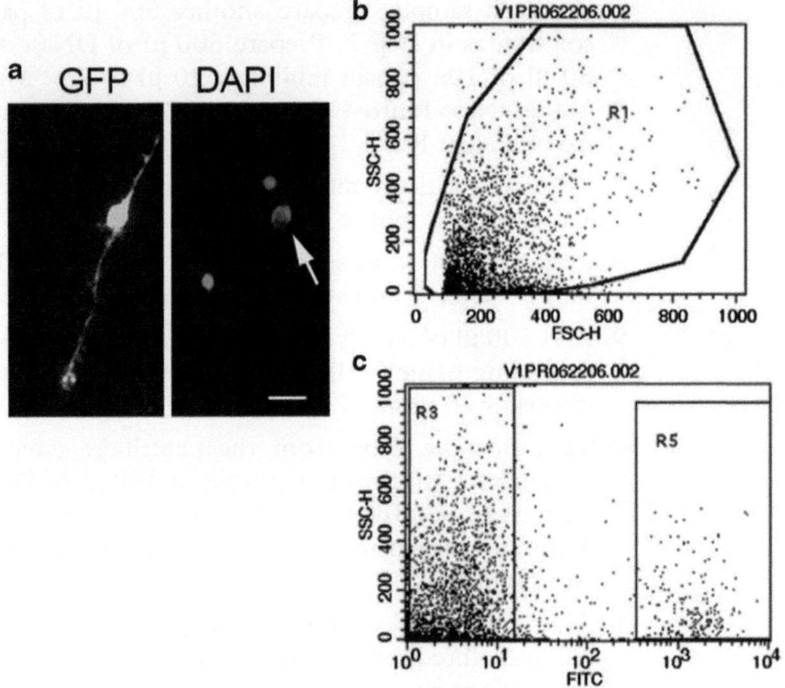

Fig. 2. Isolation of GFP+ Müller glia. (**a**) *Left panel*: dissociated GFP+ Müller glial cell. *Right panel*: same field, counterstained with DAPI. *Arrow* indicates the Müller glial cell. Scale bar: 10 μm. (**b**, **c**) Flow cytometry scatter plots; forward scatter-height (FSC-H); and side scatter-height (SSC-H). Dissociated cells from adult *Tg(gfap:GFP)mi2002* zebrafish retinas were gated by forward and side scatter (**b**). GFP+ Müller glia were isolated based on fluorescence in the FITC channel (R5 in **c**) (Modified from (2)).

2. Aliquot and freeze RNA immediately at −80°C.
3. Assess the quality and quantity of RNA using a 2100 BioAnalyzer per manufacturer's instructions.

3.5. Microarray

1. Use 20 ng of total RNA for linear amplification with Ovation Biotin Labeling System per manufacturer's instructions.
2. Hybridize 2.75 μg of biotin-labeled, fragmented cDNA to a GeneChip Zebrafish Genome Array per manufacturer's instructions (see Note 16).

4. Notes

1. Fire-polish the glass pipettes by passing the tip through the flame from a Bunsen burner a few times. This will reduce breakup of cells during trituration.
2. We always treat 5–6 adult fish in the same beaker to get consistent retinal lesions.
3. Water in the outer dish serves as a thermal buffer to avoid temperature increase from the intense light exposure.

4. Foil is used to reflect light back onto the fish in the beaker, thus enhancing the overall level of light exposure.
5. The incident light intensity at the position of the central beaker is >100,000 lux.
6. Dark adaptation promotes separation of neural retinal tissue and pigmented epithelium through the action of retinomotor movements (7).
7. Incubation pre-activates the enzymes.
8. Papain inhibitor and BSA are used to quench the enzymes. DNase reduces viscosity of the sample.
9. Tap the tube a few times to resuspend the cells.
10. DNase treatment makes the sample clear.
11. We find that tapping the tube is very efficient at breaking up the last pieces of tissue.
12. The yield of dissociated cells from 6-month-old adult zebrafish is ~2.5×10^5 cells per retina, of which ~9% are Müller glia.
13. With FACS, we recover ~2.1×10^4 GFP-labeled Müller glia per retina from 6-month-old zebrafish (~84% recovery).
14. For each sample, we combine retinas from three or four fish for cell dissociation and cell sorting. As a result, each sample contains $1–2 \times 10^5$ freshly sorted GFP^+ cells for RNA extraction.
15. The interval between retinal isolation and cell lysis is ~2.5 h.
16. We perform independent hybridizations of three biological replicates for each sampling condition.

Acknowledgment

This work was supported by NIH grant EY004318 to PAR.

References

1. Akimoto M et al (2006) Targeting of GFP to newborn rods by Nrl promoter and temporal expression profiling of flow-sorted photoreceptors. Proc Natl Acad Sci USA 103:3890–3895
2. Qin Z, Barthel LK, Raymond PA (2009) Genetic evidence for shared mechanisms of epimorphic regeneration in zebrafish. Proc Natl Acad Sci USA 106:9310–9315
3. Craig SE, Calinescu AA, Hitchcock PF (2008) Identification of the molecular signatures integral to regenerating photoreceptors in the retina of the zebra fish. J Ocul Biol Dis Inform 1:73–84
4. Von Stetina SE et al (2007) Cell-specific microarray profiling experiments reveal a comprehensive picture of gene expression in the *C. elegans* nervous system. Genome Biol 8:R135
5. Bernardos RL et al (2007) Late-stage neuronal progenitors in the retina are radial Muller glia that function as retinal stem cells. J Neurosci 27:7028–7040
6. Nelson R, Bender AM, Connaughton VP (2003) Stimulation of sodium pump restores membrane potential to neurons excited by glutamate in zebrafish distal retina. J Physiol 549: 787–800
7. Burnside B et al (1982) Induction of dark-adaptive retinomotor movement (cell elongation) in teleost retinal cones by cyclic adenosine 3′, 5-monophosphate. J Gen Physiol 79:759–774

Part V

Function/Imaging

Chapter 19

Measuring Rodent Electroretinograms to Assess Retinal Function

Molly E. Clark and Timothy W. Kraft

Abstract

Electroretinography is a noninvasive technique used to measure the electrical activity of neurons in the retina. Electroretinogram (ERG) waveforms can be used to quantify retinal function in normal and diseased rodents. In particular, the functions of rod and cone pathways can be isolated. Inner retinal neuronal functioning, such as bipolar cell activity or ganglion cell activity, can also be measured. In this chapter we describe the common full-field ERG techniques of scoptic flash, photopic flash, and flicker used to isolate and compare rod-driven and cone-driven function.

Key words: Electroretinogram, ERG, Scotopic, Photopic, Flicker, Rod, Cone, Photoreceptor, a-wave, b-wave

1. Introduction

The electroretinogram (ERG) is a measure of the massed electrical activity of retinal neurons generated in response to a light stimulus. This transretinal potential can be recorded from the eye(s) of an intact animal or from isolated retinal tissue. Light stimuli of various intensities are presented directly to the eye, and the electrical responses of the retinal neurons sum to generate the ERG waveform. In a normal functioning retina, the ERG response to a bright flash contains a negative inflection (the a-wave) followed by a positive deflection known as the b-wave (Fig. 1a) (2, 3). Bright flashes presented in the dark generate a mixed response derived from both rod and cone photoreceptors. In the presence of a modest background light the rod photoreceptors become saturated, and flashes will result in responses which are only cone-driven (Fig. 1a, *inset*).

Shu-Zhen Wang (ed.), *Retinal Development: Methods and Protocols*, Methods in Molecular Biology, vol. 884, DOI 10.1007/978-1-61779-848-1_19, © Springer Science+Business Media, LLC 2012

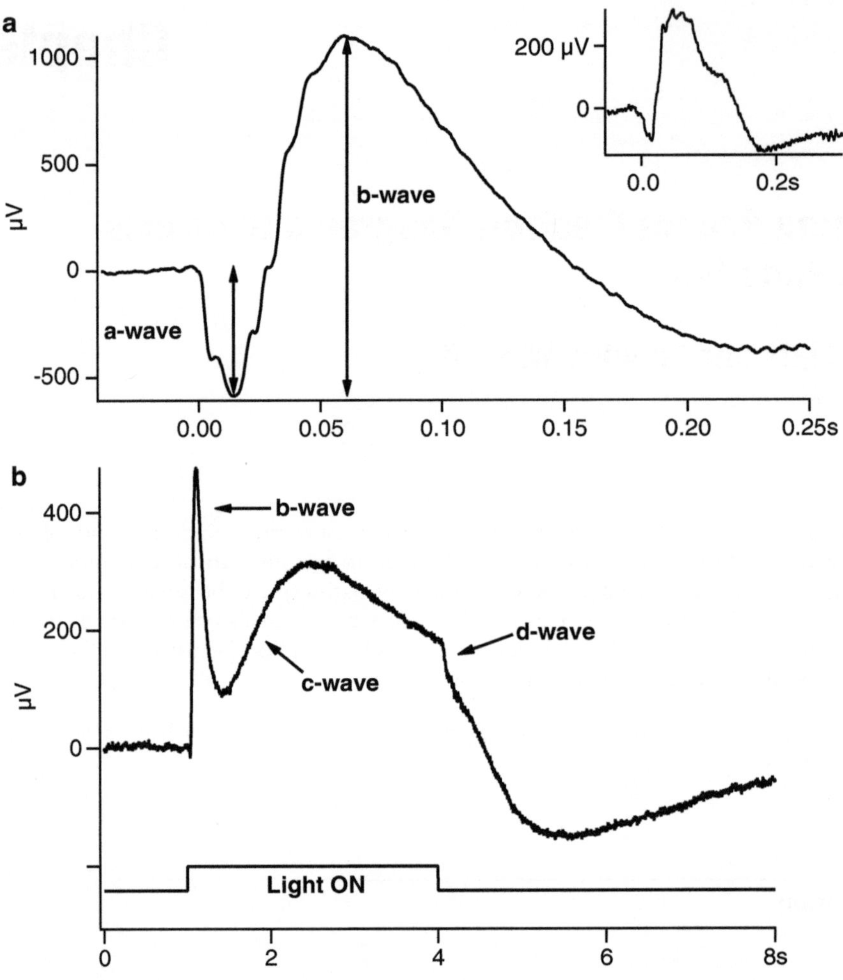

Fig. 1. (**a**) An electroretinogram (ERG) recording of a normal 43-day-old Sprague Dawley albino rat. A bright, 505-nm flash of light, 2-ms duration, was delivered at t = 0 s. The initial negative inflection is the a-wave, and the subsequent positive deflection is the b-wave. The a-wave is generated by photoreceptors, and the b-wave is derived from depolarizing ON-bipolar cells. The inset shows the light-adapted (cone-driven) response to the same bright light stimulus. (**b**) An ERG recording of a normal mouse. A longer light stimulus of 3 s duration was delivered. The positive deflection following the b-wave is called the c-wave, which corresponds to the action of RPE and Müller glial cells. When the step of light is turned off, the negative deflection is known as the d-wave. This OFF response represents the action of hyperpolarizing OFF-bipolar cells (fig. 1b adapted from ref. 1).

If the stimulus is a long step of light, then the ERG will demonstrate the c-wave—a second, slower positive deflection following the b-wave—and, at the point of light offset, a negative deflection called the d-wave (Fig. 1b).

The a-wave is derived from the function of photoreceptors (4, 5). The b-wave originates from depolarizing ON-bipolar cells (6, 7), and the d-wave, or OFF response, represents the action of hyperpolarizing or OFF-bipolar cells (8, 9). The c-wave represents the extracellular movement of potassium ions as they are transported by retinal pigmented epithelia (RPE) and Müller glial cells; the action of these cells sum to generate the c-wave (10–12). Thus, the individual waves that constitute the ERG waveform can be dissected into the physiologic function of different cell types in the retina.

The ERG can be used to characterize the physiologic state of the retina in rodents with congenital, acquired, or experimental disease. In many inherited retinal degenerative disorders, rod photoreceptor function is affected in the early stages and loss of cone photoreceptor function follows (13, 14). Other diseases may target synaptic transmission, bipolar cells, or ganglion cells (15). To help assess the state of disease in the retina using ERGs, different stimuli are presented in order to isolate rod-driven verses cone-driven function. Simple parameters that can be changed include stimulus intensity, stimulus wavelength, and background illumination. Flickering stimuli can also be used to isolate the faster cone-driven visual pathways. The standard clinical ERG for human subjects consists of only four sets of stimuli (16), but with anesthetized rodents, longer and more detailed stimulus protocols are possible.

After rodents have dark-adapted, a full-field dim flash ERG is performed first. Rods have a spectral sensitivity peak around 498 nm, so they are more sensitive to blue–green lights (17). A dim flash presented in the dark minimally activates rod photoreceptors, eliciting a purely rod response because cone photoreceptors are much less sensitive and require brighter stimuli to generate a measureable signal (18). The light intensity can be incrementally increased in order to activate cones as well, giving a mixed rod–cone response (Fig. 2, *middle* and *lower traces*). Next, a background light is turned on to saturate the rods and isolate a cone-driven response. This full-field bright flash ERG is considered a light-adapted ERG (Fig. 1a *inset*).

Another technique for isolating the cone response takes advantage of the fast response kinetics that cones exhibit. Rod photoreceptors have a low temporal resolution, and they cannot generally follow light stimuli presented at frequencies greater than 25 Hz in rodents (19). Cone photoreceptors, however, recover quickly and thus can follow a more rapidly flickering light stimulus (Fig. 3). To isolate cone function in humans, the International Standard for Clinical Electroretinography recommends using 30-Hz photopic flicker (16). Generally, as flicker frequency increases, the ERG response amplitude decreases. The critical flicker frequency (CFF) is the highest stimulus frequency for which a flickering light causes a just detectable modulated response, often threshold CFF is defined as 3 μV in ERG experiments (20). Thus, flicker and CFF can be used to detect elevated threshold in either the rod- or cone-driven pathways in the retina.

Performing a sequence of dark-adapted intensity response (IR) series, light-adapted IR series, and flicker ERG provides ample

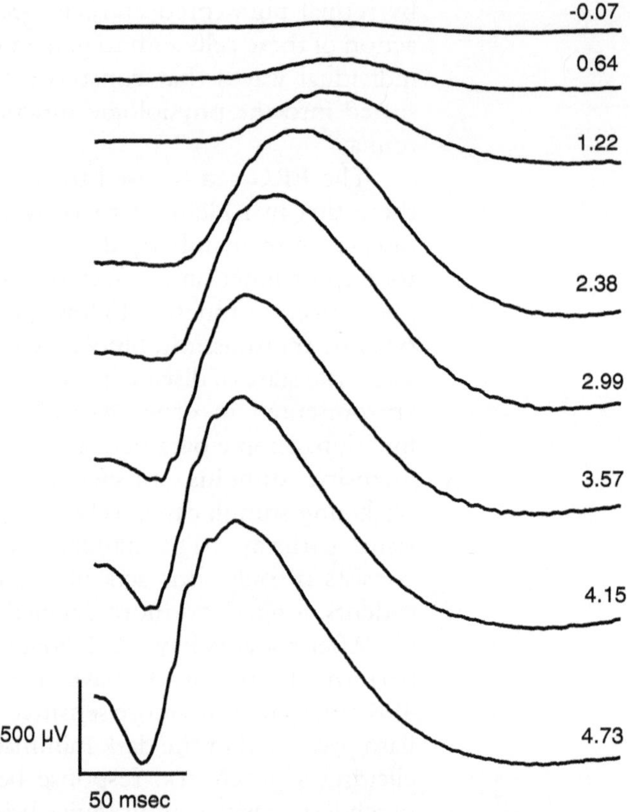

Fig. 2. Electroretinogram (ERG) responses to a series of flashes of increasing intensity recorded from a 43-day-old rat. A 505-nm, 2-ms flash was delivered to the dark-adapted rat. The light intensity was increased for each recording; the top trace represents the dimmest stimulus and the bottom trace the brightest. The values next to each trace represent light intensity (log photons/μm^2 at the cornea). Each trace is the average of 3–20 responses, AC recording.

information to assist in the assessment of the physiologic function of the retina. Review chapters and indeed volumes have been written about the interpretations and applications of ERG techniques for probing retinal health (21, 22).

2. Materials

1. Sedation agent: 3–5% isoflurane (see Note 1).
2. Systemic anesthetic: xylazine (100 mg/ml), ketamine (100 mg/ml). Combine xylazine and ketamine in a 1:10 ratio. Make an appropriate volume to inject 0.1 ml/100 g body weight (see Notes 2 and 3).

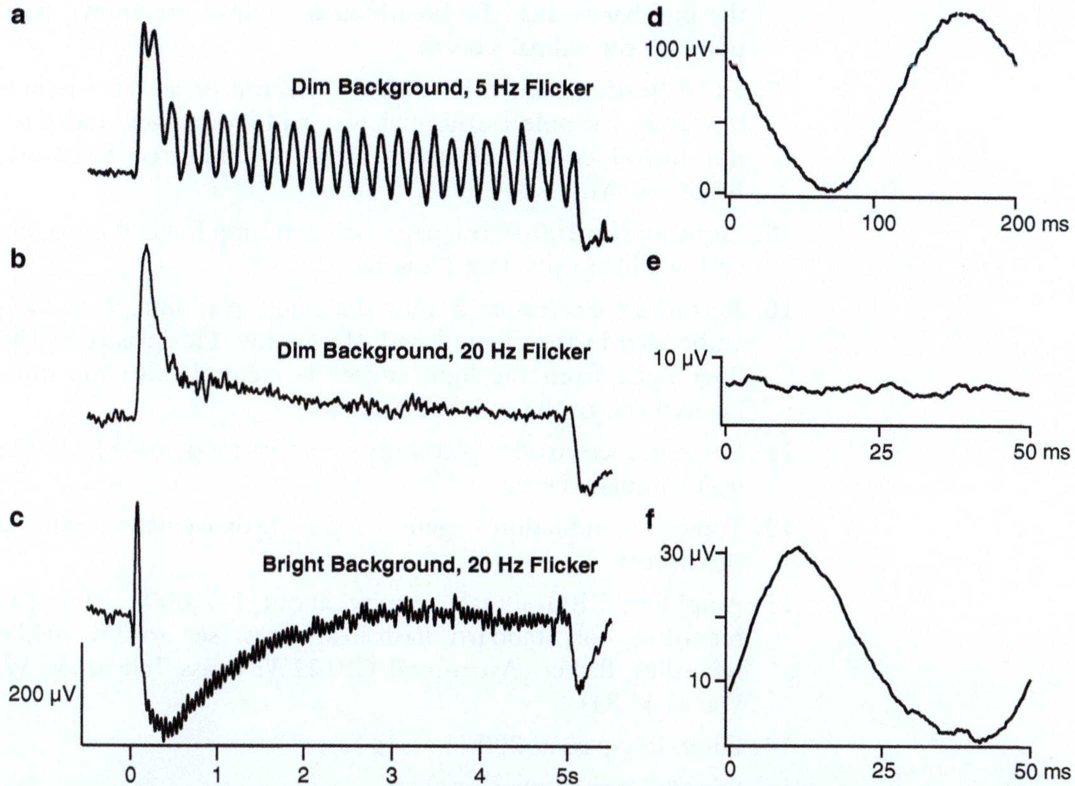

Fig. 3. An electroretinogram (ERG) recording from a normal albino Sprague Dawley rat in response to a 5-s presentation of sinusoidal flicker. (**a**) A dim stimulus evokes a large rod-driven response to 5-Hz flicker; note the on-transient in the first 0.5 s of the response. (**b**) On the same background the rod-driven flicker response decreases to nil at a stimulus frequency of 20 Hz; the on-transient and off-transients are still present. (**c**) Cone-driven responses to a bright stimulus and a fast flicker frequency of 20 Hz. The graphs **(d–f)** represent the averaged single cycle responses calculated from the raw traces to the left. (**d**) The average of 22 cycles of a 5-Hz stimulus on a dim background (312 photons/μm²/s); peak-to-peak amplitude is 127 μV. (**e**) The average of 90 cycles of a 20-Hz stimulus on a dim background, no measurable response. (**f**) The average of 90 cycles of a 20-Hz stimulus on a bright background (1.46E5 photons/μm²/s); peak to peak amplitude is 30 μV. Note the timing and voltage scale bars for *panels* (**a–c**) are shared on the *bottom left*, but individualized for *panels* (**d–f**).

3. Injection syringe: 0.5 cc insulin syringe with an attached 28-gauge needle.
4. Topical anesthetic: 0.5% proparacaine.
5. Topical dilating agents: 2.5% phenylephrine hydrochloride, and 1% tropicamide.
6. Faraday cage: any sort of a box structure wrapped in a layer of a conducting material, such as aluminum foil, in order to block external electric fields (see Note 4).
7. Rodent-size bite bar: square metal platform with a bite bar on one end. The rodent's body will lie on the platform, and the rodent's upper teeth will lock over the bar. Adjust the height of

the bite bar so that the breathing motion of the animal does not rock the animal's head.

8. Small heating pad: this could be electric or a water jacket; however, a simple isothermal wax pad works well and does not introduce any electrical artifacts (Braintree Scientific, Braintree, MA).
9. Light source: 100-W tungsten-halogen lamp focused onto one end of a fiber optic (see Note 5).
10. Recording electrode: 2 mm diameter, platinum wire loop embedded in the tapered end of a hollow Plexiglas rod. The fiber optic from the light source is secured onto the non-tapered end of the rod (see Note 6).
11. Reference electrode: platinum or silver loop placed on the non-stimulated eye.
12. Topical conduction agent: 2.5% hydroxymethylcellulose (goniosol).
13. Amplifier: ERG signals require about 1–2,000× gain, AC recording for standard flash responses, set to DC while recording flicker (Astro-med CP122W; Grass Telefactor, W. Warwick, RI).
14. Filter: low-pass at 300 Hz (see Note 7).
15. Stimulus parameters:
    (a) Light stimulus attenuation filters: calibrated neutral-density (ND) filters, allow attenuation in steps of 0.3 log units up to a maximum of 6.9 log units attenuation.
    (b) Photometer or light intensity calibrator: optical power meter, measures in units of microwatts/second. Convert to units of photons/μm^2 (Graseby Optronics, Orlando, FL).
    (c) Aperture shutter: 6 mm (Uniblitz; Vincent Associates, Rochester, NY).
    (d) Interference filter: a 40-nm bandwidth to limit the stimulus wavelength to 500 ± 20 nm (Andover Co., Salem, NH).
    (e) Flicker shutter: ferro-electric liquid crystal shutter (LV050; Displaytech Inc, Longmont, CO) (see Note 8).
16. Data acquisition:
    (a) Digitizing hardware: MIO16 (National Instruments, Austin, TX).
    (b) Digitizing software: LabView (National Instruments).
    (c) Data analysis software: IGOR PRO (Wavemetrics Inc, Lake Oswego, OR). This software can be programmed to collect and digitize data as well.

3. Methods

All steps should be performed in a dark room. Flashlights with red filters can be used sparingly for illumination during rodent preparation and adjustment. While the rodent is outside of the closed faraday cage, minimize any stray light sources in the room; for example, cover the computer monitor.

3.1. Rodent Preparation

1. Dark-adapt rodents. For wild-type animals, 1 h is sufficient; however, with albino or transgenic animals that dark-adapt slowly, longer periods may be required, up to and including overnight.
2. Place the rodent in a gas chamber, and fill the chamber with 3–5% isoflurane until the rodent is sedated. Remove the rodent from the chamber. Anesthetize the rodent with an intraperitoneal injection of xylazine (9.09 mg/kg) and ketamine (90.9 mg/kg) (see Note 9).
3. Anesthetize both corneas with a small drop of 0.5% proparacaine. Dilate the pupil of the eye to be stimulated with 2.5% topical phenylephrine HCl and 1% tropicamide (see Note 10).
4. Lay the rodent on a heating pad inside a Faraday cage. Place the upper teeth of the rodent over a bite bar in order to stabilize the head (see Note 11). Make sure that the head is positioned straight and is not tilted to one side.
5. Place a small amount of goniosol on the rodent's corneas. Wipe away excess. Place a small amount of goniosol on the reference and recording electrodes.
6. Place the recording electrode on the cornea of the eye that will be stimulated during the ERG. Place the reference electrode on the cornea of the eye that will not be stimulated during the ERG (see Note 12). Close the faraday cage with an electrically inert (blocking) curtain.
7. Perform a test flash with a 3.6 ND filter in place. If needed, readjust the electrodes and perform another test flash (see Notes 13 and 14).

3.2. Dark-Adapted Flash ERG

1. After the test flash responses have stabilized, begin ERG recording with the dimmest stimulus that generates a measurable response (ERG threshold). Sample the voltage responses for a total of 1 s at a rate of not less than 0.5 ms/pt, a minimum of 2,000 points (see Note 15).
2. Progressively increase the intensity of the 2-ms stimulus of 500-nm light by factors of two or four in order to create an IR series. As intensity increases, change the inter-stimulus interval (ISI) from 2 s up to 30 s to allow full recovery (see Note 16).

3. Perform 3–20 repeats of each stimulus depending upon the signal-to-noise ratio.
4. Analysis: average the ERG voltage responses to measure the amplitude and the (implicit) time to peak of the a- and b-wave. As a measure of sensitivity find the $I_{1/2}$, the intensity necessary to produce a half maximal response. One method is to plot the amplitude versus intensity data for both the a- and b-wave, then fit curves with a modified Michaelis function of the form:

$$R = R_{\max}\left[1-\exp\left(-\frac{i}{10^{k}}\right)\right],$$

where R = response, $R_{\max}$ = maximum response, i = log intensity, and k = (log of $I_{1/2}$) (see Note 17).

3.3. Light-Adapted Flash ERG

1. For light-adapted conditions (to isolate cone driven responses), deliver the bright flashes in the presence of an adapting (rod-saturating) background light (about 5,000 photons/μm2-s incident upon the cornea). Deliver the flashes at an ISI of 90 s.

3.4. Flicker ERG

1. Set the stimulus duration for 5 s. A ferro-electric liquid crystal shutter (e.g., LV050, Displaytech, Inc.) can produce flicker by sinusoidally modulating the stimulus intensity (see Notes 8 and 18). Set the amplifier to DC recording mode.
2. Experimental runs for each temporal frequency will contain a total of 24,000 points, at a sampling rate of 0.25 ms/pt representing the 6 s of recording time. Open the shutter after a 500-ms delay to collect baseline data. The shutter is open for 5 s of light exposure and therefore contains between 5 and 200 cycles of the sinusoidal stimulus between 1 and 40 Hz.
3. Set the stimulus frequencies to: 1, 2, 4, 5, 10, 16, 20, 25, 32, and 40 Hz. At least two mean light intensities should be used: one dim to examine rod-driven flicker responses (ex: 2.7 ND filter in place), the second brighter to isolate cone-driven retinal responses (ex: 0.3 or 0.0 ND in place).
4. Calibrate the equipment: place a 12-mm glass coverslip (in order to simulate a cornea) on the light sensor of a photometer. Place the recording electrode/light source against the coverslip, separated by a 0.5-mm thick layer of goniosol. Using the brightest light stimulus possible (unattenuated light, shutter open), move the photometer and electrode until your photometer reads the highest value possible. This value will be used for converting power or energy into stimulus strength (photons/μm^2 incident upon the cornea) (see Note 19).

5. Flicker analysis: ignore the initial on-transient of the response, and average the final 4.5 s of data into a one- or two-cycle wave to measure response amplitude. Fit the average response amplitude (μV) by a sine wave (see Note 20). Plot the $\log_{10}$ of the response amplitude at each intensity versus stimulus frequency; then fit the responses with a line to determine electrophysiological CFF using a 3 μV criterion voltage (20). The value of the CFF in dim light represents rod pathway function (~20–25 Hz), and using a brighter mean illumination, cone pathways CFF will be much higher, around 50 Hz.

4. Notes

1. For rats, isoflurane is used as a sedation agent followed by an intraperitoneal injection of an anesthetic. Because mice have small body masses, the sedation agent is not necessary; a single investigator can easily accomplish an injection of ketamine/xylazine, diluted with PBS solution. Alternatively, isoflurane can be administered as an anesthetic through a nose mask to the rodent throughout the ERG procedure. If the nose mask does not create a proper seal around the rodent's face, however, waste gas may be emitted into the room, posing a hazard to the investigators (23). Thus a vacuum-assisted scavenging device is required as well.
2. For example, if a rat weighs 300 g, combine 0.3 ml of ketamine with 0.03 ml of xylazine. Because mice have much smaller body weights (about 15 g), the injection should be diluted 5–10 times with PBS in order to reliably measure the amount of anesthetic.
3. The ketamine/xylazine combination induces lens opacities that resolve within a few hours of the animal's revival (24). The anesthetic will not affect the amplitude of the ERG responses (25).
4. The cage should also be a complete enclosure in order to block any stray light from entering and reaching the rodent's eyes.
5. LED or arrays of LEDs are commonly used as light sources. They are limited in their spectral output, but they can be turned on/off quickly, eliminating the need for a shutter.
6. This assembly ensures a constant distance between the fiber optic and the eye, and the plexiglass acts as a diffusing element (26). Other electrode material options are gold foil, silver wire, DTL microfiber electrode, silver/AgCl pellet, or cotton wicks.
7. The ERG response to a flash of light has fast (high frequency) and slow (low frequency) components. Filters can either hide

or enhance these components; they can reduce drift (AC, high-pass filters) or eliminate high frequency signals (low-pass filters). Low-pass filters reduce unwanted noise and allows a clear view of the ERG signal. For example, the typical AC filtering eliminates low frequency (drift) in the signal, but the experimenter should note this would reduce b-wave amplitude and eliminate c-wave or slow PIII response. Similarly, a 300-Hz low-pass filter reduces high-frequency noise in a signal, but if the high cutoff is placed lower, say at 100 Hz, it will diminish the a-wave and oscillatory potentials. Hardwired filters are often built into the amplifier, or they can be purchased separately; optionally signals can be filtered digitally post hoc through a variety of software applications.

8. We use variable attenuation filters to adjust the intensity of our flicker stimuli. Another option is an LED light that is simply limited by it's on and off time in order to vary the intensity (pulse width modulation). This technique can also be use to create flickering stimuli.
9. While rats generally require a sedation agent and an intraperitoneal injection, the use of both can be toxic, and care must be taken to inject the ketamine/xylazine mixture as soon as the rat is drowsy from the isoflurane. Immediately after the injection, the rat should be placed in a cage with fresh room air.
10. Apply only enough liquid to cover the cornea. A micropipetter is good for this. If the fur surrounding the eye is wet, it will short circuit the corneal surface to the skin and greatly decrease the amplitude of the ERG signal. Wait 30 s in between application of eye drops in order to ensure absorption of the drug. Dry the eyes of any excess fluid before applying the subsequent drops.
11. If performing ERG(s) over several hours, ensure that the heating pad stays warm. If the rodent's body temperature decreases below the normal 38°C, the a- and b-wave amplitudes will decrease (27).
12. Place the reference electrode tangential to the eye, separated from the corneal surface by a layer of goniosol that is less than 0.5 mm thick. The recording electrode should also be placed less than 0.5 mm away from the surface of the eye. Both electrodes should be centered over the midpoint of the cornea. Warning, do not apply pressure to the eye with the electrodes; it can disrupt blood flow to the eye, reduce ERG amplitude, and permanently damage the retina if not relieved.
13. If the test flash produces only a very small signal, no signal, or a lot of noise, the electrodes are likely placed incorrectly on the eye. The electrode may be too far away from the eye, the goniosol may have dried out, the electrode may not be

centered, or the animal may not be fully anesthetized and moving, sometimes in response to the flashes of light.

14. Two to three dim test flashes should be delivered to check for stable recording conditions. Wait 5 min in between flashes and look for growth and stabilization of the ERG amplitude. A signal that slightly increases with subsequent test flashes is ideal.
15. Sampling the ERG at high frequency (4 kHz at 0.25 ms/pt or 5 kHz at 0.20 ms/pt) will permit a more detailed view of the rising phase of the a-wave.
16. For example, the first stimulus may have 4.8 log units of ND attenuation in place with an ISI of 2 s. Then throughout the series of stimuli increasing both the stimulus intensity along with the ISI, with our strongest stimulus (0, or 0.0 ND filters in place) the ISI is 30 s.
17. Basic measures of amplitude and timing for the a- and b-wave will allow you to compare experimental animals to controls. Sensitivity measures, such as threshold and $I_{1/2}$ are also good for comparing strains of animals. Since the light stimulus illuminates the entire retina, it is considered full-field ERG, as opposed to a focal or multifocal ERG in which the stimulus only illuminates a small area of interest in the retina. The full-field ERG provides an overview of the entire retinal function, whereas the focal ERG examines a single area of interest (such as the fovea), and the multifocal ERG provides a topographical analysis of retinal activity in many small sectors (22, 28).
18. Most "all-in-one" ERG systems can deliver flashes at various frequencies. A standard of 29–31-Hz flicker isolates the cone response. Our system was designed to deliver sinusoidal stimuli, which simplifies our flicker analysis.
19. The all-in-one ERG machines [examples include: Espion E^2 (Diagnosys LLC) and the UTAS Visual Diagnostic Systems (LKC Technologies)] have an internal automatic calibration for humans. An all-in-one portable ERG system for rodents and small animals is the HMsERG (OcuScience, Kansas City, MO). With our setup of equipment using a tungsten lamp and multiple interleaved attenuation filters, we manually calibrate the maximum output of the equipment daily. Additionally, all of the attenuation filters are calibrated semiannually.
20. At lower frequencies the contribution of a second harmonic to the total response may result in a response that is not well fit by a single sine wave. In these cases, measure the peak to trough amplitude of this averaged response. An alternative to measuring the amplitude of the sine wave response generated by the retina is to do a Fourier analysis. Simply select the final 4.5 s of the response and calculate the Fourier transform. Then measure the power or magnitude at the stimulus frequency.

References

1. Rubin GR (2009) Comparisons between behavioral and electrophysiological measures of visual function in rodent models of retinal degeneration. Thesis, University of Alabama, Birmingham, p 11
2. Granit R (1963) Sensory mechanisms of the retina, with an appendix on electroretinography, vol xxiii. Oxford University Press, New York, p 412
3. Hood DC, Birch DG (1993) Human cone receptor activity: the leading edge of the a-wave and models of receptor activity. Vis Neurosci 10:857–871
4. Hood DC, Birch DG (1993) Light adaptation of human rod receptors: the leading edge of the human a-wave and models of rod receptor activity. Vision Res 33:1605–1618
5. Hood DC, Birch DG (1990) The A-wave of the human electroretinogram and rod receptor function. Invest Ophthalmol Vis Sci 31:2070–2081
6. Robson JG, Frishman LJ (1995) Response linearity and kinetics of the cat retina: the bipolar cell component of the dark-adapted electroretinogram. Vis Neurosci 12:837–850
7. Hood DC, Birch DG (1996) Beta wave of the scotopic (rod) electroretinogram as a measure of the activity of human on-bipolar cells. J Opt Soc Am A Opt Image Sci Vis 13:623–633
8. Xu X, Karwoski C (1995) Current source density analysis of the electroretinographic d wave of frog retina. J Neurophysiol 73:2459–2469
9. Sieving PA, Murayama K, Naarendorp F (1994) Push-pull model of the primate photopic electroretinogram: a role for hyperpolarizing neurons in shaping the b-wave. Vis Neurosci 11:519–532
10. Oakley B II, Green DG (1976) Correlation of light-induced changes in retinal extracellular potassium concentration with c-wave of the electroretinogram. J Neurophysiol 39:1117–1133
11. Steinberg RH, Linsenmeier RA, Griff ER (1985) Retinal pigment epithelial cell contributions to the electroretinogram. Prog Retin Res 4:33–66
12. Kofuji P, Ceelen P, Zahs KR, Surbeck LW, Lester HA, Newman EA (2000) Genetic inactivation of an inwardly rectifying potassium channel (Kir4.1 subunit) in mice: phenotypic impact in retina. J Neurosci 20:5733–5740
13. Chader GJ (2002) Animal models in research on retinal degenerations: past progress and future hope. Vision Res 42:393–399
14. Delyfer MN, Leveillar T, Mohand-Said S, Hicks D, Picaud S, Sahel JA (2004) Inherited retinal degenerations: therapeutic prospects. Biol Cell 96:261–269
15. Pardue MT et al (1998) A naturally occurring mouse model of x-linked congenital stationary night blindness. Inv Ophth Vis Sci 30(12):2443–2449
16. Marmor MF, Holder GE, Seeliger MW, Yamamoto S (2004) Standard for clinical electroretinography. Doc Ophthalmol 108: 107–114
17. Bowmaker JK, Dartnall HJ (1980) Visual pigments of rods and cones in a human retina. J Physiol 298:501–511
18. Verdon WA, Schneck ME, Haegerstrom-Portnoy G (2003) A comparison of three techniques to estimate the human dark-adapted cone electroretinogram. Vision Res 43:2089–2099
19. Bush RA, Sieving PA (1996) Inner retinal contributions to the primate photopic fast flicker electroretinogram. J Opt Soc Am A Opt Image Sci Vis 13:557–565
20. Rubin GR, Kraft TW (2007) Flicker assessment of rod and cone function in a model of retinal degeneration. Doc Ophthalmol 115:165–172
21. FishmanLJ,WangMH(2011)Electroretinogram of the human, monkey and mouse. In: Levin LA et al (eds) Adler's physiology of the eye. Edinburgh, Scotland, pp 480–501
22. Heckenlively R, Arden GB (eds) (2006) Principles and practice of clinical electrophysiology of vision, 2nd edn. Cambridge, MA
23. Smith JC, Bolon B (2006) Isoflurane leakage from non-rebreathing rodent anaestesia circuits: comparison of emissions from conventional and modified ports. Lab Animal 40:200–209
24. Calderone L, Grimes P, Shalev M (1986) Reversible cataract induced by xylazine and by ketamine–xylazine anesthesia in rats and mice. Exp Res 42(4):331–337
25. Sasovetz D (1978) Ketamine hydrochloride: an effective general anesthetic for use in electroretinography. Ann Ophthalmol 10(11):1510–1514
26. Lyubarsky AL, Pugh EN Jr (1996) Recovery phase of the murine rod photoresponse reconstructed from electroretinographic recordings. J Neurosci 16:563–571
27. Mizota A, Adachi-Usami E (2002) Effect of body temperature on electroretinogram of mice. Invest Ophthalmol Vis Sci 43(12):3754–3757
28. Bearse MA, Sutter EE (1996) Imaging localized retinal dysfunction with the multifocal electroretinogram. J Opt Soc Am A 13:634–640

Chapter 20

Functional Imaging of Retinal Photoreceptors and Inner Neurons Using Stimulus-Evoked Intrinsic Optical Signals

Xin-Cheng Yao and Yi-Chao Li

Abstract

Retinal development is a dynamic process both anatomically and functionally. High-resolution imaging and dynamic monitoring of photoreceptors and inner neurons can provide important information regarding the structure and function of the developing retina. In this chapter, we describe intrinsic optical signal (IOS) imaging as a high spatiotemporal resolution method for functional study of living retinal tissues. IOS imaging is based on near infrared (NIR) light detection of stimulus-evoked transient change of inherent optical characteristics of the cells. With no requirement for exogenous biomarkers, IOS imaging is totally noninvasive for functional mapping of stimulus-evoked spatiotemporal dynamics of the photoreceptors and inner retinal neurons.

Key words: Retinal function, Photoreceptor, Neuron, Ganglion, Electrophysiology, Optical imaging, Intrinsic optical signal

1. Introduction

As one part of the central nervous system, the retina plays a vital role in capturing photons, converting light energy to electrical signals, and several preliminary stages of visual information processing. For these, retinal photoreceptors and inner neurons form complex networks, with both feed-forward and feed-back mechanisms among different retinal layers/cells. During development, the retina displays conspicuous anatomic and functional dynamics of its photoreceptors and inner neurons (1, 2). Given the delicate structure and complex function of the retina, advanced understanding of retinal development and retinal neural information processing require the capability to simultaneously monitor dynamic activities of large populations of retinal neurons, with high-spatial and high-temporal resolution.

Shu-Zhen Wang (ed.), *Retinal Development: Methods and Protocols*, Methods in Molecular Biology, vol. 884,
DOI 10.1007/978-1-61779-848-1_20, © Springer Science+Business Media, LLC 2012

Electrophysiological methods, such as electroretinogram (ERG), have provided valuable information for functional study of the retina (3, 4), but high-resolution monitoring of multiple types of retinal cells functioning together is still challenging. In principle, optical methods can provide high-resolution imaging of the retina and other biological tissues. A variety of voltage-sensitive dyes and ion-selective indicators have been developed to allow functional imaging of neural activities. However, phototoxicity of the dyes and difficult loading procedures limit their application for functional study of the retina.

Stimulus-evoked intrinsic optical signals (IOSs) have been detected in the retina (5–7), and other neural tissues (8, 9). Fast IOSs have time courses that are comparable to stimulus-evoked electrophysiological kinetics and thus hold promise for high spatiotemporal resolution investigation of retinal neural function. Without the requirement of exogenous biomarkers, IOS imaging is totally noninvasive for dynamic monitoring of retinal neural activities. We have recently validated high-spatial (~μm) and high-temporal (ms) resolution IOS imaging of retinal neural activities in isolated, but living, retinal tissues (10–15). In principle, both flat-mounted (14) and sliced (12) retinas can be used for functional study of the retina. Flat-mounted retinas provide a simple preparation for depth-resolved mapping of neural activities at individual functional layers; while sliced retinas allow parallel monitoring of visual signal propagation from the photoreceptors to inner retinal neurons. In this article, flat-mounted Leopard frog (*Rana pipiens*) retina is taken as an example to illustrate the retinal preparation and IOS imaging procedures. The rationale of basic IOS imaging and dynamic differential IOS processing is reviewed.

2. Materials

Construct an IOS imager. Keep imaging optics clean and dry. Acquire animals from certified vendors. Prepare Ringer's solution under room temperature with purified deionized water and analytical grade reagents. Carefully follow waste disposal regulations when disposing chemical waste.

2.1. Imaging Equipment

A conventional microscope can be modified to conduct IOS imaging of living retinal tissues. Major components of the system shown in Fig. 1 are summarized as follows:

1. Optical platform: a light microscope with water dipping objective (see Note 1).
2. Light sources: the imaging system consists of two, i.e., visible and NIR, light sources. The visible light is used for retinal

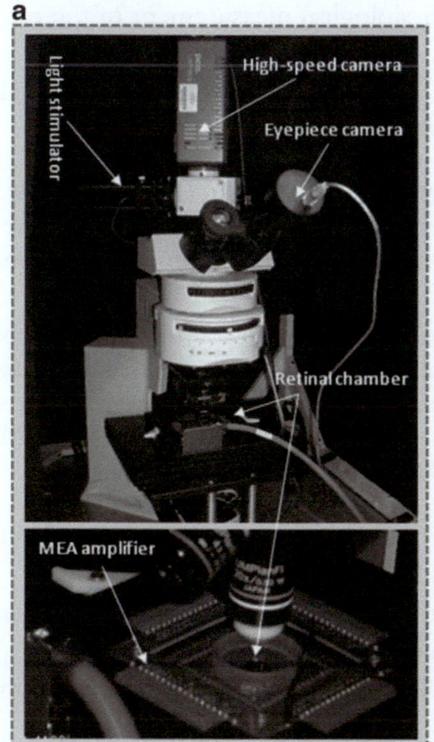

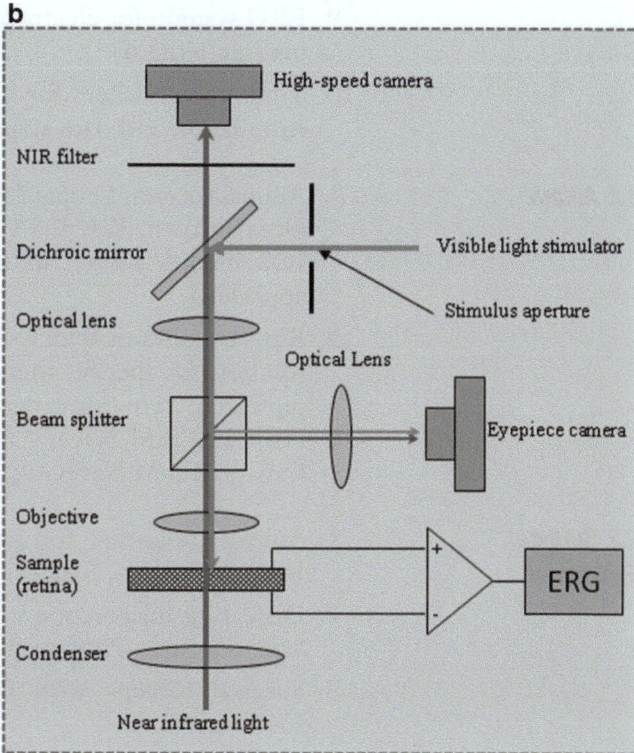

Fig. 1. Photograph (*left*) and optical diagram (*right*) of the near infrared (NIR) light microscope for intrinsic optical signal (IOS) imaging. During measurements, isolated frog retina is illuminated continuously by the NIR light. The visible light stimulator is used to produce a visible light flash for retinal stimulation. A MEA system is used for concurrent electroretinogram (ERG) measurement of retinal activation. The dichroic mirror reflects visible stimulus light and passes the NIR recording light. The eyepiece camera is used to adjust visible light stimulus aperture at the retina. In order to ensure light efficiency for intrinsic optical signal imaging, the beam splitter is removed from the optical path after the visible light stimulator is adjusted. The NIR filter before the high-speed camera is used to block visible stimulus light, and allow the NIR probe light to reach the detector for recording stimulus-evoked IOSs. (This figure is modified from Yao (13)).

stimulation, and the NIR light is used for IOS recording of retinal response (see Note 2).

3. High-speed camera: for NIR recording of stimulus-evoked IOSs in the retina (see Note 3).
4. Eyepiece camera: for test and adjustment of stimulus light patterns.
5. NIR filter: to block visible light and pass NIR light into the high-speed camera.
6. Dichroic mirror: to reflect visible stimulus light and pass the NIR recording light.
7. Beam splitter: to split the light into the high-speed camera and eyepiece camera (see Note 4).
8. Retinal chamber: to hold the sample for concurrent IOS imaging and ERG measurement (see Note 5).

9. ERG system: for electrophysiological measurement of the retina (see Note 6).
10. Timing controller: for electronic synchronization of retinal stimulation and data acquisition (see Note 7).

2.2. Animal

1. Animal species: Leopard frogs (*Rana Pipiens*) (Kons Scientific, Germantown, WI, USA) were used to collect representative IOS images shown in this article. Other animal species is applicable.
2. Ringer's solution: refer to literatures to select appropriate Ringer's solution for specific animal species. For Leopard frogs (*Rana pipiens*) used in our experiments, the Ringer's solution contains (16): 110 mM NaCl, 2.5 mM KCl, 1.6 mM $MgCl_2$, 1.0 mM $CaCl_2$, 22 mM $NaHCO_3$, and 10 mM d-glucose (see Note 8).

2.3. Surgical Equipment

1. Animal guillotine for animal decapitation (World Precision Instrument Inc., Sarasota, FL, USA).
2. Dissecting microscope with dim red light illumination (Fisher Scientific Inc., Pittsburgh, PA, USA).
3. Surgical forceps with 0.1×0.06-mm tip (World Precision Instrument).
4. McPherson-Vannas scissors with 0.1-mm straight tip (World Precision Instrument).
5. Pithing needle.

3. Methods

All animal procedures are approved by the Institutional Animal Care and Use Committee (IACUC).

3.1. Retinal Preparation

1. Dark adaptation: conduct dark adaptation as needed (see Note 9).
2. Animal euthanasia: after dark adaptation, the frog is euthanized by rapid decapitation and followed by double pithing.
3. Transfer the frog head into a Petri dish filled with frog's Ringer solution (see Note 10).
4. Eye isolation: enucleate both eyeballs from the frog head with a pair of dissecting scissors, transfer one eye into a new Petri dish filled with fresh Ringer's solution for retinal dissection, and store the other one in Ringer solution for backup experiment.
5. Retinal dissection: the procedure is performed in Ringer's solution. Trim off excess tissues around the eye ball with a pair of Vannas scissors. Hemisect the eyeball below the equator

with fine scissors to remove the lens and anterior structures. Carefully separate the retina from retinal pigment epithelium (RPE) with a pair of surgical forceps.

6. Retinal transfer: delicately transfer the isolated retina into the recording chamber, with the photoreceptor side facing towards the objective.
7. Retinal fixation: flatten out the retina in the chamber (see Note 11), cover the retina with a micromesh sheet (see Note 12), and fill the chamber with fresh Ringer's solution (see Note 13).

3.2. Retinal Imaging

1. Warm up the IOS imager (Fig. 1).
2. Test and adjust the imaging parameters, including image size, resolution, and frame rate, of the high-speed camera.
3. Test and adjust the retinal stimulator, including stimulus aperture/pattern, color, and intensity.
4. Test and adjust IOS parameters, including pre-stimulus recording phase, stimulus duration, and post-stimulus recording phase.
5. Place the retinal preparation under the IOS imager.
6. Identify a retinal area for IOS imaging.
7. Focus the NIR light to the interested retinal depth, such as photoreceptor layer (Fig. 2).
8. Record a retinal image sequence, with retinal stimulation, into the built-in RAM of the high-speed camera (PCO1200, PCO AG, Kelheim, Germany).
9. Transfer and save the image video to computer disk.
10. Change the focus plane to other interested retinal depth, such as ganglion cell layer (Fig. 3).
11. Repeat steps 8 and 9.
12. Repeat steps 6–11 for imaging other retinal areas.

3.3. Image Processing and IOS Analysis

Select one programming language, such as MATLAB or Interactive Data Language (IDL), for data processing. As shown in Figs. 2 and 3, the unit of IOS images is $\Delta I/I$, where ΔI is the stimulus-evoked dynamic optical changes and I is the background light intensity.

Basic IOS images can be constructed using the following procedure (14):

1. The raw images (Fig. 2a) from the pre-stimulus baseline recording phase is averaged, pixel by pixel, and the averaged intensity of each pixel is taken as the background intensity I of each pixel.
2. The background intensity I is subtracted from each subsequent recorded frame, pixel by pixel, to get the ΔI of each pixel.

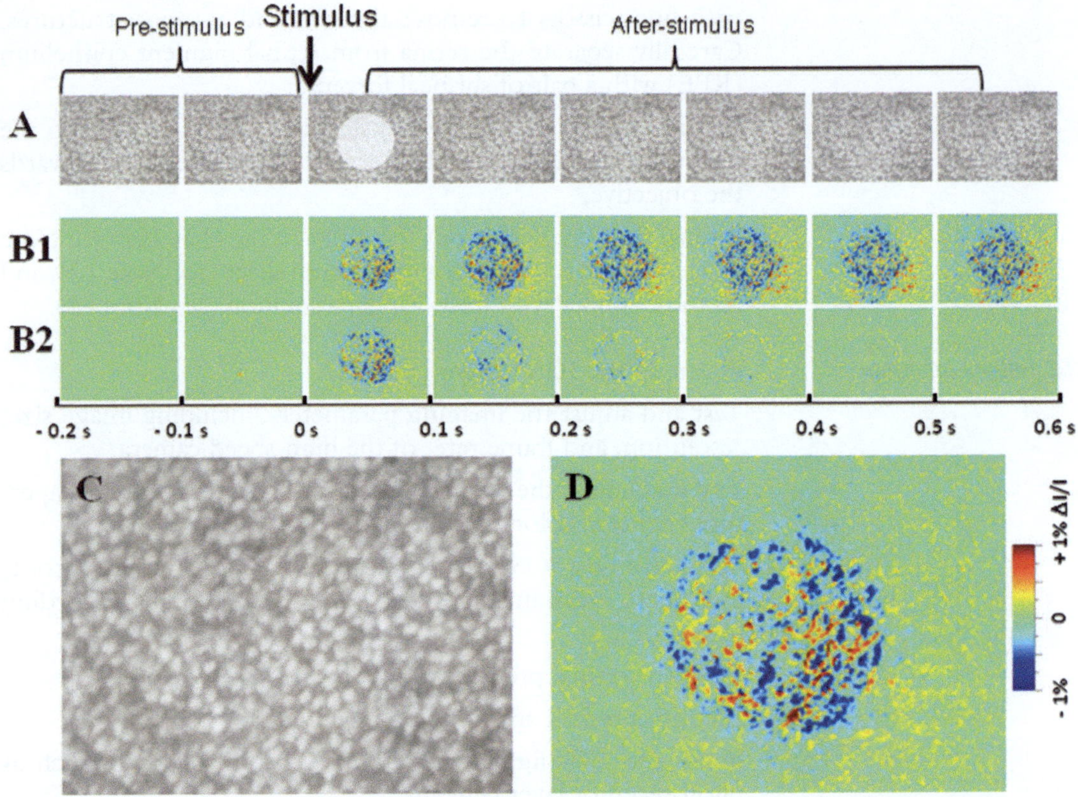

Fig. 2. Intrinsic optical signal (IOS) imaging of retinal photoreceptors. The raw images (**a**) were recorded with the CMOS camera at a speed of 1,000 frames/s. The *white spot* in the third frame of the image sequence (**a**) shows the visible stimulus pattern. (**b1**, **b2**) Reconstructed IOS and dynamic differential IOS images, respectively, based on the raw images in (**a**). Each illustrated frame is an average over 100 ms interval (100 frames); 200 ms pre-stimulus and 600 ms post-stimulus images are shown. (**c**, **d**) Enlarged images of the third frames shown in (**a**) and (**c**), respectively.

3. The $\Delta I/I$ image sequence is constructed to show the dynamic optical changes correlated with retinal activation (Fig. 2b).

Dynamic differential IOS images (Fig. 2c) can be constructed (10, 17):

$$\mathrm{IOS}_{t(x,y)} = \frac{I_{t(x,y)} - I_{\mathrm{ref}(x,y)}}{I_{\mathrm{ref}(x,y)}}, \tag{1}$$

where $I_{t(x,y)}$ is the intensity value of a pixel (x, y) at a time point t; $I_{\mathrm{ref}(x,y)}$ is the dynamic reference baseline of m consecutive frames, which can be quantified by:

$$I_{\mathrm{ref}(x,y)} = \frac{\sum_{i=t-m}^{i=t-1} I_{i(x,y)}}{m}. \tag{2}$$

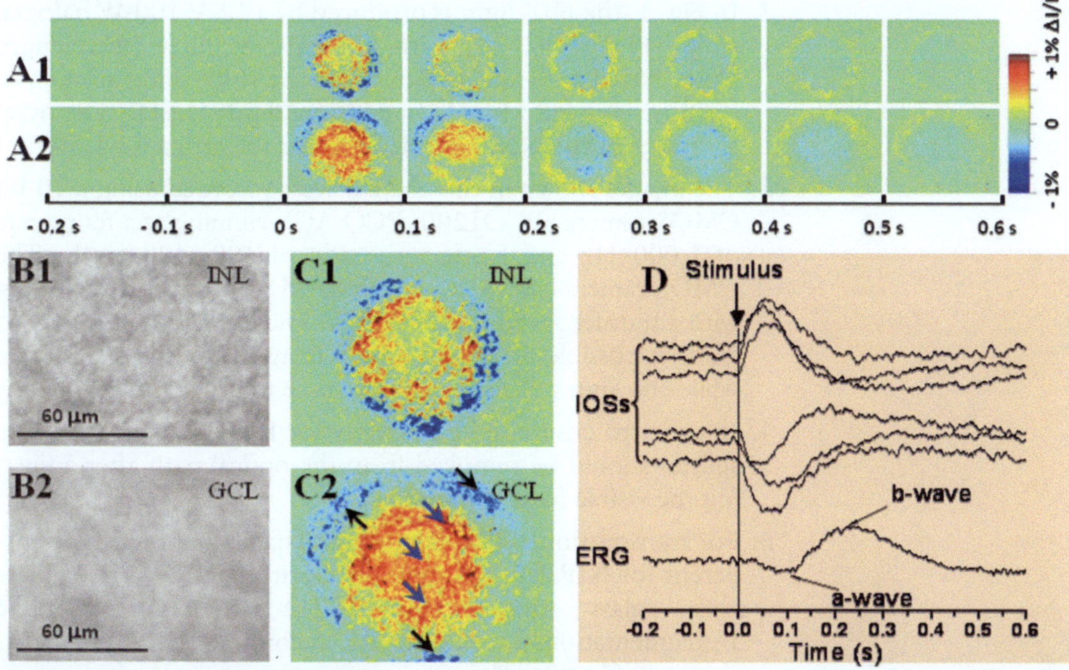

Fig. 3. Intrinsic optical signal (IOS) imaging of inner retinal neurons. (**a1**, **a2**) Dynamic differential IOS images of inner nuclear layer (INL) and ganglion cell layer (GCL). (**b1**, **b2**) Representative raw images of the INL and GCL. (**c1**, **c2**) Enlarged images of the third frames shown in (**a1**) and (**a2**), respectively. (**d**) IOS response of individual pixels pointed by arrowheads in (**c2**). Simultaneous electroretinogram (ERG) was recorded to show stimulus-evoked electrophysiological response of the retina. *Vertical line* indicates the stimulus onset and offset. The raw images were recorded with the CMOS camera at a speed of 1,000 frames/s. (This figure is modified from Yao (14)).

In other words, the averaged pixel value of m consecutive frames recorded before the time point t is used as a reference baseline to calculate the differential IOS. For the dynamic differential IOSs shown in this article, we selected $m = 100$ (i.e., images recorded over 100 ms) for the dynamic reference baseline.

Based on the IOS image sequence, stimulus-evoked retinal dynamics can be analyzed with single pixel spatial-resolution (Fig. 3b) and ms temporal-resolution (Fig. 3c). Further IOS image processing and analysis can be conducted as needed.

4. Notes

1. Both upright and inverted microscopes are applicable. For the representative IOS imager shown in Fig. 1, an upright microscope (BX51WI, Olympus America Inc., Center Valley, PA, USA) is used. In the imager, a water dipping objective is used to reduce the effect of water fluctuations.

2. In Fig. 1, the NIR light is produced by a 12-V 100-W halogen lamp (PHILIPS7724) with a band-pass filter (wavelength band: 800–1,000 nm) in front, and the visible light stimulator is a fiber-coupled white light emitting diode (LED). The overall power of the NIR light delivered at the retina is ~1 mW.
3. The images shown in this chapter were recorded using a 10-bit CMOS camera (PCO1200, PCO AG), running at a frame rate of 1,000 Hz and frame resolution of 400×400 pixels. The CMOS camera has 2 GB built-in RAM for fast image recording with a transfer speed of 820 MB/s. The ultrafast transfer speed made it possible to collect optical images at a high frame rate while allowing sufficient exposure time to ensure image quality.
4. In order to ensure light efficiency for IOS imaging, the beam splitter should be removed from the optical path after adjusting the visible light stimulator.
5. For transmission IOS imaging, the chamber should be transparent for NIR light. A multiple electrode array (MEA) plate with glass ring (100/10-ITO-gr, ALA Scientific Instrumentations) was used for recording the IOS images shown in Figs. 2 and 3.
6. Combined electrophysiological recording is helpful for the assessment of retinal viability. For the IOS imager shown in Fig. 1, a 60-channel electrophysiology recording system (MEA1060, ALA Scientific Instrumentations) is integrated.
7. Either commercial functional generator or customer-designed electronic system can be used for timing control of the retinal stimulation and date acquisition. For the system shown in Fig. 1, a commercial four-channel digital delay/pulse generator (DG535, Stanford Research Systems Inc., Sunnyvale, CA, USA) is used to synchronize the retinal stimulation and data recording.
8. Sodium bicarbonate ($NaHCO_3$) is required to adjust the pH to 7.3–7.45. Add $NaHCO_3$ after the calcium chloride ($CaCl_2$) is completely dissolved in order to prevent precipitation.
9. For our frog experiments, 1–2 h dark adaptation is typically conducted. Increased time period for dark adaptation makes it easier for separating the retina from the RPE.
10. Ice-cold solution can be helpful to slow down retinal metabolic activities during the surgery, thus to increase retinal viability for IOS imaging.
11. Slice the isolated retina radically, and thus allow it to lie flat in the recording chamber.
12. Press micromesh sheet gently to avoid retinal damage.
13. A perfusion system is useful to keep the viability of isolated retinas.

Acknowledgments

This work was supported in part by Dana Foundation (Brain and Immuno-Imaging Grant program), Eyesight Foundation of Alabama, National Institutes of Health (R21RR025788 and R21EB012264), and National Science Foundation (CBET-1055889).

References

1. Sernagor E, Eglen SJ, Wong RO (2001) Development of retinal ganglion cell structure and function. Prog Retin Eye Res 20: 139–174
2. Berardi N, Pizzorusso T, Maffei L (2000) Critical periods during sensory development. Curr Opin Neurobiol 10:138–145
3. Saszik S, Bilotta J, Givin CM (1999) ERG assessment of zebrafish retinal development. Vis Neurosci 16:881–888
4. Speer CM, Sun C, Chapman B (2011) Activity-dependent disruption of intersublaminar spaces and ABAKAN expression does not impact functional on and off organization in the ferret retinogeniculate system. Neural Dev 6:7
5. Harary HH, Brown JE, Pinto LH (1978) Rapid light-induced changes in near infrared transmission of rods in *Bufo marinus*. Science (New York, NY) 202:1083–1085
6. Pepperberg DR, Kahlert M, Krause A, Hofmann KP (1988) Photic modulation of a highly sensitive, nearinfrared light-scattering signal recorded from intact retinal photoreceptors. Proc Natl Acad Sci USA 85:5531–5535
7. Dawis SM, Rossetto M (1993) Light-evoked changes in near-infrared transmission by the ON and OFF channels of the anuran retina. Vis Neurosci 10:687–692
8. Cohen LB, Keynes RD, Hille B (1968) Light scattering and birefringence changes during nerve activity. Nature 218:438–441
9. Tasaki I, Watanabe A, Sandlin R, Carnay L (1968) Changes in fluorescence, turbidity, and birefringence associated with nerve excitation. Proc Natl Acad Sci USA 61:883–888
10. Li YG, Zhang QX, Liu L, Amthor FR, Yao XC (2010) High spatiotemporal resolution imaging of fast intrinsic optical signals activated by retinal flicker stimulation. Opt Express 18: 7210–7218
11. Li YG, Liu L, Amthor F, Yao XC (2010) High-speed line-scan confocal imaging of stimulus-evoked intrinsic optical signals in the retina. Opt Lett 35:426–428
12. Li YC, Strang C, Amthor F, Liu L, Li YG, Zhang QX, Keyser K, Yao XC (2010) Parallel optical monitoring of visual signal propagation from the photoreceptors to inner retina layers. Opt Lett 35:1810–1812
13. Yao XC (2009) Intrinsic optical signal imaging of retinal activation. Jpn J Ophthalmol 53: 327–333
14. Yao XC, Zhao YB (2008) Optical dissection of stimulus-evoked retinal activation. Opt Express 16:12446–12459
15. Zhang QX, Wang JY, Liu L, Yao XC (2010) Microlens array recording of localized retinal responses. Opt Lett 35:3838–3840
16. Sieving PA, Murayama K, Naarendorp F (1994) Push-pull model of the primate photopic electroretinogram: a role for hyperpolarizing neurons in shaping the b-wave. Vis Neurosci 11: 519–532
17. Yao XC, Liu L, Li YG (2009) Intrinsic optical signal imaging of retinal activity in frog eye. J Innov Opt Health Sci 2:201–208

Part VI

Emerging Methodologies

Chapter 21

Use of Laser Capture Microdissection for Analysis of Retinal mRNA/miRNA Expression and DNA Methylation

Laszlo Hackler Jr, Tomohiro Masuda, Verity F. Oliver, Shannath L. Merbs, and Donald J. Zack

Abstract

Laser capture microdissection (LCM) is a useful method to isolate specific cells or cell layers of interest from heterogeneous tissues, such as the retina. The collected cells can be used for DNA, RNA, or protein analysis. We have applied LCM technology to isolate cells from the outer nuclear, inner nuclear, and ganglion cell layers of the retina for mRNA and microRNA (miRNA) expression and epigenetic (DNA methylation) analysis. Here, we describe the methods we have employed for sample preparation, LCM-based isolation of retinal layers, RNA/DNA extraction, RNA quality check, microRNA analysis by quantitative PCR, and DNA methylation analysis by bisulfite sequencing.

Key words: Laser capture microdissection, Retina, Gene expression, microRNA, Epigenetics, Quantitative PCR, RNA extraction, DNA methylation, Bisulfate sequencing

1. Introduction

Laser capture microdissection (LCM) is a microscope-based technology that can be used to dissect specific cells, or groups of cells, of interest from sectioned tissues with infrared- or UV laser beam-mediated tissue cutting (1). The technology is flexible, allowing cell collection under the microscope with objectives ranging from ×4 up to ×65. This enables isolation of not only large areas of tissue sections containing mixed cell populations, but also isolation of specific cell types and even single cells. These features of the LCM are especially valuable for the isolation of specific cells from heterogeneous tissues, such as the retina. Cells can be stained to discern specific cell types, or transgenic mice carrying cell type-specific fluorescent reporters can also be used. In combination with genomic and proteomic technologies, LCM allows the study of

Shu-Zhen Wang (ed.), *Retinal Development: Methods and Protocols*, Methods in Molecular Biology, vol. 884,
DOI 10.1007/978-1-61779-848-1_21, © Springer Science+Business Media, LLC 2012

molecular mechanisms in specific cells or specific groups of cells. Fluorescence-activated cell sorting (FACS) provides another technology for isolation of specific cell types for downstream analysis, but it does not have the ability to utilize cell morphology as a section criterion, nor does it allow the selection of spatially associated clusters of cells.

The retina is composed of well-organized and stratified layers of cells and intervening synaptic areas. The cell bodies of retinal cells reside in the outer nuclear layer (ONL), inner nuclear layer (INL), and ganglion cell layer (GCL). Each layer is composed of one or more specific cell types. LCM can be a potent tool for the isolation of specific retinal cell layers and subsequent molecular analysis. Nuclear staining facilitates clear discrimination of each layer, which helps to avoid unwanted contamination of cells from other layers.

In this chapter, we describe the approaches and detailed LCM methods that we have been using to study murine retinal cell/layer-specific mRNA and microRNA (miRNA) expression and DNA methylation patterns (2, 3). We describe the methods designed to maintain RNA quality, and to prepare retinal samples, laser-microdissect the cells, and quantify expression levels using quantitative PCR (qPCR). We also describe the use of LCM for epigenetic analysis (4, 5), specifically for studying differential DNA methylation in the retinal layers. In addition to the retina-related material provided in this chapter, the reader may also find useful some of the available references on different aspects of general LCM methodology (6–8).

2. Materials

2.1. Tissue Preparation and Laser Microdissection

1. Dissecting tweezers and scissors.
2. PBS buffer: RNase free.
3. Sucrose.
4. Tissue-Tech OCT compound (Ted Pella).
5. Cryostat.
6. PEN foil slides (Leica Microsystems).
7. DEPC water.
8. Ethanol.
9. Mayer's hematoxylin.
10. 200-μl Tubes.
11. LCD system.
12. TRIzol (Invitrogen).

2.2. RNA Extraction and cDNA Synthesis

1. Chloroform.
2. Microcentrifuge.
3. Isopropyl alcohol.
4. Agilent 2100 Bioanalyzer (Agilent Technologies).
5. Agilent RNA 6000 Pico kit (Agilent Technologies).
6. NCode™ miRNA First-Strand cDNA Synthesis Kit (Invitrogen).
7. ATP.
8. Poly A Polymerase.
9. RNase-free microcentrifuge tube.
10. Water bath/heating block with temperature control.
11. miRNA-specific or control-specific reverse primer.
12. 2× SYBR® Green master mix (Bio-Rad).

2.3. DNA Methylation Analysis

1. QiaAmp DNA Micro Kit (Qiagen) or the AllPrep DNA/RNA Micro Kit (Qiagen).
2. QiaAmp DNA Micro Kit (Qiagen).
3. EZ DNA Methylation-Gold Kit (Zymo Research).
4. mM dNTPs (Fermentas, Thermo Fisher Scientific).
5. PCR: JumpStart RED*Taq* (Sigma-Aldrich).
6. MinElute Gel Extraction Kit (Qiagen).
7. ContigExpress of Vector NTI Advance 10 (Invitrogen).
8. TOPO-TA Cloning Kit for Sequencing (Invitrogen).
9. SOC medium (Invitrogen).
10. Warm LB shaker (Fisher Scientific).
11. 50 mg/ml Kanamycin (Sigma-Aldrich).
12. PyroMark Q24 pyrosequencer (Qiagen).
13. PyroMark Assay design software (Qiagen).
14. HotStar *Taq* Polymerase (Qiagen).
15. PyroMark Q24 Vacuum Prep Tool (Qiagen).
16. Streptavidin Sepharose HP beads (Amersham Biosciences).
17. Annealing Buffer (Qiagen).
18. PyroMark Q24 Pyrosequencing System (Qiagen).
19. Binding Buffer (Qiagen).

3. Methods

3.1. Sample Preparation

1. Enucleate the eye and remove the muscles with tweezers and scissors (see Note 1) (we use predominantly mouse eyes for our studies, but these methods could be adapted for use with other species).
2. Cryoprotect the eye in increasing concentration of sucrose in the following gradient: 5% sucrose/PBS, 10% sucrose/PBS, 20% sucrose/PBS, and 25% sucrose/PBS. In each condition, the eye should be immersed in the sucrose buffer at 4°C for 30 min or until the eye sinks to the bottom of the tube.
3. Immerse the eye in 25% sucrose in PBS/OCT compound mixture at the ratio of 2:1 for 1 h at 4°C.
4. Snap-freeze the eye in fresh 25% sucrose in PBS/OCT compound mixture on dry ice and store it at –80°C until use.

3.2. Laser Capture Microdissection

1. Cut frozen sections at 7–10 μm with a cryostat and mount them on PEN foil slides. Proceed to step 2 as soon as possible (see Note 2).
2. Stain the sections on ice unless otherwise mentioned as follows (see Note 3):
   (a) Fix the sections in 70% ethanol for 30 s.
   (b) Wash in DEPC water for 30 s.
   (c) Stain the sections in Mayer's hematoxylin for 10–30 s at room temperature.
   (d) Wash in DEPC water for 30 s.
   (e) Wash in 95% ethanol for 30 s.
   (f) Wash in 100% ethanol for 30 s.
   (g) Air-dry for 1–2 min.
3. During the air-dry, prepare the sample collection tubes (into the cap of a 200-μl tube, add 35 μl of lysis buffer TRIzol, see Note 4) and set them onto the microscope.
4. Microdissect the cells under the LCM microscope through provided software according to the manufacturer's instruction (see Note 5). Figure 1 shows sample images of what sections look like before and after microdissection.
5. Complete the dissection within 1 h because of the risk of RNA degradation (see Note 6).
6. Store the collected cells at –80°C until RNA extraction (see Note 7).

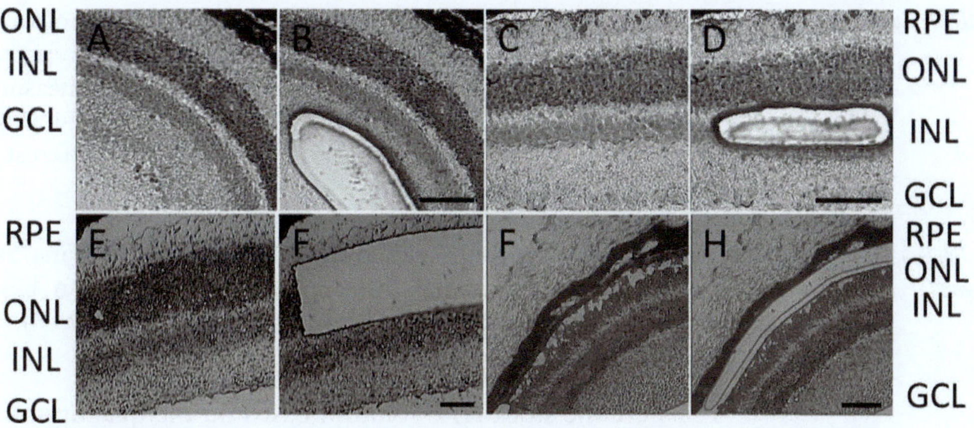

Fig. 1. Laser capture microdissection of the mouse retina. Cells from the ganglion cell layer (GCL) (**a**, **b**), inner nuclear layer (INL) (**c**, **d**), outer nuclear layer (ONL) (**e**, **f**), and retinal pigment epithelium (RPE) (**g**, **h**) were collected by LCM. Figures show section appearance before (**a**, **c**, **e**, **g**) and after (**b**, **d**, **f**, **h**) microdissection. The size bar represents 50 μm.

3.3. RNA Extraction and RNA Quality Control

1. Thaw the collected samples on ice and increase its volume to at least 100 μl with the used lysis buffer (see Note 8).
2. Add 20 μl of chloroform to separate the aqueous and organic phases (see Note 9).
3. Shake tubes vigorously by hand for 15 s and incubate them at room temperature for 2–3 min.
4. Centrifuge the samples at 12,000 × *g* for 15 min at 4°C.
5. Following centrifugation, remove the colorless upper aqueous phase. Do not take the complete colorless phase, only approximately 60–70%, and make sure not to touch the interphase.
6. Transfer the aqueous phase to a fresh tube and precipitate the RNA with 50 μl of isopropyl alcohol.
7. Mix the contents of the tube and incubate the samples for 10 min at room temperature.
8. Centrifuge the samples at 12,000 × *g* for 10 min at 4°C. The RNA precipitates to the bottom or lower side of the tube which is invisible due to its amount.
9. Remove the supernatant. Wash the RNA pellet once with 500 μl of 75% ethanol. Mix the sample by vortexing.
10. Centrifuge at 7,500 × *g* for 5 min at 4°C.
11. Remove the supernatant. Let the tube dry and dissolve the purified RNA in 10 μl of DEPC water.
12. RNA quality can be analyzed using an Agilent 2100 Bioanalyzer if an adequate amount of RNA is collected with LCM. We have used the Agilent RNA 6000 Pico kit for low RNA quantities. Typically, 1 μl of the collected samples is assessed on the chips (see Note 10).

3.4. cDNA Synthesis

There are two major strategies that are commonly used for template cDNA synthesis in the case of miRNAs. One is to prepare a "universal" cDNA sample which can be used to measure all expressed transcripts, or alternatively one can synthesize miRNA-specific cDNA which is enriched for the miRNA of interest after the reverse transcription step.

3.4.1. Universal cDNA Synthesis

The following method uses the NCode™ miRNA First-Strand cDNA Synthesis Kit with some modifications and can be used when the isolated total RNA is at least 10 ng (see Note 11).

Poly(A) Tailing

1. Dilute the stock ATP solution based on the quantity of RNA available according to the following formula: ATP dilution factor = 10,000/___ ng of total RNA.
2. Each reaction requires 0.25 μl of Poly A Polymerase. To avoid pipetting 0.25 μl of enzyme, dilute 1 μl fourfold (1 μl of enzyme with 3 μl of DEPC water) and use 1 μl in the reaction. If multiple samples are transcribed, create a master mix with the common components without diluting the enzyme.
3. Add the following at room temperature to at least 10 ng total RNA:

   2.5 μl of 5× miRNA reaction buffer

   1.25 μl of 25 mM $MnCl_2$

   1 μl of diluted ATP (from step 1)

   1 μl of Poly A Polymerase (from step 2)

   DEPC-treated water to 12.5 μl
4. Mix gently and centrifuge the tube briefly to collect the contents.
5. Incubate the tube in a heat block or water bath at 37°C for 15 min. After incubation, proceed immediately to the *first-strand cDNA synthesis.*

First-Strand cDNA Synthesis

1. Add the following to an RNase-free microcentrifuge tube:

   4 μl of polyadenylated RNA

   1 μl of annealing buffer

   3 μl of universal RT primer (25 μM)
2. Incubate the tube at 65°C for 5 min.
3. Place the tube on ice for 1 min.
4. Add the following to the tube, for a final volume of 20 μl:

   10 μl of 2× First-Strand Reaction Mix

   2 μl of SuperScript™ III RT/RNaseOUT™ Enzyme Mix

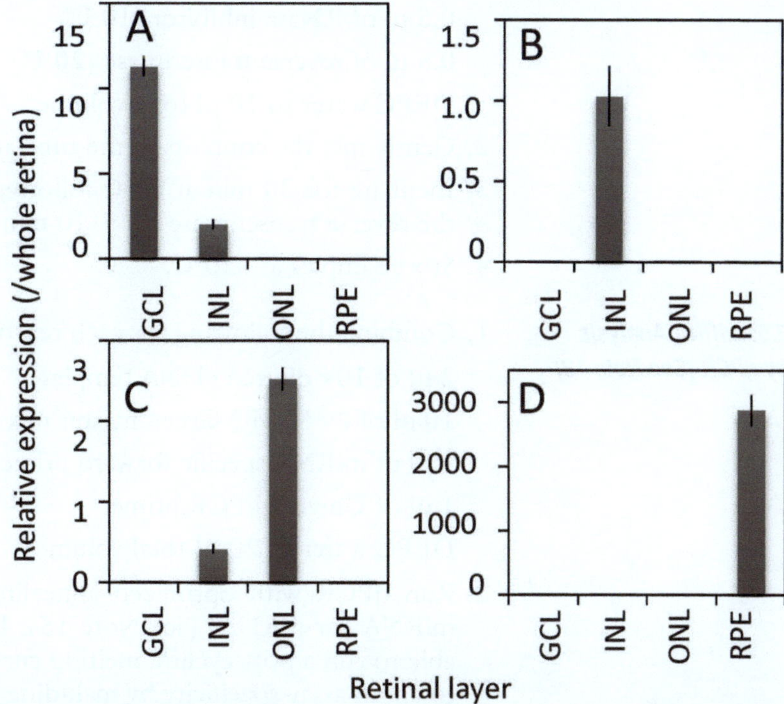

Fig. 2. QPCR analysis of marker gene expressions in microdissected retinal layers. To evaluate cross-contamination, qPCR was carried out for marker genes known to be preferentially expressed in the GCL (**a**, *Thy1*), INL (**b**, *Chx10*), ONL (**c**, *Crx*), and RPE (**d**, *RPE65*) of microdissected mouse retina. Expression of each gene was normalized to Gapdh and then expressed as relative expression compared to whole retina. Note that *Crx*, although predominantly expressed in photoreceptors, is also expressed in bipolar cells of the INL.

5. Spin the tube briefly to collect the contents.
6. Transfer the tube to a thermal cycler preheated to 50°C and incubate for 50 min.
7. Incubate at 85°C for 5 min to stop the reaction.
8. Chill the reaction on ice. Store samples at −20°C or proceed directly to qPCR. Figure 2 shows an example of qPCR analysis with marker genes for the different retinal cell layers, demonstrating that strong layer-specific signals can be obtained with minimal contamination from other layers.

3.4.2. Transcript-Specific cDNA Synthesis

This method is the appropriate choice in most cases when the isolated RNA is below 10 ng (see Note 12).

1. In an RNase-free tube, combine:

   1–10 ng of template total RNA

   2 μl of 5× RT Reaction buffer

   2 μl of miRNA-specific or control-specific reverse primer

   0.5 μl of 10 mM dNTP mix

0.5 μl of RNase inhibitor (10 U)

0.5 μl of reverse transcriptase (20 U)

DEPC water to 10 μl total volume

2. Gently mix the contents of the tube and spin down.
3. Incubate for 30 min at 50°C followed by heat inactivation of the reverse transcriptase for 5–10 min at 85°C.
4. Store samples at -20°C.

3.5. miRNA Analysis by qPCR (See Note 13)

1. Combine the following for each reaction:

2 μl of 10× diluted cDNA template

10 μl of 2× SYBR® Green master mix

1 μl of miRNA specific forward primer (see Note 14)

1 μl of Universal PCR primer

DEPC water to 20-μl total volume

2. Run qPCR with optimized annealing temperature for each miRNA (or mRNA) (see Note 15). If the instrument used is able to run a post-cycling melting curve analysis, it is desirable to check assay specificity by including this step.

3.6. Bisulfite Sequencing for DNA Methylation Analysis

Genomic DNA for bisulfite sequencing is isolated using either the QiaAmp DNA Micro Kit or the AllPrep DNA/RNA Micro Kit.

3.6.1. Isolate DNA Using the QiaAmp DNA Micro Kit

1. Microdissect the desired cells into 15 μl of Buffer ATL in a 0.2-ml tube (see Notes 16 and 17). The cells can be stored in the buffer at -80°C until you are ready to isolate the DNA.
2. Follow the steps as listed for "Isolation of Genomic DNA from Laser-Microdissected Tissues" in the kit manual. Elute the DNA with 20 μl of Buffer AE. As suggested in the last step, incubate the column with Buffer AE for 5 min at room temperature before centrifugation to increase DNA yield.

3.6.2. Isolate DNA Using the AllPrep DNA/RNA Micro Kit

1. Microdissect the desired cells into 28 μl of Buffer RLT Plus. The cells can be stored in the buffer at -80°C until you are ready to isolate the RNA and DNA. Multiple dissections can be combined prior to isolation of nucleic acid up to 350 μl.
2. Follow the steps as listed for "Simultaneous Purification of Genomic DNA and Total RNA from Microdissected Cryosections" in the manufacturer's instructions. Elute the DNA with 20 μl of Buffer EB (preheated to 70°C).

3.6.3. Bisulfite Conversion of DNA

Bisulfite modification of DNA creates sequence differences by converting unmethylated cytosines to uracil, leaving methylated cytosines unchanged (9). Methylation state is proportionately represented by presence of C versus T residues, as determined by DNA sequencing, after PCR amplification.

1. Bisulfite modification of genomic DNA is performed using the EZ DNA Methylation-Gold Kit, according to the manufacturer's instructions (see Note 18) with the modifications described below.
2. After the 64°C step is complete and the temperature has reached 4°C, complete the rest of the protocol immediately. Do not store the sample at 4°C for up to 20 h as suggested by the manufacturer.
3. Elute the column twice with 10 μl of M-Elution buffer instead of the suggested 10 μl. For each elution, the column is incubated for 10 min at room temperature prior to centrifugation. Pool and divide the 20 μl into four aliquots and freeze at −80°C. Bisulfite treatment is harsh, and converted DNA is prone to degradation, which is accelerated by freeze–thaw cycles. Use 1 μl per reaction below.

3.6.4. Bisulfite Sequencing by Direct Sequencing or Cloning

1. Identify a 200–300-bp region to be sequenced. In our experience, regions of more than 350 bp do not amplify efficiently and are more prone to PCR bias.
2. Copy and paste the genomic sequence (with 100 bp on either side of your identified region) into the Web-based MethPrimer program that identifies CpG sites and CpG islands (10). This program will design primers designed to specifically amplify bisulfite-modified genomic DNA and will include sequence that contains C not followed by G (represented by T in the primers to selectively amplify only bisulfite-modified DNA) and avoid regions that contain CpG (see Note 19).
3. For PCR amplification using the primers from MethPrimer, combine the following for each reaction:

   3 μl of 10× PCR buffer

   0.6 μl of 10 mM dNTPs 1.2 μl of Forward Primer (10 μM)

   1.2 μl of Reverse Primer (10 μM)

   1.2 μl of JumpStart RED*Taq*

   21.8 μl of DEPC water

   1 μl of bisulfite-converted genomic DNA
4. Run the PCR amplification with the optimized annealing temperature (*X*°C) with the following cycling conditions: 94°C for 1 min for initial denaturation followed by 40 cycles of 94°C for 30 s, *X*°C for 30 s, 72°C for 1 min, and a final elongation of 72°C for 10 min, followed by incubation at 4°C (short term) or 20°C (long term).
5. Visualize 10–20 μl of each reaction on a 2% agarose gel to confirm a single product of the appropriate size.

6. For direct sequencing, agarose gel-purify the PCR product (MinElute Gel Extraction Kit) and directly sequence using the forward primer. Using the ContigExpress function of Vector NTI Advance 10, the chromatograph peak heights for T and C at each CpG site can be measured and the $\log_{10}$ [T peak height (unmethylated)/C peak height (methylated)] versus CpG position with respect to the transcription start site was plotted. CpG sites with $\log_{10}$ (U/M) > 1 have a relative lack of methylation and CpG sites with $\log_{10}$ (U/M) < 1 are relatively methylated.
7. Alternatively, sequencing of individual clones after PCR amplification can be performed using the TOPO-TA Cloning Kit for Sequencing.
   (a) Combine the following (see Note 20):

   0.5–2 μl of fresh (see Note 21) PCR product, depending on the band intensity

   0.5 μl of salt solution

   0–1.5 μl of DEPC water

   0.5 μl of pCR™4-TOPO® TA vector in a total volume of 3 μl

   (b) Incubate at room temperature for 5 min and then place the reaction on ice.

   (c) Transform One Shot® TOP10 Chemically Competent *Escherichia coli* as follows:

   Prechill microfuge tubes and thaw competent cells on ice.

   Transfer 25 μl of competent cells (half of each tube of cells) to the prechilled microfuge tube.

   Add 2 μl of this ligation mix to a tube of competent cells, and flick gently by hand (do not mix by pipette).

   Incubate on ice for 5 min.

   Heat shock cells at 42°C for 30 s and immediately transfer to ice.

   Add 125 μl of SOC medium and shake at 200 rpm at 37°C for 1 h.

   Warm LB + 50 mg/ml kanamycin (Sigma-Aldrich) plates for 30 min before plating cells.

   Plate 50–100 μl of each transformation and let the plate dry on the bench.

   Incubate overnight at 37°C.

   (d) The following day, pick colonies for sequencing and analyze by colony PCR, if necessary, prior to sequencing using the M13 reverse primer. Typically, ten clones per sample per sequence are averaged. BiQ Analyzer is a good software tool for analysis, visualization, and quality control of bisulfite sequencing data (11).

3.6.5. Bisulfite Sequencing by Pyrosequencing

We use the PyroMark Q24 pyrosequencer, which is based upon the principle of sequencing by synthesis and provides quantitative methylation data at each CpG site (12). Up to 24 samples can be analyzed in parallel.

1. Using the PyroMark Assay design software, paste in the sequence of interest and highlight the target region. The chosen target region should only encompass about 20 bases for the easiest assay design (see Note 22).
2. Perform PCR optimization with the forward (biotinylated, HPLC purified) and reverse primers, with HotStar *Taq* Polymerase and additional $MgCl_2$ when necessary according to the manufacturer's protocol. For each PCR, 1 μl of bisulfite-converted DNA should be sufficient.
3. Run 5 μl of product from a 30-μl PCR reaction on a 2% agarose gel to ensure that a single product of the expected size is present.
4. If a reasonable intensity band is visualized on the gel, use 10 μl of PCR product in the pyrosequencing reaction. In the case of inefficient PCR amplification, up to 20 μl of product can be used.
5. Purify the biotinylated PCR product and make it single stranded with PyroMark Q24 Vacuum Prep Tool, following manufacturer's direction. During this step, the PCR product is bound to Streptavidin Sepharose HP beads, purified, washed, denatured using a 0.2 M NaOH solution, and washed again as per the manufacturer's protocol.
6. Anneal 0.3 μM of pyrosequencing primer to the purified single-stranded PCR product in Annealing Buffer.
7. Perform pyrosequencing using the PyroMark Q24 Pyrosequencing System.
8. If needed, design the additional sequencing primers using the PyroMark Assay design software, fix the original forward and reverse primers, set a new target region, and repeat the assay design. With a single sequencing primer, 40–60 bp can usually be sequenced. If more sequence is desired, a serial pyrosequencing assay can be performed (13). In this case, the same PCR product is sequenced with different sequencing primers to cover a longer region of interest.
9. Run serial pyrosequencing reactions by adding 20 μl of Binding Buffer to each well of a plate immediately after the initial run has been performed, followed by the 10-min incubation with shaking and the usual vacuum prep workstation protocol (see Note 23).

4. Notes

1. For the mouse pups, earlier than postnatal day 10, we generally cryoprotect the eyes as part of the whole head. The head is decapitated right below the upper jaw, and the eyelids are removed to directly expose the eyes to the buffer.
2. Once slides are prepared, they should be used as soon as possible. However, in our lab, we have tested the RNA quality obtained from fresh versus stored slides (–80°C in sealed box with desiccant), and did not observe a significant difference in quality.
3. One can wash for a shorter period of time in each step, but this is a trade-off between RNA quality and stress-free laser dissection. One should proceed to the next step as soon as possible due to the risk of RNA degradation, but inadequate washing of the section can lead to residual OCT compound on the section, which causes significant slowing of the LCM cutting process, and hence increases the risk of reduced RNA quality and quantity.
4. The method of RNA extraction must be decided prior to microdissection. In our LCM experiments, we collect samples directly into tubes containing lysis buffer (35 μl). For miRNA analysis, we collect into TRIzol. It is useful to check the evaporation rate of one's chosen lysis buffer before use. Extended cutting times (over 1 h) may reduce the volume of the buffer, changing its properties, which may result in degraded samples or low yield.
5. For fine cutting, one should adjust the laser power according to the section thickness and the characteristics of the layer being cut. High-power laser output burns the cells along the edge of the cut.
6. Longer exposure of sections to ambient temperature causes significant loss of intact RNA. First, check how long you take to finish one section to estimate how many sections you can put on one slide. For qPCR using cDNA templates synthesized with random hexamers, you can extend the dissection time somewhat as partial degradation of mRNA length is less of a problem than with oligo-dT primed cDNA.
7. Samples may be stored in TRIzol for a few days at –80°C, but it is recommended to process them as soon after cutting as possible.
8. We decided to use TRIzol for isolation because it captures all types of RNA from the samples, including the miRNA and mRNA fractions. We found this useful when expression of intronic miRNAs was compared to that of their host genes,

which we could achieve from the same sample (2). Also the lack of a miRNA enrichment step helps ensure that a complete miRNA fraction is collected. If small RNA enrichment is necessary, it can follow the TRIzol-based purification.

9. The manufacturer's protocol uses 200 μl of chloroform for phase separation for each milliliter of TRIzol used; we use 20 for 100 μl of TRIzol and 1/10 of other volumes indicated (except for the washing of the precipitated RNA pellet, for which we use 500 μl of 75% ethanol).
10. Generally, it is difficult to assess RNA quality from LCM samples due to the limited amount of RNA that is collected. In our experience, a BioAnalyzer run is useful with samples that contain at least 1 ng of total RNA. In that case, visual inspection of the 5S, 18S, and 28S ribosomal RNA peaks or use of the RNA integrity number (RIN) gives an estimation of RNA integrity (14). With LCM samples containing less than 1 ng of total RNA, we generally do not assess RNA integrity. As one rough method to assess cDNA quality when amounts are limiting, a qPCR run with known primers for a highly and constitutively expressed gene can be used. In case of miRNAs, it is safe to assume that RNases would digest larger transcripts with greater probability (e.g., mRNA) than mature miRNAs with 18–22-nucleotide length. Hence, moderately degraded samples should still yield acceptable miRNA expression results (see http://www.abrf.org/Other/ABRFMeetings/ABRF2011/RG%20Presentions/SChittur-STighe_R5a.pdf).
11. Due to the short length of mature miRNAs and the lack of a polyadenylated tail, the priming step of the reverse transcription step provides a challenge. The former prevents random priming while the latter prevents the use of oligo-dT to initiate the reaction. For universal cDNA synthesis, in the first step, a polyadenylated tail is generated on the 3′ end of each transcript by addition of poly A polymerase and ATP. In the second step, the resulting tailed transcript is primed with a modified oligo-dT primer that carries a specific overhang that is a unique sequence that will serve as a general priming site in the following qPCR reaction. The amplification is then carried out with a miRNA-specific forward primer and a universal reverse primer. Despite the short length of miRNAs, the resulting amplicon is in the range of 80–90 bp, which is close to the preferred amplicon length for qPCR. The major advantage of this method is that from one cDNA sample multiple miRNAs can be assayed. In this approach, the major limit is the sample quantity itself (preferably the input total RNA should be above 10 ng). One but significant drawback of the universal cDNA method, however, is that its sensitivity is not always adequate for LCM samples. Only abundant miRNAs can be measured

with great confidence. This approach is often the method of choice when dozens of miRNAs have to be investigated from a limited amount of material. When needed, sensitivity can be improved through the use of locked nucleic acid (LNA)-enhanced oligonucleotides in a slightly different method when not only the forward but also the reverse primer contain a short miRNA-specific sequence (see Note 12).

12. Alternatively, a transcript-specific cDNA can be generated by miRNA-specific RT primers. One method is to use the miRCURY LNA™ microRNA PCR System from Exiqon. Generally, 1 ng of total RNA should be used, but it may work with less (10–100 pg) in the case of abundant miRNAs (see Note 11). The primers used contain the reverse complement of the last few nucleotides at the 3′ end of the given miRNA, reducing significantly the subset of transcripts that are copied, and thereby enriching for the desired transcripts. The RT primer also contains a unique sequence for priming during amplification, while an miRNA-specific oligo (which usually covers the full length of the mature miRNA) serves as the forward primer. The major advantage of this approach is improved sensitivity over the universal cDNA method, while the most significant disadvantages are the increased labor, sample, and reagent requirements. For each investigated miRNA, a different RT reaction is needed which can also lead to increased technical variation, which may be decreased through the use of master mixes where possible and precise pipetting. This method may be preferred in the case of a rare target and when the number of miRNAs to investigate is low. It can complement the universal cDNA method or serve as a backup when it fails.

13. There are two choices of sample preparation before template synthesis for qPCR. One is to enrich the small RNAs from the collected samples, and the other is to measure miRNA expression from total RNA. Depending on the downstream protocol used, enrichment may be necessary. qPCR analysis of miRNAs follows the same steps as when one measures mRNA levels in a sample. After reverse transcription to generate template cDNA, specific forward and reverse primers are used for amplification. The same simple method of SybrGreen detection can be applied to follow the amplification, and depending on the qPCR instrument, reaction quality can be assessed by melting curve analysis.

14. miRNA-specific forward primers have to be designed and ordered separately if one uses the universal cDNA method. Usually, it is the same sequence as the mature sequence of the miRNA of interest. In case of the miRNA-specific cDNA method, these primers are included in pairs with the appropriate primers for the reverse transcription step.

15. Once the template cDNA is synthesized, it is useful to test the resulting cDNA with a primer set for a transcript that could later serve as a normalizing control (in the case of the universal cDNA method). Usually, 5S rRNA or U6 snRNA can be used for these purposes. It is recommended to use more than one "housekeeping" gene to avoid normalization problems associated with differential expression of a particular "housekeeping" gene. miRNA qPCR assays should be optimized similarly to regular qPCR assays. Optimal primer concentration, annealing temperature, and cDNA dilution should be determined. It is also recommended to check assay linearity and efficiency by obtaining a standard curve with each assayed primer set. Data analysis should proceed as with regular qPCR.
16. If you are just dissecting cells for DNA, you can dissect for longer than 1 h because of the greater stability of DNA. However, depending on the humidity of the room, you will periodically need to add 15 μl of nuclease-free water to the cap as the fluid will slowly evaporate.
17. The number of sections you dissect depends upon the desired cell type and species. Typically, 14–16 cryosections of a mouse eye will fit on one PEN foil slide.
18. The CT Conversion Reagent is designed for ten reactions. Although once prepared, the instructions say that the solution can be stored for 1 week at 4°C or up to 1 month at -20°C, we use the reagent the same day and discard any leftover.
19. As a first pass, try the default parameters for primer selection. However, by manipulating the target, excluded region, product size, and product CpGs, you can often direct primer design to a particular region. Occasionally, in a very dense CpG island or in a region that results in long strings of T's after bisulfite conversion, primer design or efficient sequencing is impossible.
20. The TOPO TA Cloning manual (Invitrogen) recommends setting up a 6-μl final volume cloning reaction. We have found that we get satisfactory results using a reaction volume of 3 μl, which enables twice the number of cloning reactions from each kit.
21. The TOPO TA Cloning kit relies on a standard PCR product having single 3′ adenine overhangs. If your Taq polymerase is a proofreading polymerase, if you plan to store your samples before cloning, or if you need to gel-purify your PCR product after amplification, you will need to add 3′ adenines as described in the manual under "Addition of 3′ A-Overhangs Post-Amplification."
22. The analysis target region can be expanded after suitable primers have been found. To avoid incorporating CpG sites in primer sequences, ensure that the "allow primer over variable

position" is turned off under "Assay Settings." Bisulfite PCR products should be no larger than 350 bp to enable efficient amplification. In our experience, assays with scores of 65 or greater as determined by the PyroMark Assay design software work effectively. If primers cannot be designed in your target region, try using the reverse strand, shortening the target region (design a serial pyrosequencing reaction to compensate) or, if the sequencing primer indicates that self-annealing is possible, the addition of a ~2 mismatched bases at the 5′ end may prevent self-annealing.

23. The overall signal may be lower in a serial pyrosequencing reaction.

Acknowledgments

The work described in this chapter was supported by grants from the National Eye Institute, NIH (5R01EY009769, 5P30EY001765, and 5R01EY020406), and by generous gifts from the Guerrieri Family Foundation and from Mr. and Mrs. Robert and Clarice Smith.

References

1. Curran S, McKay JA, McLeod HL, Murray GI (2000) Laser capture microscopy. Mol Pathol 53(2):64–68
2. Hackler L Jr, Wan J, Swaroop A, Qian J, Zack DJ (2010) MicroRNA profile of the developing mouse retina. Invest Ophthalmol Vis Sci 51(4):1823–1831
3. Wahlin KJ, Hackler L Jr, Adler R, Zack DJ (2010) Alternative splicing of neuroligin and its protein distribution in the outer plexiform layer of the chicken retina. J Comp Neurol 518(24):4938–4962
4. Kalantari M et al (2009) Laser capture microdissection of cervical human papillomavirus infections: copy number of the virus in cancerous and normal tissue and heterogeneous DNA methylation. Virology 390(2):261–267
5. Wu Y, Strawn E, Basir Z, Halverson G, Guo SW (2006) Promoter hypermethylation of progesterone receptor isoform B (PR-B) in endometriosis. Epigenetics 1(2):106–111
6. Arin MJ, Roop DR (2002) Use of laser capture microscopy in the analysis of mouse models of human diseases. Methods Enzymol 356:207–215
7. Espina V et al (2006) Laser-capture microdissection. Nat Protoc 1(2):586–603
8. Neira M, Azen E (2002) Gene discovery with laser capture microscopy. Methods Enzymol 356:282–289
9. Frommer M et al (1992) A genomic sequencing protocol that yields a positive display of 5-methylcytosine residues in individual DNA strands. Proc Natl Acad Sci USA 89(5): 1827–1831
10. Li LC, Dahiya R (2002) MethPrimer: designing primers for methylation PCRs. Bioinformatics 18(11):1427–1431
11. Bock C et al (2005) BiQ analyzer: visualization and quality control for DNA methylation data from bisulfite sequencing. Bioinformatics 21(21):4067–4068
12. Dupont JM, Tost J, Jammes H, Gut IG (2004) De novo quantitative bisulfite sequencing using the pyrosequencing technology. Anal Biochem 333(1):119–127
13. Tost J, El abdalaoui H, Gut IG (2006) Serial pyrosequencing for quantitative DNA methylation analysis. Biotechniques 40(6):721–722, 724, 726
14. Schroeder A et al (2006) The RIN: an RNA integrity number for assigning integrity values to RNA measurements. BMC Mol Biol 7:3

Chapter 22

Revealing Looping Organization of Mammalian Photoreceptor Genes Using Chromosome Conformation Capture (3C) Assays

Guang-Hua Peng and Shiming Chen

Abstract

Chromosome conformation capture (3C) is a biochemical assay to reveal higher order chromosomal organizations mediated by physical contact between discrete DNA segments in vivo. Chromosomal organizations are involved in transcriptional regulation of a number of genes in various cell types. We have adapted 3C for analyzing the intrachromosomal looping organization of rod and cone photoreceptor genes in the mammalian retina. Here, we describe a detailed protocol for 3C assays on whole mouse retinas. Using the M-cone *opsin* gene as an example, we demonstrate how to genetically distinguish 3C signals from cones versus rods in retinal 3C assays. We also describe the challenges and key points of 3C design and performance as well as appropriate controls and result interpretations.

Key words: Chromosomal conformation capture (3C), Intrachromosomal loops, Retina, Photoreceptors, *Opsin* genes, Transcriptional regulation

1. Introduction

The chromosomal conformation capture (3C) assay (1) is a biochemical method to reveal physical contact between distant chromatin segments that occurs in vivo. Briefly, chromatin of living cells is fixed with formaldehyde to cross-link the interacting sites, digested with a restriction enzyme that cleaves the intervening DNA sequence, and ligated in situ to join the cross-linked chromatin segments, which are then analyzed and quantified by PCR (2, 3). This method successfully revealed looping organization of the *β-globin* locus and showed that its shared enhancer *LCR* acts by making physical contacts with the promoter of actively transcribed *β-globin* genes in an erythroid cell-specific manner (4–6). Chromosomal looping has been detected for several other genomic loci, including *interferon gamma* (*Ifng*) (7, 8), *immunoglobulin*

Shu-Zhen Wang (ed.), *Retinal Development: Methods and Protocols*, Methods in Molecular Biology, vol. 884, DOI 10.1007/978-1-61779-848-1_22, © Springer Science+Business Media, LLC 2012

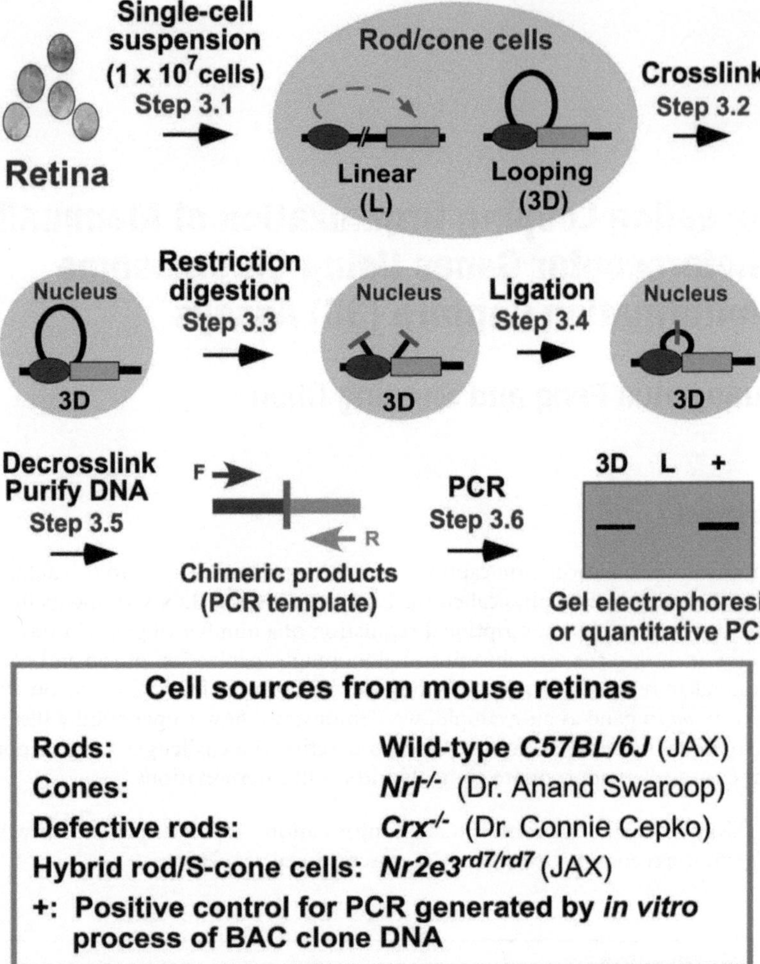

Fig. 1. Diagram of 3C assays using mammalian retinas, illustrating the key steps and genetic dissection of photoreceptor subtypes in mouse retina.

kappa (*Igkappa*) (9, 10), *insulin-like growth-factor 2* (*Igf2*) (11), *T-helper type 2* (*T(H)2*) *cytokine* (12), *α-globin* (13), and distal-less homeobox gene 5 and 6 (Dlx5–Dlx6) (14).

Most reported 3C assays were performed using relatively pure cell populations, such as various cell types derived from the hematopoietic system. 3C assays using multicellular fresh tissues, like the retina, pose many challenges. In order to appropriately perform 3C assays on multicellular tissues and identify 3C signals from a specific cell type of interest, one needs to collect $>1\times10^7$ cells with 70–80% enrichment of that cell type. This can be achieved by either fluorescence-activated cell sorting (FACS) (15) or immunomagnetic enrichment (16). Both methods require labeling of the desired cell type. Our laboratory is studying transcriptional regulation of the rod- and cone-specific genes in retinal photoreceptors using *opsin* loci as models. To develop 3C specifically for rod/cone photoreceptors, we need to achieve ≥70% rods or cones in retinal samples. Ordinarily, rod photoreceptors represent approximately 70% of all cells in normal mouse or human retinas. Thus, it is feasible to use whole retinas of normal mice in 3C assays to assess rod/cone gene chromosomal organization in rods without specific rod enrichment procedures. However, cones in mouse retinas represent only 3% of retinal cells. To achieve cone-rich 3C preparations, we have used whole retinas of *Nrl*$^{-/-}$ mice, which lack the rod-specific transcription factor NRL, therefore converting all the photoreceptors to cones. We have also used several other mutant strains of mice for 3C assays, including *Crx*$^{-/-}$ (17) to provide defective rods and *Nr2e3*$^{rd7/rd7}$ to provide immature rod/S-cone hybrid cells (18–20). For human retinas, we manually dissect the *Macula* and periphery subregions to achieve cone-rich and rod-rich populations, respectively.

We have performed 3C using mouse and human retinas following a protocol described by Hagege et al. (2) with modifications. Figure 1 shows a diagram of the main steps in 3C with retinal samples. In the following sections, we describe detailed protocols and important points for each of the main steps using murine *M-cone opsin* (*Mop*) locus as an example. We showed that *Mop* has a linear organization in rods but adopts a looping (3D) conformation in cones, corresponding to cone-specific *Mop* expression.

2. Materials

2.1. Equipment

Orbital shaker/rocker: e.g., VSN-5 Variable Speed Nutating Mixer (PRO Scientific).

Incubator shaker or shaking water bath: e.g., SWB25 shaking water bath (Thermo Scientific).

PCR machine: e.g., CFX96™ Real-Time PCR System (Bio-Rad).

Centrifuge and microcentrifuge: e.g., Eppendorf centrifuges for 1.5- and 15–50-ml tubes.

2.2. Mouse Strains

C57BL/6J wild-type (WT) (Stock no. 0664) and *Nr2e3*$^{rd7/rd7}$ mutant (Stock no. 004643) mice were purchased from the Jackson Laboratory. *Nrl*$^{-/-}$ and *Crx*$^{-/-}$ knockout mice were obtained from Dr. Anand Swaroop (NIH, NEI) and Dr. Constance Cepko (Harvard University), respectively.

2.3. Bacterial Artificial Chromosome DNA for Positive Controls

Bacterial artificial chromosome (BAC) clones containing genes of interest can be identified from the UCSC Web site (http://genome.ucsc.edu/cgi-bin/hgGateway) and purchased from

BACPAC Resources (http://bacpac.chori.org). BAC clone DNA was prepared using QIAGEN Large-Construct Kit, following the manufacturer's protocol. The BAC 3C templates were prepared in parallel with retinal samples, beginning at the step of restriction digestion in Subheading 3.3, step 1 (see Note 1).

2.4. Reagents and Solutions for Preparing Single-Cell Suspension from Retinas

10× Phosphate-buffered saline (PBS) without calcium and magnesium, pH 7.5.

1× PBS/10% (v/v) fetal bovine serum (FBS).

RPMI1640 (Invitrogen)/10% (v/v) FBS.

DMEM (Invitrogen)/10% (v/v) FBS.

Collagenase Type I.

TESCA buffer: 50 mM *N*-[Tris(hydroxymethyl)methyl]-2-aminoethanesulfonic acid (TES) (pH 7.4 at 37°C); 0.36 mM $CaCl_2$.

2.5. Cross-Link and Cell Lysis Solutions

1× PBS with 10% (v/v) FBS (1× PBS/10% FBS).

37% (v/v) Formaldehyde, diluted to 2% in 1× PBS/10% FBS.

1 M glycine.

Cell lysis buffer: 10 mM Tris–HCl (pH 7.5), 10 mM NaCl, 5 mM $MgCl_2$, 0.1 mM EGTA, and 1× protease inhibitor cocktail (Roche).

2.6. Restriction Enzymes and Buffers

Restriction enzyme(s) of choice: for selecting appropriate enzymes (see Note 2). We have used Bgl II (New England BioLabs) in this study (Fig. 2a) and Bpml (New England BioLabs) in another study (21).

Appropriate restriction enzyme buffer diluted to 1.2×.

10× Digestion buffer (dependent on specific enzyme(s) of choice): for BglII or Bpml, use NEB buffer 3: 1 M NaCl; 0.5 M Tris–HCl (pH 7.9); 0.1 M $MgCl_2$; and 10 mM DTT.

20% SDS (*w*/*v*).

20% Triton X-100 (v/v).

2.7. Enzyme and Buffers for In Situ Ligation

20% SDS (*w*/*v*).

20% Triton X-100 (v/v).

T4 DNA ligase (New England BioLabs).

10× Ligation buffer: 0.5 M Tris–HCl (pH 7.5), 0.1 M $MgCl_2$, 0.1 M DTT, and 10 mM ATP (diluted to 1.15× for Subheading 3.4, step 4).

2.8. De-cross-link and DNA Purification Reagents

Proteinase K, 10 mg/ml (Research Products International Corp).

5× Proteinase K buffer: 50 mM Tris–HCl (pH 8.0), 25 mM EDTA (pH 8.0), and 2.5% SDS.

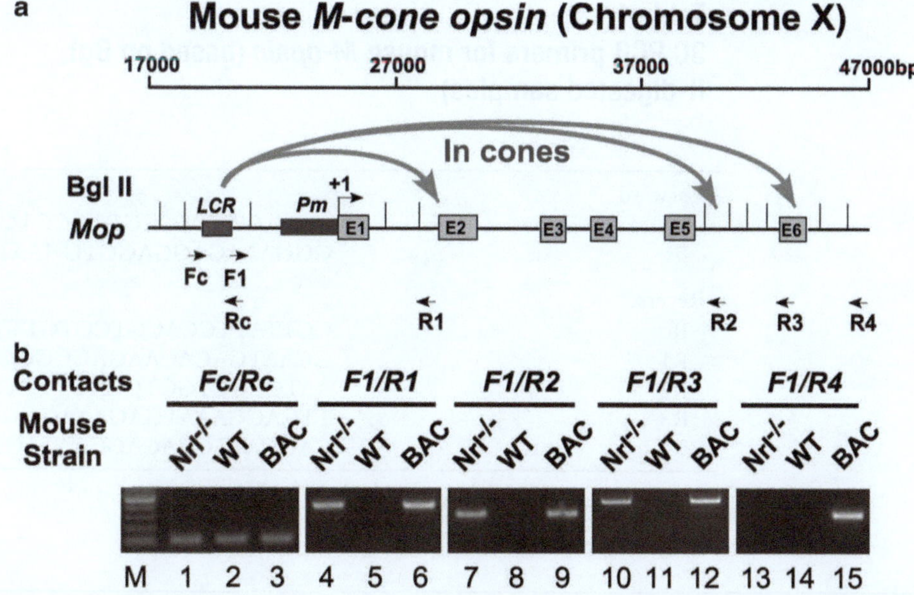

Fig. 2. Examples of 3C assays on mouse M-cone *opsin* (*Mop*) locus. (**a**) Diagram of mouse *Mop* gene structure showing the enhancer (*LCR*), promoter (*Pm*), and six exons (*E1–E6*). The Bgl II restriction sites are shown with *vertical lines* on *top* of the gene frame. 3C primers are shown in *small arrows* (see Table 1 for sequences). (**b**) 3C results presented as *gel images*, which demonstrate that *Mop* adopts intrachromosomal looping conformation in cones (*Nrl*$^{-/-}$) but has a linear configuration in rods (*WT*).

Ribonuclease A (RNase A), 1 mg/ml.

Phenol:chloroform:isoamyl alcohol 25:24:1, saturated with 10 mM Tris–HCl (pH 8.0) and 1 mM EDTA.

3 M sodium acetate (pH 5.2).

Ethanol (100%).

10 mM Tris–HCl (pH 7.5).

2.9. PCR Analysis Reagents

Primers: for appropriate design of 3C PCR primers (see Note 3). Table 1 lists the primers for analyzing mouse *M-opsin* conformation as shown in Fig. 2a.

Deoxynucleotide (dNTP) set, 100 mM (Sigma).

10× Taq polymerase buffer (Sigma).

JumpStart REDTaq DNA polymerase (Sigma).

SYBR Green JumpStart Taq ReadyMix for QPCR (Sigma).

Agarose (Bioline): 1% gel in 1× Tris–Acetate–EDTA (TAE) buffer (Sigma).

Table 1
3C PCR primers for mouse *M-opsin* (based on Bgl II-digested samples)

Forward	
Fc	CTCAGTTTTGCCCTGCCTG
F1	GGGAAAGAGGAGGTGGAATAAG
Reverse	
Rc	CCTTATTCCACCTCCTCTTTCC
R1	GCAATGACACAAGAGCGGC
R2	AATCCCCTGCTCCAGTTTTC
R3	TTCAGCAAATGAGAAAGGGC
R4	CAGCATTCCACAGCAATAGTCTG

3. Methods

3C design and protocol are described in detail by Hagege et al. (2). We have modified this protocol as described below for analyzing photoreceptor genes using mammalian retinas. The entire procedure takes about 2 weeks to complete (Fig. 1).

3.1. Preparation of Single-Cell Suspension (Time Required: 30 min to 1 h)

Single-cell suspensions are prepared from mouse or human retinas (Subheading 3.1.1) or cultured cells (Subheading 3.1.2).

3.1.1. Single-Cell Preparation from Mouse or Human Retinas

1×10^7 Cells are required for each 3C sample. One mouse retina contains about 10^6 cells. To achieve sufficient number of photoreceptor cells, we used 16 mouse retinas for each 3C sample. Human retina is >10-fold bigger than the mouse retina. We found that two human retinas manually dissected into *Macula* and periphery subregions are sufficient to achieve reliable 3C results.

1. For each 3C sample, quickly dissect 16 mouse retinas (or two human retinas) and transfer them into a 1.5-ml tube.
2. Wash retinas with 1× PBS, spin 400 × *g* for 1 min at room temperature, and remove all PBS.
3. Incubate in 250 μl of TESCA buffer supplemented with 0.5–2% (*w*/*v*) collagenase at 37°C with gentle shaking on an orbital shaker/rocker [e.g., Speed setting 4, or 30 rpm, on VSN-5 Variable Speed Nutating Mixer) for 15–18 min (see Note 4).

4. Stop the collagenase treatment by addition of 500 μl of DMEM/10% FBS. Resuspend the cells by gently pipetting up and down through a large orifice tip (Fisher, 02-707-134).
5. Centrifuge the cell suspension for 1 min at $400 \times g$ at room temperature.
6. Discard the supernatant and resuspend the pellet in 500 μl of 1× PBS/10% FBS.
7. Filter through the 40-μm cell strainer (BD Falcon, 352340) to make a single-cell suspension. Aliquot 1×10^7 cells in 500 μl of 1× PBS/10% FBS. The cells are ready for cross-link (Subheading 3.2.1).

3.1.2. Single-Cell Preparation from Suspended Cell Cultures (e.g., Y79 Retinoblastoma Cells)

1. Transfer cultured cell suspension to a 14-ml centrifuge tube. Centrifuge for 1 min at $400 \times g$ at room temperature.
2. Discard the supernatant and resuspend the pellet in 500 μl of RPMI1640/10% FBS. Spin $400 \times g$ for 1 min at room temperature.
3. Discard the supernatant and resuspend the pellet in 500 μl of 1× PBS/10% FBS.
4. Filter through the 40-μm cell strainer to make a single-cell suspension. Aliquot 1×10^7 cells in 500 μl of 1× PBS/10% FBS. The cells are ready for cross-link (Subheading 3.2.1).

3.2. Cross-Link and Cell Lysis (Time Required: 40–50 min)

3.2.1. Cross-Link

1. Dilute 37% formaldehyde to 2% in 1× PBS/10% FBS just before use.
2. Add 9.5 ml of 2% formaldehyde in 1× PBS/10% FBS per 1×10^7 cells and incubate in a 14-ml tube with shaking on the PRO orbital shaker (Setting 4, or 30 rpm) for 10 min at room temperature (see Note 5).
3. Transfer the reaction tubes to an ice bath, add 1.425 ml of ice-cold 1 M glycine, and chill for 5 min to stop the cross-link reaction.
4. Spin for 8 min at $250 \times g$ at 4°C and carefully remove all the supernatant.

3.2.2. Cell Lysis and Nuclei Collection

1. Resuspend the cell pellet in 5 ml of cold lysis buffer and incubate for 10 min on ice. Gently pipette up and down the mixture through a large orifice tip to prepare a homogeneous nuclei suspension.
2. Centrifuge at 4°C for 5 min at $400 \times g$ and remove the supernatant containing cell debris. The pellet now contains the intact nuclei. The pelleted nuclei can be frozen in liquid nitrogen or stored at –80°C for several months.

3.3. Restriction Enzyme Digestion in Intact Nuclei (Time Required: 16 h)

1. Resuspend the nuclei (and the desired BAC clone DNA) in 0.5 ml of 1.2× restriction enzyme buffer and transfer to a 17 mm×100 mm Round-Bottom tube (BD Falcon, 352059).
2. Place the tube at 37°C and add 7.5 μl of 20% (*w*/*v*) SDS (final: 0.3% SDS).
3. Incubate for 1 h at 37°C while shaking on the PRO orbital shaker (Setting 4, or 30 rpm).
4. Add 50 μl of 20% (v/v) Triton X-100 (final: 2% Triton X-100).
5. Incubate for 1 h at 37°C while shaking on the PRO orbital shaker (Setting 4, or 30 rpm).
6. Remove a 5-μl aliquot of the sample and label as an undigested genomic DNA control (UND). This sample may be stored at −20°C until it is needed to determine the digestion efficiency (Subheading 3.3, step 8; see Note 6).
7. Add 400 U of the selected restriction enzyme (e.g., Bgl II in this study) to the remaining sample and incubate overnight at 37°C while shaking on the PRO orbital shaker (Setting 4, or 30 rpm).
8. Take a 5-μl aliquot of the sample and label as a digested genomic DNA control (D). To process the remaining sample, proceed to Subheading 3.4, step 1. To determine the digestion efficiency, analyze the aliquots from Subheading 3.3, step 6 (undigested control), and 3.3, step 8 (digested sample), as described in Note 6. If desired, this efficiency analysis can be carried out in parallel with the following steps (Subheadings 3.4–3.6). However, it is highly recommended to complete the digestion efficiency assessment before moving to the next step.

3.4. Ligation and De-cross-link (Time Required: 18–20 h)

1. Add 40 μl of 20% (*w*/*v*) SDS (final: 1.6%) to the digested sample from Subheading 3.3, step 7.
2. Incubate for 30 min at 65°C in a water bath with shaking to stop restriction digestion.
3. Transfer the digested nuclei to a 50-ml tube.
4. Add 6.125 ml of 1.15× ligation buffer.
5. Add 375 μl of 20% (v/v) Triton X-100 (final: 1% Triton X-100).
6. Incubate for 1 h at 37°C while shaking gently.
7. Add 5 μl ligase (100 U total) and incubate for 4 h at 16°C followed by 30 min at room temperature.
8. Add 30 μl of 10 mg/ml Proteinase K (final: 300 μg).
9. Incubate at 65°C overnight to de-cross-link the sample.

3.5. DNA Purification (Time Required: 6–7 h)

1. Add 30 μl of 10 mg/ml RNase (final: 300 μg) to each sample from Subheading 3.4, step 9.
2. Incubate for 45 min at 37°C.
3. Add 7 ml of phenol:chloroform:isoamyl alcohol and mix vigorously.
4. Centrifuge for 15 min at 2,200 × *g* at room temperature.
5. Transfer the top phase into a new 50-ml tube and add 7 ml of distilled water and 1.5 ml of 2 M sodium acetate (pH 5.2). Mix, and then add 35 ml of ethanol (100%).
6. Mix and place at −80°C for 2 h.
7. Centrifuge at 4°C for 45 min at 2,200 × *g*.
8. Remove the supernatant and wash the pellet with 10 ml of 70% (v/v) ethanol.
9. Centrifuge at 4°C for 15 min at 2,200 × *g*.
10. Remove the supernatant and air-dry the pellet at room temperature.
11. Dissolve the DNA pellet in 150 μl of 10 mM Tris–HCl, pH 7.5. The 3C DNA template is ready for PCR analysis. The 3C template DNA may be kept at −20°C for several months.

3.6. Agarose Gel Analysis and Real-Time PCR Quantifications of Ligation Products (Time Required: 3–5 days)

3.6.1. Agarose Gel Analysis

1. Prepare serial dilutions, such as 10-, 20-, and 40-fold dilutions, for each 3C sample and BAC control.
2. Perform regular PCR for each dilution using a primer pair (e.g., *Fc/Rc* in Fig. 2) designed to generate a PCR product within the same restriction fragment of each candidate gene. This product serves as a linear input control for all 3C products. Run 1% agarose gel to determine an optimal dilution for further PCR analyses with other 3C primer pairs.
3. Take one best dilution of each 3C sample and BAC control to complete PCR analyses with the remaining 3C primer pairs and visualize the results on 1% agarose gel. An example of such analyses is shown in Fig. 2b.

3.6.2. Quantitative Real-Time PCR Analyses of Ligation Products Using SYBR Green Kits

1. Make serial dilutions in H_2O of the original 3C template (from Subheading 3.5, step 11) (e.g., 10-, 20-, and 40-fold dilution). For each candidate gene, generate a standard curve using serial dilutions of the template with the linear control primer pair *Fc/Rc*, and choose the best dilution for quantitative PCR (qPCR) analyses of the ligation products. We have used SYBR Green Jumpstart Taq Readymix qPCR Kit (Sigma, S4438) and CFX96™ Real-Time PCR System (Bio-Rad) to perform these assays. However, any other regular or fast SYBR green kits and real-time PCR machines should work as well. Quantifications are based on the cycle thresholds (Ct values) and calculated using the comparative Ct ($2^{-\Delta\Delta Ct}$) method (22).

2. Perform qPCR using other 3C primer pairs: Looping frequencies were calculated as ratios (×100) relative to the constant DNA control using the *Fc/Rc* primer pair (% Fc/Rc). Error bars represent standard error of mean from several independent experiments ($n \geq 3$).

4. Notes

1. PCR control 3C template: Digested/ligated BAC DNA containing the gene of interest is used as a positive control for all primer pairs used in 3C PCR reactions. Six to ten micrograms of each BAC clone DNA is cut with 100–400 U of the same restriction enzyme and ligated with 100 U T4 DNA ligase under the same conditions as the experimental samples (from Subheading 3.3, step 1, and thereafter). The optimal dilution of the BAC control 3C template for PCR needs to be determined experimentally as described in Subheading 3.6.1 and may be different from that of the retina sample DNA.
2. Choice of restriction enzyme(s): A restriction enzyme selected for 3C assays should cut the locus in appropriate frequency to allow for separate analyses of the relevant regulatory elements (e.g., enhancer, promoter, exon, 5′ or 3′ regions, etc.). For long-range looping contacts (>20 kbp), 6-bp cutters such as BglII (Fig. 2), BamHI, or EcoRI are good choices. For short-range contacts (5–20 kbp), more frequent cutters should be used, such as *Bpm*1 or some 4-bp cutters.
3. PCR primer design: 3C PCR primers are typically 20–22 oligomers with the following parameters: Tm: 52–60°C with 2°C maximum difference between primers, percent G+C: 45–58, and product size: 150–500 bp. Positions: Forward primers should reside near the 3′ end of candidate restriction fragments, while the reverse primer should reside near the 5′ end of the target restriction fragment (Fig. 2a). The primer should be close to but not crossing the desired restriction enzyme-cutting site. The primers should not form self 3′-dimer, hairpin, or self-duplex, and should not bind anywhere else in the genome besides the target gene sequence. Primer pairs should have high amplification efficiency (>90% in qPCR analysis) and generate a single PCR product in the BAC control lane of agarose gel analysis or a single melting curve in qPCR analysis. We have used the MacVector software (MacVector, Inc.) to design all our primers. However, several online primer design programs, such as Primer3 (http://frodo.wi.mit.edu/primer3/), are also suitable for this purpose.

Table 2
Collagenase digestion parameters for single-cell preparation of mouse and human retinas

Species	Age (day)	Concentration (w/v %)	Time (min)
Mouse (16 retinas)	P0	0.50	15
	P3	0.75	15
	P5	1.00	16
	P7	1.25	16
	P10	1.50	18
	P14	2.00	18
Human (2 retinas)	Adult	2.00	18

4. Single-cell preparation: The optimal concentration and reaction time for collagenase digestion depend on the age and number of the retinas. These should be determined experimentally. Table 2 lists suggested ranges for different samples. The digestion effectiveness (>80% single cells) should be verified using light microscopy. Under- or over-digestion could affect 3C results.
5. Cross-link: It is important to use formaldehyde freshly diluted to 2% in 1× PBS/10% FBS. For example, dilute 541 μl of stock formaldehyde (37%) to a final volume of 10 ml.
6. Determine the digestion efficiency: Digestion efficiencies should be carefully assessed for each selected restriction enzyme as follows (time required: 8 h).
   - Add 500 μl of 1× Proteinase K buffer and 1 μl of 20 mg/ml Proteinase K (final: 20 μg) to the aliquots saved in Subheading 3.3, step 6 (UND), and 3.3, step 8 (D).
   - Incubate for 60 min at 65°C.
   - Equilibrate for a few minutes at 37°C, then add 1 μl of 1 mg/ml RNase A (final: 1 μg), and incubate for 2 h at 37°C.
   - Add 500 μl of phenol:chloroform:isoamyl alcohol, and mix vigorously.
   - Centrifuge at room temperature for 5 min at 16,000 ×*g*.
   - Transfer the top phase into a new tube, add 50 μl of 2 M sodium acetate, pH 5.2, mix, and then add 1.5 ml of ethanol to precipitate DNA.
   - Mix well and place at −80°C until frozen (about 60 min).
   - Centrifuge at 4°C for 20 min at 16,000 ×*g*.

- Remove the supernatant and add 500 μl of 70% ethanol.
- Centrifuge for 10 min at 16,000 × *g* at room temperature.
- Remove the supernatant and air-dry the pellet at room temperature.
- Resuspend the DNA in 60 μl of water.
- Analyzing digestion efficiencies:
  - Run 0.8% agarose gel with 10 μl per sample. Well-digested samples should lose the high-molecular-weight bands and appear as a smear on agarose gel (see Fig. 3 for an example with Bgl II digestion).
  - One could also quantify the digestion efficiency using the method described by Hagege et al. (2). The efficiency should be above 60–70%, ideally >80%. Samples with lower digestion efficiencies should be digested longer with more units of the enzyme.

In summary, 3C is a challenging multistep technique. Failure in any of the key steps described in Notes 1–6 above can lead to negative results. Thus, a negative result for a particular PCR reaction is not meaningful unless it is accompanied by positive results using the same 3C preparation with a different primer pair to verify the acceptability of the sample. Positive results also need careful negative controls for specificity. In our experience, the best controls for positive results are the following two: (1) analyzing the same candidate gene using 3C preparations from different cell types. In the example shown in Fig. 2, we analyzed *M-opsin* conformation using both rod-rich and cone-rich 3C samples. (2) Analyzing multiple candidate genes with both similar and opposite

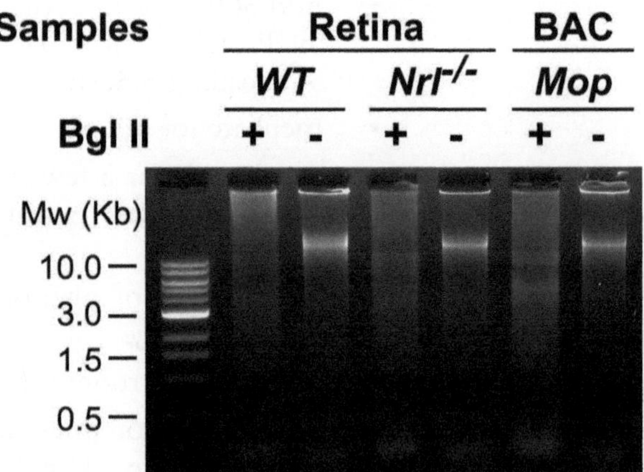

Fig. 3. Example of examining restriction digestion efficiency described in Note 6. Bgl II-digested (+) and -undigested (–) samples are shown for the retina of the indicated strains and a *Mop*-containing BAC control DNA. Note the disappearance of the high-molecular-weight band in the digested lanes, demonstrating good digestion efficiency.

cell-type expression patterns. For example, the same *WT* (rod rich) and *Nrl*$^{-/-}$ (cone rich) 3C preparations described in Fig. 2 should also be analyzed for a rod-specific gene, such as *rhodopsin*. If *rhodospin* shows looping conformation in *WT* but not in *Nrl*$^{-/-}$ samples, it will provide a good confirmation for the Fig. 2 results showing that *M-opsin* adopts looping organization in cones but not in rods. Using *WT* and *Nrl*$^{-/-}$ retinas, we have performed these types of assays for three rod- and cone-opsin genes as well as two other rod and cone genes and demonstrated that, indeed, rod and cone genes adopt different conformations in rods vs. cones (21).

Acknowledgment

We thank Dr. Anand Swaroop and Dr. Connie Cepko for providing *Nrl*$^{-/-}$ and *Crx*$^{-/-}$ mice, Hui Wang for technical assistance, and Anne Hennig for critical reading of the manuscript. This work was supported by NIH EY012543 (to SC), NIH EY02687 (to WU-DOVS), Lew Wasserman Merit Award (to SC), and unrestricted fund from Research to Prevent Blindness (to WU-DOVS).

References

1. Dekker J, Rippe K, Dekker M, Kleckner N (2002) Capturing chromosome conformation. Science 295:1306–1311
2. Hagege H et al (2007) Quantitative analysis of chromosome conformation capture assays (3C-qPCR). Nat Protoc 2:1722–1733
3. Vassetzky Y et al (2009) Chromosome conformation capture (from 3C to 5C) and its ChIP-based modification. Methods Mol Biol 567:171–188
4. Tolhuis B, Palstra RJ, Splinter E, Grosveld F, de Laat W (2002) Looping and interaction between hypersensitive sites in the active beta-globin locus. Mol Cell 10:1453–1465
5. Carter D, Chakalova L, Osborne CS, Dai YF, Fraser P (2002) Long-range chromatin regulatory interactions in vivo. Nat Genet 32:623–626
6. Palstra RJ et al (2003) The beta-globin nuclear compartment in development and erythroid differentiation. Nat Genet 35:190–194
7. Eivazova ER, Aune TM (2004) Dynamic alterations in the conformation of the Ifng gene region during T helper cell differentiation. Proc Natl Acad Sci USA 101:251–256
8. Eivazova ER, Vassetzky YS, Aune TM (2007) Selective matrix attachment regions in T helper cell subsets support loop conformation in the Ifng gene. Genes Immun 8:35–43
9. Liu Z, Garrard WT (2005) Long-range interactions between three transcriptional enhancers, active Vkappa gene promoters, and a 3 boundary sequence spanning 46 kilobases. Mol Cell Biol 25:3220–3231
10. Liu Z, Ma Z, Terada LS, Garrard WT (2009) Divergent roles of RelA and c-Rel in establishing chromosomal loops upon activation of the Igkappa gene. J Immunol 183:3819–3830
11. Murrell A, Heeson S, Reik W (2004) Interaction between differentially methylated regions partitions the imprinted genes Igf2 and H19 into parent-specific chromatin loops. Nat Genet 36:889–893
12. Spilianakis CG, Flavell RA (2004) Long-range intrachromosomal interactions in the T helper type 2 cytokine locus. Nat Immunol 5:1017–1027
13. Zhou GL et al (2006) Active chromatin hub of the mouse alpha-globin locus forms in a transcription factory of clustered housekeeping genes. Mol Cell Biol 26:5096–5105
14. Horike S, Cai S, Miyano M, Cheng JF, Kohwi-Shigematsu T (2005) Loss of silent-chromatin looping and impaired imprinting of DLX5 in Rett syndrome. Nat Genet 37:31–40

15. Akimoto M et al (2006) Targeting of GFP to newborn rods by Nrl promoter and temporal expression profiling of flow-sorted photoreceptors. Proc Natl Acad Sci USA 103: 3890–3895
16. van Beijnum JR, Rousch M, Castermans K, van der Linden E, Griffioen AW (2008) Isolation of endothelial cells from fresh tissues. Nat Protoc 3:1085–1091
17. Furukawa T, Morrow EM, Li T, Davis FC, Cepko CL (1999) Retinopathy and attenuated circadian entrainment in Crx-deficient mice. Nat Genet 23:466–470
18. Chen J, Rattner A, Nathans J (2005) The rod photoreceptor-specific nuclear receptor Nr2e3 represses transcription of multiple cone-specific genes. J Neurosci 25:118–129
19. Corbo JC, Cepko CL (2005) A hybrid photoreceptor expressing both rod and cone genes in a mouse model of enhanced S-cone syndrome. PLoS Genet 1:e11
20. Peng GH, Ahmad O, Ahmad F, Liu J, Chen S (2005) The photoreceptor-specific nuclear receptor Nr2e3 interacts with Crx and exerts opposing effects on the transcription of rod versus cone genes. Hum Mol Genet 14:747–764
21. Peng GH, Chen S (2011) Active opsin loci adopt intrachromosomal loops that depend on the photoreceptor transcription factor network. Proc Natl Acad Sci USA. 108: 17821–17826
22. Schmittgen TD, Livak KJ (2008) Analyzing real-time PCR data by the comparative C(T) method. Nat Protoc 3:1101–1108

Chapter 23

Retinal Transcriptome Profiling by Directional Next-Generation Sequencing Using 100 ng of Total RNA

Matthew J. Brooks, Harsha Karur Rajasimha, and Anand Swaroop

Abstract

RNA expression profiles produced by next-generation sequencing (NGS) technology (RNA-seq) allow comprehensive investigation of transcribed sequences within a cell or tissue. RNA-seq is rapidly becoming more cost-effective for transcriptome profiling. However, its usage will expand dramatically if one starts with low amount of RNA and obtains transcript directionality during the analysis. Here, we describe a detailed protocol for the creation of a directional RNA-seq library from 100 ng of starting total RNA.

Key words: RNA Sequencing, Next-generation Sequencing, Massively Parallel Sequencing, Directional RNA-seq, Low input RNA

1. Introduction

Next-generation sequencing (NGS) is transforming the study of transcriptome profiling as new methodologies are making it more comprehensive, quantitative, accessible, and affordable (1). With the expanding availability and throughput of NGS systems, the cost is becoming similar to current methods of transcriptome analysis, including microarrays and serial analysis of gene expression (SAGE). NGS of RNA molecules (RNA-seq) provides a comprehensive and unbiased analysis of the trancsriptome as it does not rely on probe hybridization or suffers from design limitations (2). A simultaneous detection of transcripts at low levels (3) and identification of novel transcripts and splice variants (4, 5) is now possible with RNA-seq. A recent investigation has demonstrated its value for the analysis of retinal transcriptome profiling in adult wild-type and Nrl$^{-/-}$ mice (6).

RNA-seq is currently limited by the amount of starting RNA and the ability of the library construction methods to retain mRNA strand polarity. As the use of RNA-seq becomes more widespread,

Shu-Zhen Wang (ed.), *Retinal Development: Methods and Protocols*, Methods in Molecular Biology, vol. 884, DOI 10.1007/978-1-61779-848-1_23, © Springer Science+Business Media, LLC 2012

the types and amounts of tissues and/or cells being analyzed will become more diverse. This is especially true for eye tissues as one would like to obtain expression profiles from single cell types or from a clinically affected part. In addition to the need for protocols using low starting material, retaining strand polarity during RNA-seq will permit proper quantification of expression, particularly for overlapping genes. Several protocols have attempted to retain the mRNA polarity to varying success (7). One of the protocols utilizes the labeling of the second strand of cDNA with uracil, which allows the degradation of this strand prior to sequencing (8). We have modified this protocol and used Illumina TruSeq RNA Sample Prepartion kit to obtain directional RNA sequence using 100 ng of total RNA. We have successfully used this protocol to generate libraries from wild-type mouse retina at postnatal day 21 comparing input amounts at 100 ng and 1 μg of total RNA. Results from this experiment can be observed in Table 1 and Fig. 1.

2. Materials

2.1. Reagents

Agencourt AMPure XP Kit (Beckman Coulter Genomics).

RNA Nano 6000 Kit (Agilent).

DNA 1000 Kit (Agilent).

UltraPure™ DNase/RNase-Free Distilled Water (Invitrogen).

100% Ethanol (Sigma-Aldrich).

SuperScript II (Invitrogen).

dUTP (Epicentre).

Uracil-DNA Glycosylase (UDG) (New England Biolabs).

TruSeq RNA Sample Prep Kit (Illumina).

TruSeq PE Cluster Kit v5–CS–GA (Illumina).

TruSeq SBS Kit v5-GA (Illumina).

2.2. Equipment

Agilent 2100 Bioanalyzer (Agilent).

Thermal cycler (Eppendorf Mastercycler or equivalent).

Microcentrifuge (Eppendorf or equivalent).

Magnetic Stand-96 (Invitrogen).

P10, P20, P200, and P1000 pipettes (Pipetman or equivalent).

Real-Time PCR System (The 7900HT Fast Real-Time PCR System or equivalent).

Ice bucket.

Powder-free gloves.

Table 1
FPKM values for selected genes in 100 ng and 1 μg adult retina libraries

Transcript ID	Gene symbol	Gene description	100 ng	1 μg
NM_133205	Arr3	Arrestin 3, retinal	59.20	63.95
NM_008553	Ascl1	Achaete–scute complex homolog 1	0.37	0.36
NM_016864	Atoh7	Atonal homolog 7	0.15	0.00
NM_009788	Calb1	Calbindin 1	19.30	18.21
NM_007770	Crx	Cone-rod homeobox containing gene	133.39	132.12
NM_001113330	Crx	Cone-rod homeobox containing gene	168.72	163.69
NM_001109752	Dlg4	Discs, large homolog 4	43.05	43.02
NM_007864	Dlg4	Discs, large homolog 4	40.62	40.73
NM_010277	Gfap	Glial fibrillary acidic protein	5.44	5.93
NM_001131020	Gfap	Glial fibrillary acidic protein	0.27	0.47
NM_008140	Gnat1	Guanine nucleotide binding protein, alpha transducing 1	2662.64	2830.72
NM_008141	Gnat2	Guanine nucleotide binding protein, alpha transducing 2	59.89	59.33
NM_001160017	Gnb1	Guanine nucleotide binding protein (G protein), beta 1	433.06	427.54
NM_001160016	Gnb1	Guanine nucleotide binding protein (G protein), beta 1	311.57	307.60
NM_008142	Gnb1	Guanine nucleotide binding protein (G protein), beta 1	311.22	307.25
NM_010314	Gngt1	Guanine nucleotide binding protein (G protein), gamma transducing activity polypeptide 1	376.79	396.45
NM_173372	Grm6	Glutamate receptor, metabotropic 6	23.99	23.76
NM_010710	Lhx2	LIM homeobox protein 2	7.74	7.40
NM_010710	Lhx2	LIM homeobox protein 2	7.74	7.40
NM_010712	Lhx4	LIM homeobox protein 4	5.49	6.33
NM_001170537	Mef2c	Myocyte enhancer factor 2C	7.75	7.37
NM_025282	Mef2c	Myocyte enhancer factor 2C	5.25	4.79
NM_001178049	Mitf	Microphthalmia-associated transcription factor	0.21	0.37
NM_008601	Mitf	Microphthalmia-associated transcription factor	0.49	0.68

(continued)

Table 1 (continued)

Transcript ID	Gene symbol	Gene description	100 ng	1 µg
NM_001113198	Mitf	Microphthalmia-associated transcription factor	0.55	0.23
NM_010894	Neurod1	Neurogenic differentiation 1	166.95	155.00
NM_007501	Neurod4	Neurogenic differentiation 4	66.22	62.74
NM_013708	Nr2e3	Nuclear receptor subfamily 2, group E, member 3	167.71	175.67
NM_008736	Nrl	Neural retina leucine zipper gene	17.19	18.82
NM_001136074	Nrl	Neural retina leucine zipper gene	459.51	464.32
NM_008106	Opn1mw	Opsin 1 (cone pigments), medium-wave-sensitive (color blindness, deutan)	85.12	80.88
NM_007538	Opn1sw	Opsin 1 (cone pigments), short-wave-sensitive (color blindness, tritan)	136.43	135.00
NM_001128599	Opn4	Opsin 4 (melanopsin)	0.48	0.76
NM_013887	Opn4	Opsin 4 (melanopsin)	0.41	0.15
NM_144841	Otx2	Orthodenticle homolog 2 (Drosophila)	71.23	69.07
NM_001159925	Pax4	Paired box gene 4	0.00	0.00
NM_001159926	Pax4	Paired box gene 4	0.00	0.00
NM_011038	Pax4	Paired box gene 4	0.00	0.00
NM_013627	Pax6	Paired box gene 6	58.83	62.55
NM_024458	Pdc	Phosducin	606.10	636.29
NM_001159730	Pdc	Phosducin	502.98	526.22
NM_146086	Pde6a	Phosphodiesterase 6A, cGMP-specific, rod, alpha	336.89	357.72
NM_008806	Pde6b	Phosphodiesterase 6B, cGMP, rod receptor, beta polypeptide	261.59	274.74
NM_001170959	Pde6c	Phosphodiesterase 6C, cGMP-specific, cone, alpha prime	12.06	12.39
NM_033614	Pde6c	Phosphodiesterase 6C, cGMP-specific, cone, alpha prime	8.94	9.60
NM_012065	Pde6g	Phosphodiesterase 6G, cGMP-specific, rod, gamma	941.28	951.44
NM_023898	Pde6h	Phosphodiesterase 6H, cGMP-specific, cone, gamma	70.97	71.98

(continued)

Table 1 (continued)

Transcript ID	Gene symbol	Gene description	100 ng	1 μg
NM_011143	Pou4f1	POU domain, class 4, transcription factor 1	5.62	5.31
NM_138944	Pou4f2	POU domain, class 4, transcription factor 2	2.91	2.57
NM_138945	Pou4f3	POU domain, class 4, transcription factor 3	1.14	0.98
NM_011101	Prkca	Protein kinase C, alpha	43.88	41.90
NM_013833	Rax	Retina and anterior neural fold homeobox	22.32	18.43
NM_145383	Rho	Rhodopsin	5517.13	5671.94
NM_146095	Rorb	RAR-related orphan receptor beta	23.74	22.42
NM_001043354	Rorb	RAR-related orphan receptor beta	5.77	6.29
NM_009118	Sag	Retinal S-antigen	529.33	549.90
NM_011381	Six3	Sine oculis-related homeobox 3 homolog (Drosophila)	16.70	15.21
NM_011384	Six6	sine oculis-related homeobox 6 homolog (Drosophila)	13.25	12.69
NM_148938	Slc1a3	Solute carrier family 1 (glial high affinity glutamate transporter), member 3	34.10	34.17
NM_001113417	Thrb	Thyroid hormone receptor beta	9.50	9.46
NM_009380	Thrb	Thyroid hormone receptor beta	3.64	3.73
NM_007701	Vsx2	Visual system homeobox 2	49.81	48.31

FPKM values reflect the abundance of specific transcripts. The rod phototransduction genes—rhodopsin (*Rho*), rod transducin (*Gnat1*), and rod phoshodiesterase gamma (*Pde6g*)—are the top three highly expressed genes in the mature mouse retina. As predicted, the genes expressed in early stages of retinal or neuronal development (e.g., *Ascl1*, *Pax4*, and *Atoh7*) show extremely low FPKM values in the adult retina. *Pax6*, *Rax* (or *Rx*) and *Otx2* demonstrate reasonable expression, whereas *Nrl* and *Crx* are highly expressed

PCR plate.

PCR plate seals.

1.7-ml Microcentrifuge tube.

Sterile, nuclease-free aerosol barrier pipette tips.

Timer.

Vortex mixer.

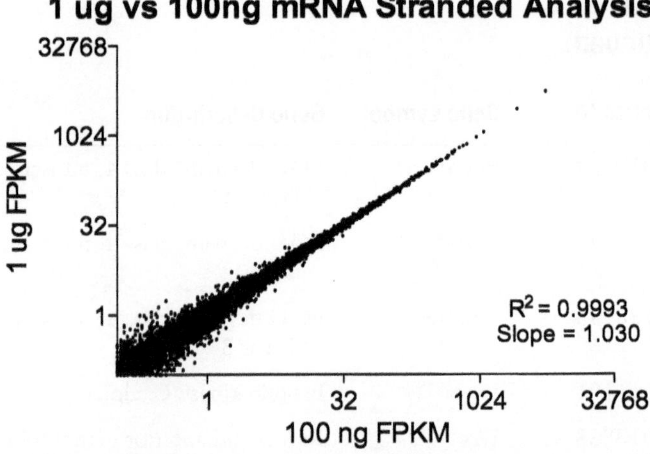

Fig. 1. FPKM correlation between libraries made from 100 ng and 1 μg total RNA. Each library was constructed from the same RNA sample from a wild-type postnatal day 21 mouse retina and sequenced on an individual lane of the GAIIx. Correlation of 18,572 coding transcripts between both libraries shows extremely consistent FPKM values between both samples ($R^2 = 0.9993$ and slope = 1.030).

3. Methods

To avoid cross-contamination between samples, perform all dilutions and reactions in a dedicated clean area or PCR hood with UV sterilization and positive airflow. The use of filter tips is highly recommended. The whole procedure takes about 2 days (Fig. 2). However, the protocol can be stopped after any AMPure XP beads purification step (unless otherwise specified), and the samples can be stored at –20°C until required.

3.1. mRNA Purification and Fragmentation

1. Allow RNA Purification Beads to reach room temperature.
2. Thaw bead binding buffer, bead washing buffer, elution buffer and elute, prime, fragmentation mix at room temperature.
3. Dilute 100 ng of high quality total RNA (see Note 1) in 50 μl of UltraPure™ water in a PCR plate.
4. Vigorously vortex RNA Purification Beads to allow complete resuspension.
5. Add 50 μl of RNA Purification Beads to the diluted RNA. This allows binding of the poly-A RNA to Oligo-dT beads. Pipette entire volume up and down six times to mix thoroughly.
6. Seal plate and incubate in a preprogrammed thermocycler at 65°C for 5 min followed by a 4°C hold.
7. When PCR plate has reached 4°C, remove it and incubate at room temperature for 5 min.

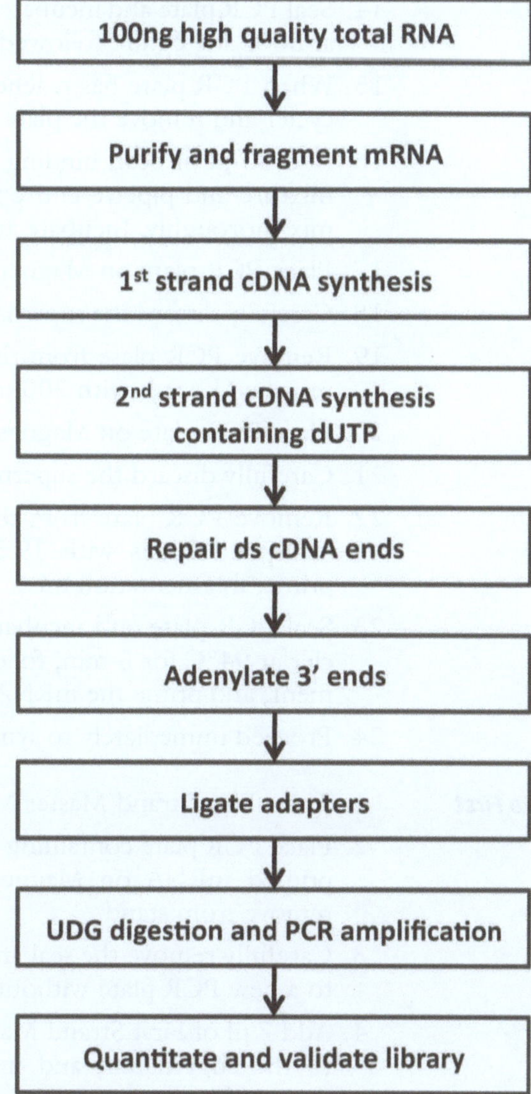

Fig. 2. Directional RNA-seq library preparation workflow.

8. Place PCR plate on Magnetic Stand-96 for 5 min to separate the poly-A RNA bound beads from the solution.
9. Carefully discard the supernatant. Do not disturb the beads.
10. Remove PCR plate from magnetic stand and thoroughly resuspend beads with 200 μl of bead washing buffer at room temperature.
11. Place PCR plate on Magnetic Stand-96 for 5 min.
12. Carefully discard the supernatant, making sure that the beads are not disturbed.
13. Remove PCR plate from the magnetic stand and thoroughly resuspend beads with 50 μl of room temperature elution buffer.

14. Seal PCR plate and incubate in a preprogrammed thermocycler at 80°C for 2 min, followed by a 25°C hold.
15. When PCR plate has reached 25°C, take it out from thermocycler and remove the plate seal.
16. Add 50 μl of bead binding buffer to the eluted mRNA/bead mixture and pipette entire volume up and down six times to mix thoroughly. Incubate at room temperature for 5 min.
17. Place PCR plate on Magnetic Stand-96 for 5 min.
18. Carefully discard the supernatant without disturbing the beads.
19. Remove PCR plate from the magnetic stand and thoroughly resuspend beads with 200 μl of bead washing buffer.
20. Place PCR plate on Magnetic Stand-96 for 5 min.
21. Carefully discard the supernatant. Do not disturb the beads.
22. Remove PCR plate from the magnetic stand and thoroughly resuspend beads with 19.5 μl of room temperature elute, prime, fragmentation mix.
23. Seal PCR plate and incubate in a preprogrammed thermocycler at 94°C for 8 min, followed by a 4°C hold to elute, fragment, and prime the mRNA.
24. Proceed immediately to synthesize first strand cDNA.

3.2. Synthesize First Strand cDNA

1. Thaw First Strand Master Mix at room temperature.
2. Place PCR plate containing beads and eluted, fragmented, and primed mRNA on Magnetic Stand-96 for 5 min. Do not remove from stand.
3. Carefully remove the seal and transfer 17 μl of the supernatant to a new PCR plate without disturbing the beads.
4. Add 7 μl of First Strand Master Mix and 1 μl of SuperScript II to the supernatant and mix the entire volume gently, but thoroughly, six times.
5. Seal PCR plate and run the program listed in Table 2.
6. When the thermocycler has reached 4°C, remove PCR plate, and proceed immediately to synthesize second strand cDNA.

Table 2
Thermocycler settings for 1st Strand cDNA synthesis

Step	Temperature (°C)	Time
1	95	10 min
2	42	50 min
3	70	15 min
4	4	Hold

3.3. Synthesize Second Strand cDNA Containing UTP

1. Thaw Second Strand Master Mix and Resuspension Buffer at room temperature.
2. Remove seal and add 25 μl of Second Strand Master Mix and 1 μl of dUTP to the PCR plate containing the first strand cDNA. Mix the entire volume up and down six times.
3. Seal PCR plate and incubate in a preprogrammed thermocycler (no heated lid) at 16°C for 60 min.
4. Remove PCR plate and allow it to reach room temperature.

3.4. AMPure XP Bead Purification

1. Allow AMPure XP Beads to reach room temperature.
2. Vigorously vortex AMPure XP beads until well dispersed.
3. Add 90 μl of well-mixed AMPure XP beads to 50 μl of ds cDNA and pipette entire volume up and down ten times.
4. Incubate PCR plate at room temperature for 15 min.
5. Place PCR plate on Magnetic Stand-96 for 5 min.
6. Remove supernatant, without disturbing the beads.
7. With PCR plate on the Magnetic Stand-96, add 200 μl of freshly made 80% EtOH without disturbing the beads.
8. Incubate for 30 s and completely remove supernatant.
9. Repeat steps 7 and 8 for a total of two 80% EtOH washes.
10. Let PCR plate stand for 15 min to dry the wells and remove from Magnetic Stand-96.
11. Completely resuspend beads with 52.5 μl of Resuspension Buffer and incubate at room temperature for 2 min.
12. Place the PCR plate on Magnetic Stand-96 for 5 min.
13. Transfer 50 μl of the supernatant to a new PCR plate.
14. If not planning to proceed right away, seal the PCR plate and store at −20°C for up to 7 days.

3.5. End Repair

1. Thaw End Repair Mix and Resuspension Buffer at room temperature.
2. Add 10 μl of Resuspension Buffer and 40 μl of End Repair Mix to the ds cDNA and pipette the entire volume up and down ten times to mix thoroughly.
3. Seal PCR plate and incubate at 30°C for 30 min.
4. Purify the end repaired ds cDNA using 160 μl of AMPure XP beads, following the instruction from Subheading 3.4.
5. Elute in 17.5 μl of Resuspension buffer, bind beads to the Magnetic Stand-96, and transfer 15 μl of the supernatant to a new PCR plate.
6. If not planning to proceed right away, seal the PCR plate and store at −20°C for up to 7 days.

Table 3
Adapter ligation reaction set-up

Reagent	Volume (μl)
A-tailed cDNA sample	30
DNA ligase mix	2.5
Resuspension buffer	2.5
RNA Adapter Index (AR001–AR012)	2.5

3.6. A-Tailing

1. Thaw A-Tailing Mix and Resuspension Buffer at room temperature.
2. Add 2.5 μl of Resuspension Buffer and 12.5 μl of A-Tailing Mix to the end repaired ds cDNA and pipette the entire volume up and down ten times to mix thoroughly.
3. Seal PCR plate and incubate at 37°C for 30 min.
4. Proceed immediately to Ligate Adapters.

3.7. Ligate Adapters

1. Thaw RNA Adapter Index tubes (AR001–AR012, see Note 2), Stop Ligase Mix, and Resuspension Buffer at room temperature.
2. Remove seal from PCR plate, add reagents in the order as described in Table 3, and mix entire volume up and down ten times to mix thoroughly.
3. Seal PCR plate and incubate at 30°C for 10 min.
4. Remove plate seal, add 5 μl of Ligation Stop Mix, and pipette entire volume up and down ten times to mix thoroughly.
5. Purify the ligated cDNA using 42 μl of AMPure XP beads, following the instruction from Subheading 3.4.
6. Elute in 52.5 μl of Resuspension buffer, bind beads to the Magnetic Stand-96, and transfer 50 μl of the supernatant to a new PCR plate.
7. Again purify the ligated cDNA using 50 μl of AMPure XP beads, following the instructions from Subheading 3.4.
8. Elute in 22.5 μl of Resuspension buffer, bind beads to the Magnetic Stand-96, and transfer 20 μl of the supernatant to a new PCR plate.
9. If not planning to proceed immediately, seal the PCR plate and store at −20°C for up to 7 days.

3.8. Enrich DNA Fragments

1. Thaw PCR Master Mix, PCR Primer Coctail, and Resuspension Buffer at room temperature.
2. Remove seal from PCR plate, add reagents in the order as described in Table 4, and mix entire volume up and down ten times to mix thoroughly.

Table 4
PCR and UDG reagent set-up

Reagent	Volume (μl)
Ligated cDNA sample	20
PCR Primer Cocktail	5
UDG	1
PCR Master Mix	25

Table 5
Thermocycler settings for UDG treatment and PCR enrichment of cDNA library

Step	Temperature (°C)	Time
1	37	15 min
2	98	30 s
3	98	10 s
4	65	30 s
5	72	30 s
7		Repeat steps 3 through 5 for a total of 15 times
8	72	5 min
9	4	Hold

3. Seal PCR plate and run the program listed in Table 5.
4. Purify the Enriched cDNA using 50 μl of AMPure XP beads, following the instructions from Subheading 3.4.
5. Elute in 32.5 μl of Resuspension buffer, bind beads to the Magnetic Stand-96, and transfer 30 μl of the supernatant to a new 1.7-ml tube for storage at −20°C.

3.9. Library Assessment and Quantitation

1. Assess quality and quantity of cDNA libraries with the Agilent 2100 Bioanalyzer DNA 1000 assay.
2. Determine the concentration of the sample by integration under the peak.
3. The electropherogram should show a peak in the size range of approximately 250–400 nucleotides (Fig. 3).
4. Do additional quantification by real-time PCR (see Note 3).

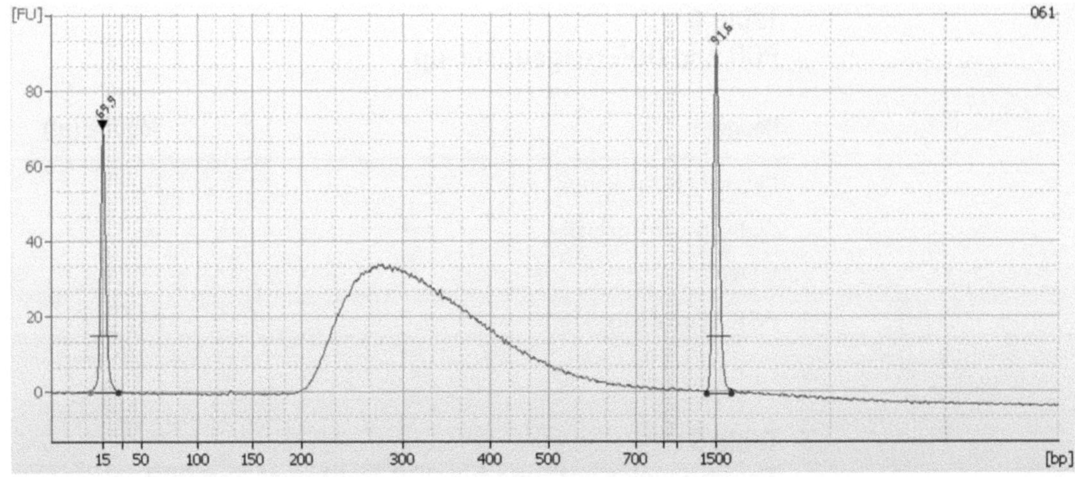

Fig. 3. Assessing quality and size distribution of ds cDNA library prior to sequencing with Agilent 2100 Bioanalyzer using a DNA 1000 chip. The ds cDNA library should range in size from 200 to 700 bp long with a peak at approximately 300 bp.

3.10. Cluster Generation

Cluster generation protocol turns captured libraries into clonal clusters of ~1,000 copies each, ready for sequencing on the Genome Analyzer. During this step, samples are denatured and hybridized to a flowcell. Captured DNA is used as template for second strand synthesis and is amplified into a clonal cluster. Subsequently, clusters are linearized, active sites are blocked, and sequencing primer provides a site for sequencing by synthesis.

1. Dilute and denature the libraries for a 10 pM load onto the flow cell (see Note 4).
2. Thaw and prepare the reagents following the instruction of Illumina Cluster Generation Kit.
3. Open and run appropriate recipe on the cluster station.
4. Follow recipe prompts to load flow cell.
5. Follow recipe prompts to load reagents.
6. Complete the cluster generation steps: hybridization, amplification, linearization, blocking, and primer hybridization.
7. Take the flowcell for sequencing (see Note 5).

3.11. Sequencing

1. Perform a prerun wash step on sequencer.
2. Thaw and prepare sequencing reagent following the instruction from Sequencing Kit Sequencing Kit.
3. Load sequencing reagent.
4. Prime positions on the genome analyzer.
5. Clean and install prism and flow cell.
6. Check for proper reagent delivery and apply oil.
7. Perform read 1 first-base incorporation and auto-calibration.

8. Check quality metrics.
9. Continue the run for desired number of cycles.

3.12. Data Analysis

Analysis of directional RNA sequencing data involves the following steps (Fig. 4):

1. Convert the raw reads data file to standard fastq format for downstream analysis.
2. Align pass filter reads to the mouse reference genome build mm9 or later. Among alignment algorithms that are available (BWA (9), Bowtie (10), ELAND (11), and TopHat (12)). TopHat is recommended due to its ability to map across splice junctions.

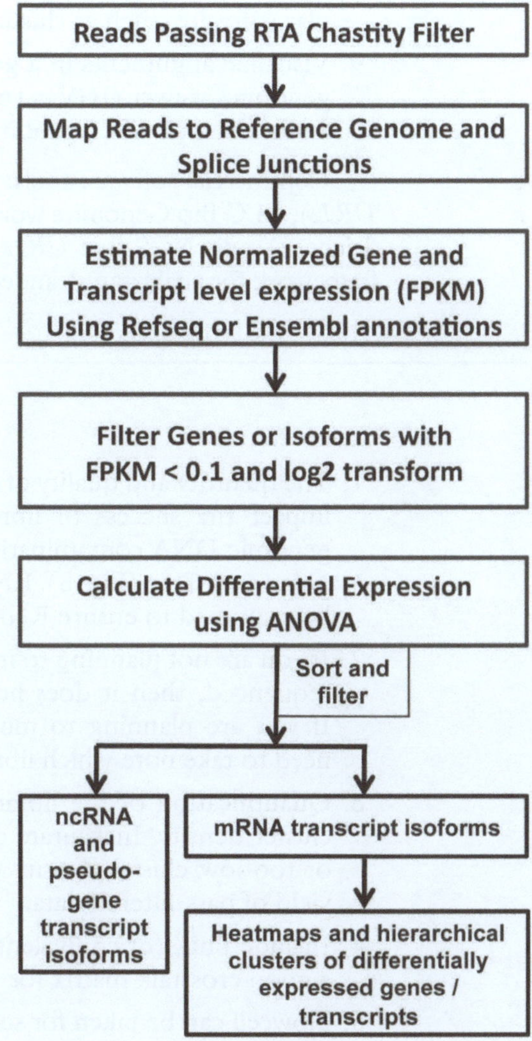

Fig. 4. Directional RNA-seq data analysis workflow.

3. Reverse the reported strand in the resulting Sequence Alignment/Map (SAM) format (13) file. This is a necessary step as sequencing was done on the antisense strand.
4. Estimate transcript abundance in each sample using Cufflinks (4).
5. Filter out transcripts that are expressed at very low levels (FPKM < 0.1) in all samples
6. Compute differential transcript/gene expression using CuffDiff (4) or ANOVA.
7. Sort the resulting list of differentially expressed transcripts into categories, such as mRNA transcripts, pseudogenes, ncRNA transcripts and filter based on statistical significance (e.g., fold change ≥1.5 and *p*-value <0.05).
8. Draw heatmaps and dendrograms for clusters of genes obtained by performing unsupervised hierarchical clustering using popular software, such as cluster3.0 (14) and Java Treeview (15).
9. Visualize alignments in a genome browser, such as integrated genomics viewer (IGV) (16) or the University of California at Santa Cruz (UCSC) genome browser (17).

Commercial software tools, such as Partek Genomics Suite (see *URLs*), CLC Bio Genomics workbench (see *URLs*), and Strand Life Science Avadis NGS (see *URLs*), provide a one-stop start to finish framework for analyzing stranded and unstranded RNA-seq data.

4. Notes

1. The quantity and quality of starting RNA material can critically impact the success of library preparation. RNA, free from genomic DNA contamination, should be assessed on an RNA Nano 600 Chip (Fig. 5). RNA Integrity Number (RIN) should be examined to ensure RNA quality is above 7.0.
2. If you are not planning to multiplex the libraries when they are sequenced, then it does not matter which adapters you use. If you are planning to multiplex the libraries, then you will need to take note which library is associated with which index.
3. Quantification of the library is critical for obtaining optimal cluster density. Inaccurate quantification can lead to too high or too low cluster density on the flowcell, resulting in lower yield of pass-filtered data.
4. Include PhiX (or equivalent) as a control sample as it provides a good crosstalk matrix for use in base-biased samples.
5. Flowcell can be taken for sequencing directly or stored at 4°C. It is, however, advisable to sequence the flowcell within 24 h.

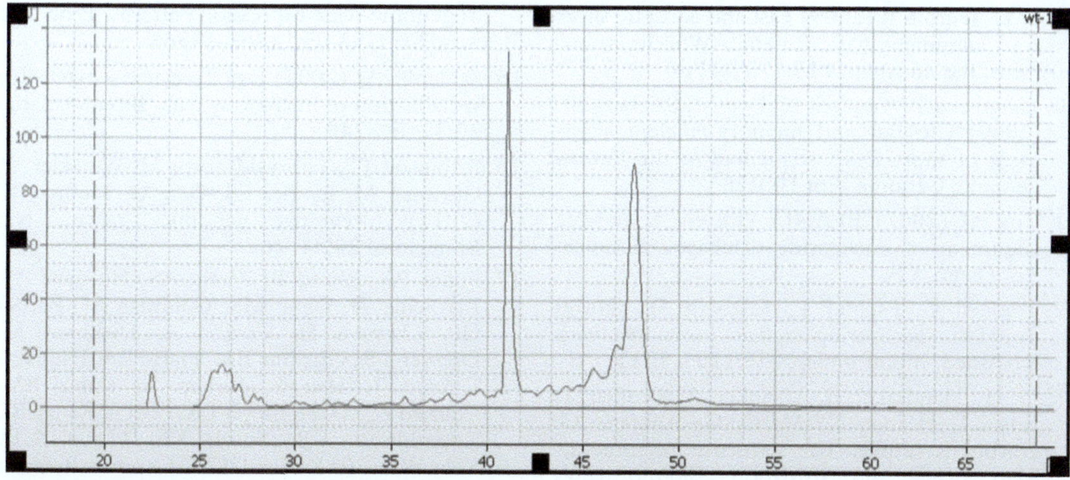

Fig. 5. Assessing quality and quantity of initial total RNA with Agilent 2100 Bioanalyzer using a Nano 6000 RNA chip. RNA isolated using Trizol reagent from a postnatal day 21 mouse retina was used to generate the ds cDNA libraries from 100 ng and 1 μg initial RNA amounts. The four prominent peaks from left to right are: spike-in marker, miRNA/tRNA/5S rRNA, 18S rRNA, and 28S rRNA.

Acknowledgment

The authors are supported by Intramural Research Program of the National Eye Institute, National Institutes of Health, Bethesda, MD, USA.

References

1. Mortazavi A, Williams BA, McCue K, Schaeffer L, Wold B (2008) Mapping and quantifying mammalian transcriptomes by RNA-Seq. Nat Methods 5:621–628
2. Bradford JR, Hey Y, Yates T, Li Y, Pepper SD, Miller CJ (2010) A comparison of massively parallel nucleotide sequencing with oligonucleotide microarrays for global transcription profiling. BMC Genomics 11:282
3. Wilhelm BT, Marguerat S, Watt S, Schubert F, Wood V, Goodhead I, Penkett CJ, Rogers J, Bahler J (2008) Dynamic repertoire of a eukaryotic transcriptome surveyed at single-nucleotide resolution. Nature 453:1239–1243
4. Trapnell C, Williams BA, Pertea G, Mortazavi A, Kwan G, van Baren MJ, Salzberg SL, Wold BJ, Pachter L (2010) Transcript assembly and quantification by RNA-Seq reveals unannotated transcripts and isoform switching during cell differentiation. Nat Biotechnol 28:511–515
5. Pal S, Gupta R, Kim H, Wickramasinghe P, Baubet V, Showe LC, Dahmane N, Davuluri RV (2011) Alternative transcription exceeds alternative splicing in generating the transcriptome diversity of cerebellar development. Genome Res 21:1260–1272
6. Brooks MJ, Rajasimha HK, Roger JE, Swaroop A (2011) Next generation sequencing facilitates quantitative analysis of wild type and Nrl−/− retinal transcriptomes. Mol Vis 17:3034–3054. Epub 2011 Nov 23
7. Levin JZ, Yassour M, Adiconis X, Nusbaum C, Thompson DA, Friedman N, Gnirke A, Regev A (2010) Comprehensive comparative analysis of strand-specific RNA sequencing methods. Nat Methods 7:709–715
8. Parkhomchuk D, Borodina T, Amstislavskiy V, Banaru M, Hallen L, Krobitsch S, Lehrach H, Soldatov A (2009) Transcriptome analysis by strand-specific sequencing of complementary DNA. Nucleic Acids Res 37:e123

9. Li H, Durbin R (2009) Fast and accurate short read alignment with Burrows–Wheeler transform. Bioinformatics 25:1754–1760
10. Langmead B, Trapnell C, Pop M, Salzberg SL (2009) Ultrafast and memory-efficient alignment of short DNA sequences to the human genome. Genome Biol 10:R25
11. Cox AJ (2007) ELAND: efficient large-scale alignment of nucleiotide databases. Illumina, San Diego, CA
12. Trapnell C, Pachter L, Salzberg SL (2009) TopHat: discovering splice junctions with RNA-Seq. Bioinformatics 25:1105–1111
13. Li H, Handsaker B, Wysoker A, Fennell T, Ruan J, Homer N, Marth G, Abecasis G, Durbin R (2009) The sequence Alignment/Map format and SAMtools. Bioinformatics 25:2078–2079
14. Eisen MB, Spellman PT, Brown PO, Botstein D (1998) Cluster analysis and display of genome-wide expression patterns. Proc Natl Acad Sci USA 95:14863–14868
15. Saldanha AJ (2004) Java Treeview – extensible visualization of microarray data. Bioinformatics 20:3246–3248
16. Robinson JT, Thorvaldsdottir H, Winckler W, Guttman M, Lander ES, Getz G, Mesirov JP (2011) Integrative genomics viewer. Nat Biotechnol 29:24–26
17. Fujita PA, Rhead B, Zweig AS, Hinrichs AS, Karolchik D, Cline MS, Goldman M, Barber GP, Clawson H, Coelho A, Diekhans M, Dreszer TR, Giardine BM, Harte RA, Hillman-Jackson J, Hsu F, Kirkup V, Kuhn RM, Learned K, Li CH, Meyer LR, Pohl A, Raney BJ, Rosenbloom KR, Smith KE, Haussler D, Kent WJ. (2011) The UCSC Genome Browser database: update 2011. Nucleic Acids Res 39(Database Issue):D876–D882.

Web Resources/URLs

RefSeq: http://www.ncbi.nlm.nih.gov/RefSeq/

Ensembl: http://www.ensembl.org

UCSC Genome Browser: http://www.genome.ucsc.edu

Illumina, Inc: http://illumina.com

Rfam: http://rfam.sanger.ac.uk/

Bowtie: http://bowtie-bio.sourceforge.net/index.shtml

Burrows-Wheeler Aligner (BWA): http://bio-bwa.sourceforge.net/

TopHat: http://tophat.cbcb.umd.edu

Cufflinks: http://cufflinks.cbcb.umd.edu

NovoAlign: http://www.novocraft.com/main/index.php

Piccard: http://picard.sourceforge.net/index.shtml

Annovar: http://www.openbioinformatics.org/annovar/

CLC Bio Genomics Workbench: http://www.clcbio.com/

Partek: http://www.partek.com/

Strand Life Sciences Avadis NGS: http://www.strandls.com/

Chapter 24

Exome Sequencing: Capture and Sequencing of All Human Coding Regions for Disease Gene Discovery

Rinki Ratna Priya, Harsha Karur Rajasimha, Matthew J. Brooks, and Anand Swaroop

Abstract

In humans, protein-coding exons constitute 1.5–1.7% of the human genome. Targeted sequencing of all coding exons is termed as exome sequencing. This method enriches for coding sequences at a genome-wide scale from 3 μg of DNA in a hybridization capture. Exome analysis provides an excellent opportunity for high-throughput identification of disease-causing variations without the prior knowledge of linkage or association. A comprehensive landscape of coding variants could also offer valuable mechanistic insights into phenotypic heterogeneity and genetic epistasis.

Key words: Targeted sequencing, Next-generation sequencing, Massively parallel sequencing, Genetic variation, Mutation, Inherited retinal disease, Neurodegeneration

1. Introduction

Recent advent of next-generation sequencing technology permits the examination of whole genome in a comprehensive and unbiased manner. Genetic variations in the coding exons (the "exome") are of significant interest for researchers as a majority of Mendelian disorders are caused by mutations in the protein-coding sequences (1). A significant decrease in the cost and time associated with exome sequencing compared to whole genome sequencing has made exome capture a preferred method for disease gene identification. Since the first "proof of principle" report demonstrated the efficacy of exome sequencing for finding a disease mutation in Freeman–Sheldon syndrome (2), this method has uncovered novel causative mutations in several Mendelian diseases (3), and more recently even for the identification of de novo mutations in sporadic cases of autism (4).

Shu-Zhen Wang (ed.), *Retinal Development: Methods and Protocols*, Methods in Molecular Biology, vol. 884, DOI 10.1007/978-1-61779-848-1_24, © Springer Science+Business Media, LLC 2012

Inherited forms of retinal degeneration commonly display monogenic inheritance, and hence positional cloning has been highly successful for mutation identification with almost 150 genes implicated so far (RetNet data base; see *URLs*). Exome capture method is promising for the gene hunt in small families or simplex cases, which are not amenable to the traditional method of positional cloning. The analysis of all coding exons will also be more productive compared to candidate gene screening, where the chromosomal location of the disease has been deciphered. Recently, exome sequencing identified the causative gene, *DHDDS*, in a single-generation family with retinitis pigmentosa (5). However, a major challenge is to assign the causality among many variants that are identified through this approach. Inheritance pattern (if even small families are available), combined with expression and functional data, should enable exome sequencing to reach its full potential as a tool for gene discovery.

Among the methods of target capture, array-based (2) and liquid-based (6) hybridization have been extended to target the human exome. At present, commercial targeted sequence capture kits are available from Agilent, Illumina, and Nimblegen. In this chapter, we describe a liquid-based hybridization capture adapted from the SureSelect Human All Exon 50 Mb Kit (see *URLs*), which includes all coding exons annotated by the GENCODE (see *URLs*) project as well as CCDS (see *URLs*) and RefSeq databases (see *URLs*). The kit also contains ten bases of flanking sequence for each targeted region and small noncoding RNAs from miRBase v.13 (see *URLs*) and Rfam (see *URLs*). The captured exon library is then subjected to cluster generation and sequencing protocols, which have been described here for Illumina platform GAIIx.

2. Materials

2.1. Reagents

1× Low TE Buffer (10 mM Tris–HCl, pH 8.0, 0.1 mM EDTA). Agencourt AMPure XP Kit (Beckman Coulter Genomics). DNA 1000 Kit (Agilent).

High Sensitivity DNA Kit (Agilent).

UltraPure™ DNase/RNase-Free Distilled Water (Invitrogen).

100% Ethanol (Sigma-Aldrich).

T4 DNA ligase buffer with 10 mM ATP (New England BioLabs or equivalent).

10 mM dNTP mix (New England BioLabs or equivalent).

T4 DNA polymerase (New England BioLabs or equivalent).

Klenow DNA polymerase (New England BioLabs or equivalent).

T4 Polynucleotide Kinase (New England BioLabs or equivalent).

10× Klenow buffer (New England BioLabs or equivalent).

Klenow fragment (3′–5′ exonuclease minus) (New England BioLabs or equivalent).

1 mM dATP (New England BioLabs or equivalent).

2× Rapid DNA ligase buffer (Enzymatics or equivalent).

Rapid DNA ligase (enzymatics or equivalent).

Adapter oligo mix.

Adapter PCR Primer 1.

Adapter PCR Primer 2.

5× Phusion HF buffer (New England BioLabs or equivalent).

Phusion high-fidelity DNA polymerase (New England BioLabs or equivalent).

SureSelect Human All Exon Plus Kit (Agilent).

Dynabeads MyOne Streptavidin T1 (Invitrogen).

Paired-End Genomic DNA Sample Prep Kit (Illumina or equivalent).

TruSeq PE Cluster Kit v5–CS–GA (Illumina or equivalent).

TruSeq SBS Kit v5-GA (Illumina or equivalent).

2.2. Equipment

Covaris S-series Single Tube Sample Preparation System, Model S2 (Covaris).

Covaris microTUBE with AFA fiber and snap cap (Covaris).

Agilent 2100 Bioanalyzer (Agilent).

Thermal cycler (Eppendorf Mastercycler or equivalent).

Microcentrifuge (Eppendorf or equivalent).

DNA LoBind Tubes, 1.5-ml (Eppendorf or equivalent).

Magna-Sep™ Magnetic Particle Separator (Invitrogen).

P10, P20, P200, and P1000 pipettes (Pipetman or equivalent).

Vacuum concentrator (Savant SpeedVac or equivalent).

Real-Time PCR System (The 7900HT Fast Real-Time PCR System or equivalent).

Nutator (BD Diagnostics or equivalent).

Ice bucket.

Powder-free gloves.

PCR tubes, strips, or plates.

Sterile, nuclease-free aerosol barrier pipette tips.

Timer.

Vortex mixer.

Heat block at 37 and 65°C.

3. Methods

To avoid cross contamination between samples, perform all dilutions and reactions in a dedicated clean area or PCR hood with UV sterilization and positive airflow. The use of filter tips is highly recommended. The completion of whole procedure takes 4–5 days (Fig. 1). However, the protocol can be stopped after any AMPure XP beads purification step (unless otherwise specified), and the samples can be stored in low-bind tube at −20°C until required.

3.1. Fragmentation

1. Dilute 3 μg of high-quality genomic DNA (see Note 1) in 120 μl of 1× Low TE Buffer in a 1.5-ml LoBind tube. Set up Covaris instrument with the settings listed in Table 1.
2. Use a tapered pipette tip to slowly transfer 120 μl DNA sample through the presplit septa (see Note 2).
3. Transfer the DNA into a fresh LoBind tube.

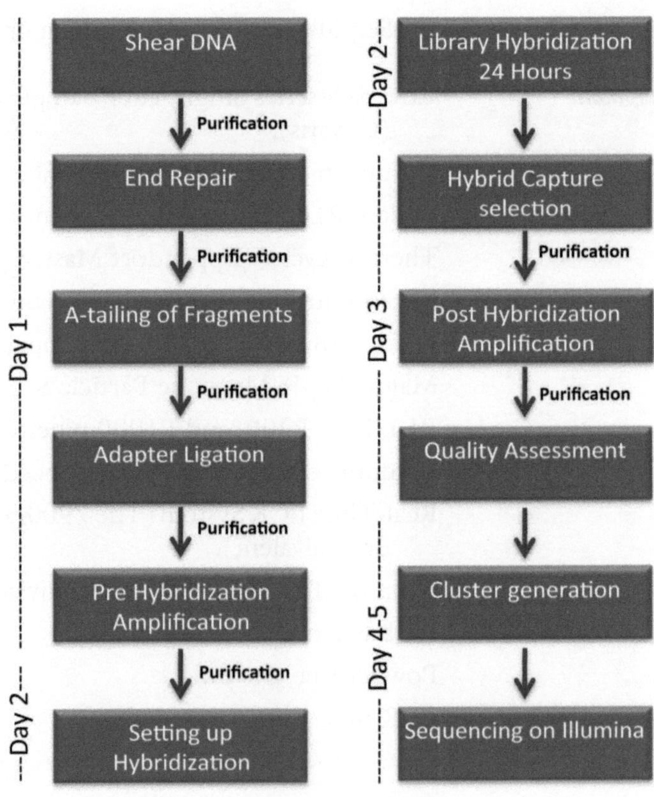

Fig. 1. Exome library preparation and sequencing workflow.

Table 1
Covaris DNA shear settings

Setting	Value
Duty cycle	10%
Intensity	5
Cycles per burst	200
Time	Six cycles of 60 s each
Set mode	Frequency sweeping
Temperature	4–7°C

3.2. Purification

1. Let AMPure XP beads come to room temperature for at least 30 min.
2. Mix the reagent well on a vortex mixer till it appears homogenous. Add 180 μl of homogenous AMPure XP beads to a 1.5-ml LoBind tube, and add the sheared DNA library (~120 μl).
3. Mix well on a vortex mixer and incubate for 5 min at room temperature.
4. Put the tube in a magnetic stand and wait for the solution to clear (3–5 min).
5. Keep the tube in the magnetic stand. Discard the cleared solution from the tube without disturbing the beads.
6. Continue to keep the tube in the magnetic stand while you dispense 500 μl of 70% ethanol in each tube.
7. Let the tube sit for 1 min to allow any disturbed beads to settle, and then remove ethanol. Repeat this step once.
8. Dry the samples on a 37°C heat block for 5 min or until the residual ethanol completely evaporates.
9. Add 30 μl nuclease-free water, mix well on a vortex mixer, and incubate for 2 min at room temperature.
10. Place the tube in the magnetic stand and leave for 2–3 min or until the solution is clear.
11. Remove approximately 30 μl of the supernatant to a fresh 1.5-ml LoBind tube.

3.3. Assess Quality with the Agilent 2100 Bioanalyzer

1. Use Agilent DNA 1000 chip and reagent kit for preparing the chip with samples and ladder.
2. Load the chip into the 2100 Bioanalyzer and start the run within 5 min after preparation.
3. Check that the electropherogram shows a distribution with a peak size of ~190 nucleotides (Fig. 2a; see Note 3).

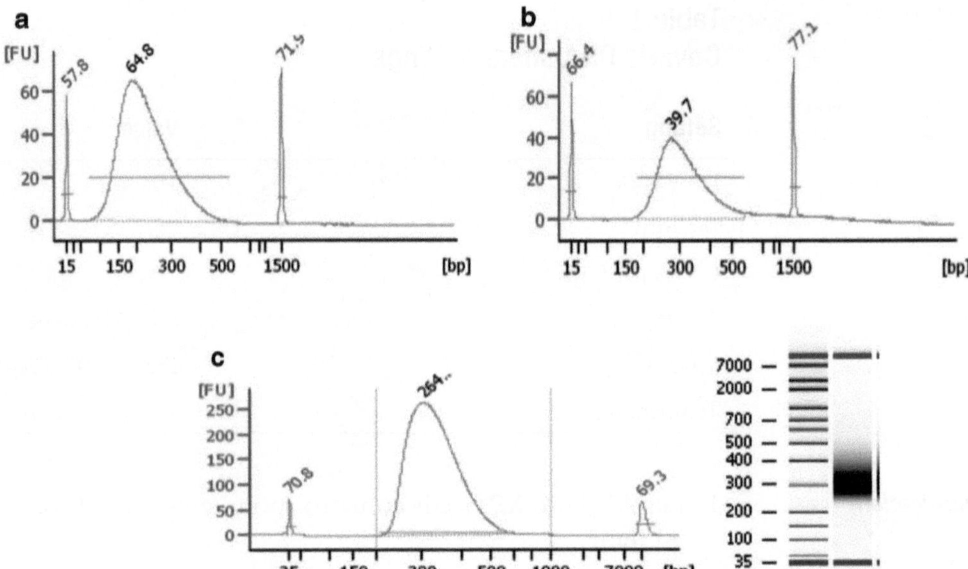

Fig. 2. Assessing quality, quantity, and size distribution with Agilent 2100 Bioanalyzer. (**a**) Analysis of sheared DNA using a DNA 1000 Bioanalyzer assay. (**b**) Analysis of amplified adapter-ligated library (pre-hybridization) using DNA 1000 Bioanalyzer assay. (**c**) Analysis of amplified capture DNA library (post-hybridization) using the High Sensitivity DNA Kit.

3.4. End Repair

1. Prepare the reaction mix, according to Table 2, on ice. Mix well on a vortex mixer. Add 71 μl of the reaction mix and 29 μl of DNA sample to each tube. Mix by pipetting (see Note 4).
2. Incubate tube in a thermal cycler for 30 min at 20°C. Do not use heated lid.
3. Purify the sample using 90 μl of homogenous AMPure XP beads, following the instruction from Subheading 3.2. Elute in 32 μl nuclease-free water.

3.5. A-Tailing

1. Prepare the reaction mix as in Table 3. Add 32 μl of each DNA sample to each well or tube. Mix by pipetting.
2. Incubate in a thermal cycler for 30 min at 37°C with the lid temperature not exceeding 50°C.
3. Purify the sample using 90 μl of homogenous AMPure XP beads, following the instruction from Subheading 3.2, and elute in 15 μl nuclease-free water.
4. Proceed immediately to adapter ligation step.

3.6. Adapter Ligation

1. Anneal the adapters (see Note 5).
2. Prepare the reaction mix as in Table 4.
3. Add 36 μl of the reaction mix and 14 μl of DNA sample to each tube. Mix by pipetting.
4. Incubate for 15 min at 20°C on a thermal cycler. Do not use a heated lid.

Table 2
End Repair Mix

Reagent	Volume (μl)
DNA sample	~29
Nuclease-free water	46
T4 DNA ligase buffer with 10 mM ATP	10
10 mM dNTPmix	4
T4 DNA polymerase	5
Klenow DNA polymerase	1
T4 polynucleotide kinase	5

Table 3
Adding "A" Bases

Reagent	Volume (μl)
DNA sample	~32
10× Klenow buffer	5
1 mM dATP	10
Klenow fragment (3′–5′ exo minus)	3

Table 4
Ligation mix

Reagent	Volume (μl)
DNA sample	14
2× Rapid ligation buffer	25
Adapter oligo mix	6
Rapid T4 DNA ligase	5

5. Purify the sample using 90 μl of homogenous AMPure XP beads, following the instruction from Subheading 3.2. Elute in 52 μl nuclease-free water.

3.7. Pre-hybridization Amplification

1. Use half of the adapter-ligated fragments for amplification and save the other half at 20°C for future use, if needed.

2. Prepare the reaction mix (25 μl) as instructed in Table 5 and add 25 μl of each DNA sample to each tube (see Note 6).
3. Mix by pipetting and run the program listed in Table 6 (see Note 7).
4. Purify the sample using 90 μl of homogenous AMPure XP beads and elute in 30 μl nuclease-free water (follow the instruction from Subheading 3.2).

3.8. Assess Quality and Quantity with Agilent 2100 Bioanalyzer

1. Use Agilent DNA 1000 chip and reagent kit for preparing the chip with samples and ladder.
2. Load the chip into the 2100 Bioanalyzer and start the run within 5 min after preparation.
3. After the run, determine the concentration of the sample by integration under the peak (see Note 8).

Table 5
Components for PCR mix

Reagent	Volume (μl)
Adapter-ligated library	25
Nuclease-free water	11.5
Adapter PCR primer 1	1
Adapter PCR primer 2	1
5× Phusion HF buffer	10
25 mM dNTP mix	0.5
Phusion high-fidelity DNA polymerase	1

Table 6
PCR program

Step	Temperature (°C)	Time
1	98	30 s
2	98	10 s
3	65	30 s
4	72	30 s
5		Repeat steps 2 through 4 for a total of 4–6 times
6	72	5 min
7	4	Hold

4. Check that the electropherogram shows a distribution with a peak size of 250 ± 10% bp (Fig. 2b).

3.9. Hybridization

1. Concentrate a 500-ng aliquot of library to 3.4 μl using a vacuum concentrator at ≤45°C (see Note 9).
2. Mix the components in Table 7 to prepare block mix in a PCR tube (see Note 10).
3. Add 3.4 μl of 147 ng/μl prepped library with 5.6 μl of SureSelect Block Mix (Tube A).
4. Mix the components in Table 8 at room temperature to prepare the hybridization buffer (Tube B) (see Note 11).
5. Dilute 1 μl RNase Block with 2 μl nuclease-free water.
6. In a separate tube, mix 5 μl of SureSelect capture library with 2 μl of diluted RNase Block (Tube C).
7. Place the tube "A" in thermal cycler with the following program in Table 9.

Table 7
Block mix

Reagent	Volume (μl)
SureSelect Block #1	2.5
SureSelect Block #2	2.5
SureSelect Block #3	0.6

Table 8
Hybridization Buffer

Reagent	Volume (μl)
SureSelect Hyb # 1	25
SureSelect Hyb # 2	1
SureSelect Hyb # 3	10
SureSelect Hyb # 4	13

Table 9
Exon-capture PCR program

Step	Temperature (°C)	Time
1	95	5 min
2	65	Hold

8. Use a heated lid on the thermal cycler at 105°C throughout.
9. Maintain the cycler at 65°C while you add 40 μl of hybridization buffer (Tube B).
10. Incubate both tubes for a minimum of 5 min at 65°C.
11. Place the capture library mix (Tube C) in cycler.
12. Incubate the samples at 65°C for 2 min.
13. Maintain the cycler at 65°C while using pipette to take 13 μl of Hybridization Buffer from the tube "B" and add it to SureSelect capture library mix tube "C."
14. Maintain the cycler at 65°C and transfer the entire content of prepped library mix in tube "A" to the hybridization solution in tube "C."
15. Seal the tube "C" carefully and incubate the hybridization mixture for 24 h at 65°C with a heated lid at (see Note 12).

3.10. Magnetic Beads' Preparation

1. Keep SureSelect Wash Buffer #2 at 65°C water bath for Subheading 3.11.
2. Let the Dynal MyOne Streptavidin T1 (Invitrogen) beads come to room temperature for at least 30 min.
3. Vigorously resuspend on a vortex mixer and take 50 μl of magnetic beads to a 1.5-ml microfuge tube.
4. Wash the beads:
   (a) Add 200 μl of SureSelect Binding buffer.
   (b) Mix the beads on a vortex mixer for 5 s.
   (c) Place the tube into a magnetic stand and wait till solution becomes clear (3–5 min).
   (d) Carefully remove and discard the supernatant.
   (e) Repeat steps (a through d) for a total of three washes.
5. Resuspend the beads in 200 μl of SureSelect Binding buffer.

3.11. Hybrid Capture Selection

1. Estimate and record the volume of hybridization that remained after 24 h incubation (see Note 13).
2. Add the hybridization mixture directly from the thermal cycler to the bead solution, and mix by inverting three to five times.
3. Incubate the hybrid-capture/bead solution on a Nutator or equivalent for 30 min at room temperature.
4. Make sure that the content is mixing properly.
5. Briefly spin in a centrifuge and separate the beads and buffer on a magnetic separator.
6. Keep the tube in the magnetic stand for 3–5 min.

7. Discard the cleared solution from the tube without disturbing the beads.
8. Resuspend the beads in 500 μl of SureSelect Wash Buffer #1 by mixing on a vortex mixer for 5 s.
9. Incubate the samples for 15 min at room temperature. Occasionally mix on a vortex mixer.
10. Briefly spin in a centrifuge, separate the beads and buffer on a magnetic separator, and remove the supernatant.
11. Wash the beads:
    (a) Resuspend the beads in 500 μl of 65°C pre-warmed SureSelect Wash Buffer #2 and mix on a vortex mixer for 5 s to resuspend the beads.
    (b) Incubate the samples for 10 min at 65°C in a heat block or equivalent. Occasionally mix on a vortex mixer.
    (c) Briefly spin in a centrifuge.
    (d) Separate the beads and buffer on a Dynal magnetic separator and remove the supernatant.
    (e) Repeat steps (a through d) for a total of three washes. Remove all traces of wash buffer before proceeding to step 12.
12. Mix the beads in 50 μl of SureSelect Elution Buffer on a vortex mixer for 5 s to resuspend the beads.
13. Incubate the samples for 10 min at room temperature. Occasionally mix on a vortex mixer.
14. Briefly spin in a centrifuge, and separate the beads and buffer on a magnetic stand.
15. Transfer the supernatant (captured DNA) to a new 1.5-ml microfuge tube.
16. Add 50 μl of SureSelect Neutralization Buffer to the captured DNA and mix briefly.
17. Purify the sample using 180 μl of homogenous AMPure XP beads following the instruction from Subheading 3.2 and elute in 30 μl nuclease-free water.

3.12. Post-hybridization Amplification

1. Prepare the reaction mix, as described in Table 10, on ice. Mix well on a vortex mixer.
2. Add 15 μl of DNA sample to 35 μl of the reaction mix.
3. Mix by pipetting, put the tube in a thermal cycler, and run the program in Table 11 (see Note 14).
4. Purify the sample using 90 μl of homogenous AMPure XP beads and elute in 30 μl nuclease-free water following the instruction from Subheading 3.2.

Table 10
Post-capture PCR master mix

Reagent	Volume (μl)
Captured DNA	15
Nuclease-free water	22.5
5× Phusion HF buffer	10
Phusion high-fidelity DNA polymerase	1
25 mM dNTP mix	0.5
SureSelect GA PCR primers	1

Table 11
Post-capture PCR program

Step	Temperature (°C)	Time
1	98	2 min
2	98	20 s
3	60	30 s
4	72	30 s
5		Repeat steps 2 through 4 for a total of 10–12 cycles
6	72	5 min
7	4	Hold

5. Assess quality and quantity with the Agilent 2100 Bioanalyzer High Sensitivity DNA assay.
6. Determine the concentration of the sample by integration under the peak.
7. The electropherogram should show a peak in the size range of approximately 300–400 nucleotides (Fig. 2c).
8. Do additional quantification by real-time PCR (see Note 15).

3.13. Cluster Generation

Cluster generation turns captured libraries into clonal clusters of ~1,000 copies each, ready for sequencing on the Genome Analyzer. During this step, samples are denatured and hybridized to flow cell. Captured DNA is used as template for second-strand synthesis and is amplified into a clonal cluster. Subsequently, clusters are linearized and active sites are blocked and sequencing primer provides a site for sequence by synthesis.

1. Dilute and denature the libraries at 8 pM (~1.6 pg/ul of 300 bp DNA) using 0.1 N NaOH (see Note 16).
2. Thaw and prepare the reagents following the instruction of Illumina Cluster Generation Kit.
3. Open and run appropriate recipe on the cluster station.
4. Follow recipe prompts to load flow cell.
5. Follow recipe prompts to load reagents.
6. Complete the cluster generation steps: hybridization, amplification, linearization, blocking, and primer hybridization.
7. Take the flow cell for sequencing (see Note 17).

3.14. Sequencing

1. Perform a pre-run wash step on sequencer.
2. Thaw and prepare sequencing reagent following the instruction from Sequencing Kit.
3. Load sequencing reagent.
4. Prime positions on the genome analyzer.
5. Clean and install prism and flow cell.
6. Check for proper reagent delivery and apply oil.
7. Perform read 1 first-base incorporation and auto-calibration.
8. Check quality metrics.
9. Continue the run for desired number of cycles.

3.15. Data Analysis

Analysis of exome sequencing data involves the following steps (Fig. 3).

1. Convert the raw reads data-file to standard fastq format for downstream analysis.
2. Align pass filter reads to the human reference genome build hg19 (see Note 18). Among alignment algorithms that are available [BWA (7), Bowtie (8), ELAND (9), and NovoAlign (see *URLs*)], BWA is the preferred choice because of speed and accuracy.
3. Estimate the depth of coverage of the target regions to ascertain capture efficiency. BEDTools (10) is useful in estimating the depth of coverage of the target.
4. Realign reads around indels, filter duplicate reads (PCR bias) using Piccard, and recalibrate base qualities.
5. Variant calling (single nucleotide variations, insertions, deletions, and structural variations) can be performed by Samtools (11), Umake (12), GATK (13, 14), MAQ (15), CLC Bio (see *URLs*), Partek genomics suite (see *URLs*), or Avadis NGS (see *URLs*). A combination of Samtools for likelihood calculation and Umake for variant calling and filtration may provide an optimal balance between false discovery rate and false negatives.

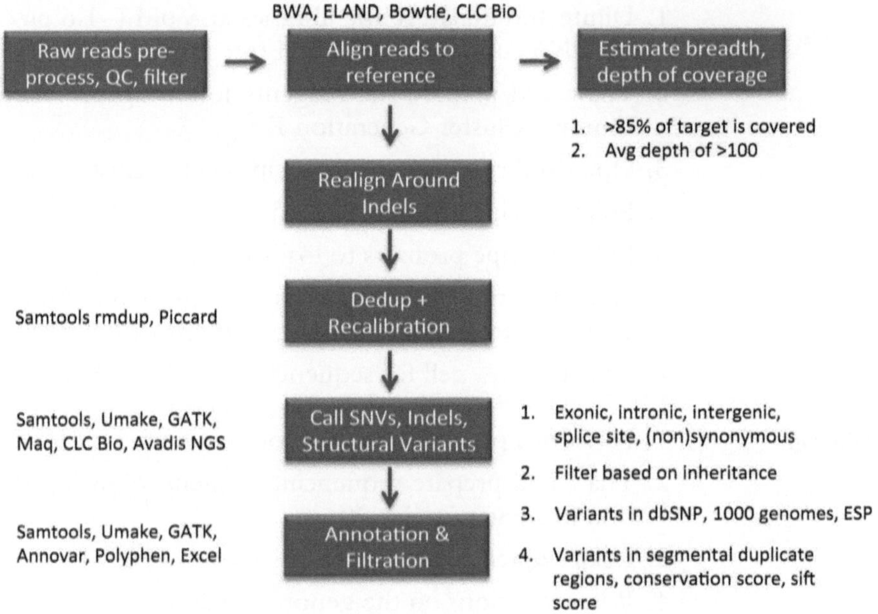

Fig. 3. Exome sequencing data analysis workflow.

6. Filter the variants, based on technical parameters such as minimum and maximum depth of coverage, base quality, mapping quality, and allelic imbalance. Umake may offer robust variant filtering options for SNVs.
7. Functionally annotate the variants based on gene function (e.g., synonymous, non-synonymous, stopgain, stoploss, homozygous or heterzygous, and splice site). Annovar (16) can provide functional annotation of genetic variants predicted from deep sequencing data.
8. Filter the variants based on family pedigree (in the case of related individuals) and phenotype (clinical) information (if available).
9. Visualize variants in a genome browser, such as integrated genomics viewer (IGV) (17) or University of California at Santa Cruz (UCSC) genome browser (18).

4. Notes

1. The quantity and quality of starting DNA material can critically impact the success of library preparation. Quantification of genomic DNA can be unreliable; thus, SYBRGreen-based assays are recommended in addition to spectrophotometric methods. Measuring concentration after diluting the sample in

TE is also a good practice. If the starting material is limited, one can work with whole genome-amplified (WGA) DNA. However, WGA can cause drop in capture efficiency.

2. Do not introduce any air bubbles as they interfere with the shearing process.
3. Unexpected size distribution after shearing can be observed at times because of inappropriate settings used on Covaris or using the wrong kind of vials. Also temperature increases when shearing multiple samples. Make sure that the water temperature is between 4 and 7°C.
4. Subheadings 3.4 through 3.7 (end repair, A-tailing, adapter ligation, and pre-hybridization amplification) can be performed using homemade recipes. Alternatively, Paired-End Genomic DNA Sample Prep Kit (Illumina or equivalent) includes all the reagents, adapters, and primers that are needed for the above-mentioned steps.
5. The choice of adapters depends on the platform used for sequencing of the library. For sequencing on Illumina platform, the adaptors comprised an alternative set of two oligonucleotides: 5′-ACACTCTTTCCCTACACGACGCTCTTC CGATCxT (x = phosphorothioate bond) and 5′-phosphate-GATCGGAAGAGCGGTTCAGCAGGAATGCCGAG (19).
6. Use primers that are appropriate for your adapter sequence. For sequencing paired-read libraries on Illumina platform, the primer sequences are: 5′-CAAGCAGAAGACGGCATACGA GATCGGTCTCGGCATTCCTGCTGAACCGCTCT TCCGATCxT and 5′-AATGAT ACGGCGACCACCGA GATCTACACTCTTTCCCTACACGACGCTCTT CCGATCxT (x = phosphorothioate bond resistant to excision by 3′–5′ exonucleases) (19). The adapters and primers should be PAGE purified.
7. Keep the amplification to four cycles to minimize PCR-related bias or nonspecific amplification. However, if the yield is too low, the number of cycles can be increased to a maximum of six cycles.
8. A minimum of 500 ng of library is required for hybridization. Low recovery can be caused by PCR failure or inefficient cleanup. Repeating PCR with fresh reagents can help.
9. Keep checking the volume in the tube while using vacuum concentrator. If it completely dries off, add 3.4 μl nuclease-free water, mix on a vortex mixer, and spin in a microfuge for 1 min.
10. Use a good-quality tube/plate for performing hybridization. Test for evaporation by a mock hybridization using 27 μl of hybridization buffer, with 24 h incubation at 65°C with a

heated lid. Do not use the tube/plate if less than 20 μl of reaction mixture remains after the hybridization.

11. If precipitate forms, warm the hybridization buffer at 65°C for 5 min.
12. If the lid temperature can be adjusted in the thermal cycler (e.g., Eppendorf Mastercycler), set it to 75°C for minimum evaporation.
13. Check the volume left after 24 h of hybridization. Volume <20 μl can result in suboptimal capture performance.
14. Subheading 3.12 uses half the volume of the capture library for amplification. Keep the amplification to ten cycles in order to minimize PCR-related bias or nonspecific amplification. Check the yield on a bioanalyzer. If the yield is too low, cycles can be adjusted to a maximum of 12 cycles.
15. Quantification of the library is critical for obtaining optimal cluster density. Inaccurate quantification can lead to too high or too low cluster density on the flow cell, resulting in lower yield of pass-filtered data.
16. Include PhiX (or equivalent) as a control sample as it provides a good cross-talk matrix for use in base-biased samples.
17. Flow cell can be taken for sequencing directly or stored at 4°C. It is, however, advisable to sequence the flow cell within 24 h.
18. Paired-end sequencing is recommended for better alignment of sequencing reads.

Acknowledgment

The authors are supported by Intramural Research Program of the National Eye Institute, National Institutes of Health, Bethesda, MD, USA.

References

1. Stenson PD et al (2009) The human gene mutation database: 2008 update. Genome Med 1:13
2. Ng SB et al (2009) Targeted capture and massively parallel sequencing of 12 human exomes. Nature 461:272–276
3. Ku CS, Naidoo N, Pawitan Y (2010) Revisiting Mendelian disorders through exome sequencing. Hum Genet 129:351–370
4. O'Roak BJ et al (2011) Exome sequencing in sporadic autism spectrum disorders identifies severe de novo mutations. Nat Genet 43: 585–589
5. Züchner S et al (2011) Whole-exome sequencing links a variant in DHDDS to retinitis pigmentosa. Am J Hum Genet 88:201–206
6. Bainbridge MN et al (2010) Whole exome capture in solution with 3 Gbp of data. Genome Biol 11:R62
7. Li H, Durbin R (2009) Fast and accurate short read alignment with Burrows-Wheeler transform. Bioinformatics 25:1754–1760

8. Langmead B et al (2009) Ultrafast and memory-efficient alignment of short DNA sequences to the human genome. Genome Biol 10:R25
9. Cox AJ (2007) ELAND: efficient large-scale alignment of nucleotide databases. Illumina, San Diego
10. Quinlan AR, Hall IM (2010) BEDTools: a flexible suite of utilities for comparing genomic features. Bioinformatics 26:841–842
11. Li H et al (2009) The sequence alignment/map (SAM) format and SAMtools. Bioinformatics 25:2078–2079
12. Li Y et al (2011) Low-coverage sequencing: implications for design of complex trait association studies. Genome Res 21:940–951
13. McKenna A et al (2010) The Genome Analysis Toolkit: a MapReduce framework for analyzing next-generation DNA sequencing data. Genome Res 20:1297–1303
14. DePristo M et al (2011) A framework for variation discovery and genotyping using next-generation DNA sequencing data. Na Genet 43:491–498
15. Li H, Ruan J, Durbin R (2008) Mapping short DNA sequencing reads and calling variants using mapping quality scores. Genome Res 18:1851–1858
16. Wang K, Li M, Hakonarson H (2010) ANNOVAR: functional annotation of genetic variants from next-generation sequencing data. Nucleic Acids Res 38:e164
17. Robinson JT et al (2011) Integrative genomics viewer. Nat Biotechnol 29:24–26
18. Fujita PA et al (2011) The UCSC genome browser database: update 2011. Nucleic Acids Res 39:D876–D882
19. Bentley DR et al (2008) Accurate whole human genome sequencing using reversible terminator chemistry. Nature 456:53–59

Web Resources/URLs

RetNet: http://www.sph.uth.tmc.edu/retnet/

GENCODE: http://genecodes.com/

CCDS: http://www.ncbi.nlm.nih.gov/CCDS/

RefSeq: http://www.ncbi.nlm.nih.gov/RefSeq/

SureSelect Human All Exon 50 Mb kit: http://www.home.agilent.com/

miRBase: http://www.mirbase.org/

Rfam: http://rfam.sanger.ac.uk/

Bowtie: http://bowtie-bio.sourceforge.net/index.shtml

Burrows-Wheeler Aligner (BWA): http://bio-bwa.sourceforge.net/

NovoAlign: http://www.novocraft.com/main/index.php

Genome Analysis Toolkit (GATK): http://www.broadinstitute.org/gsa/wiki/index.php/Main_Page

Umake: http://genome.sph.umich.edu/wiki/UMAKE

Piccard: http://picard.sourceforge.net/index.shtml

Annovar: http://www.openbioinformatics.org/annovar/

CLC Bio Genomics Workbench: http://www.clcbio.com/

Partek: http://www.partek.com/

Strand Life Sciences Avadis NGS: http://www.strandls.com/

Chapter 25

Determination of Posttranslational Modifications of Photoreceptor Differentiation Factor NRL: Focus on SUMOylation

Jerome E. Roger, Jacob Nellissery, and Anand Swaroop

Abstract

Conjugation of SUMO (small ubiquitin-related modifier 1) is a critical posttranslational modification, with significant impact on protein function/activity. Here, we describe direct SUMOylation of GST (glutathione S-transferase)-fusion protein and immunoprecipitation assays for investigating SUMOylation of any protein of interest. We have employed these methods to examine SUMOylation of the basic-motif leucine zipper transcription factor NRL.

Key words: Retina, Photoreceptor, Vision, Transcription factor, Protein modification, Protein function, SUMO1, Immunoprecipitation, Immunoblotting, GST

1. Introduction

Posttranslational modifications (PTMs), such as phosphorylation, acetylation, ubiquitination, and SUMOylation, are a rapid and transient way for regulating protein activity/function (1–3). They represent an adaptive, quick, and crucial response to changes in the cellular environment. PTMs can also affect DNA/protein or protein–protein interaction and subcellular localization, providing a high level of complexity in modulating cellular physiology (4–6).

SUMOylation is a protein-based modification similar to ubiquitination, controlled by an enzymatic cascade (7). Even though small ubiquitin-like modifier (SUMO) and ubiquitin are structurally close and target lysine residues, their conjugation to the same target protein may have different functional consequences (8). SUMO is synthesized as a precursor, which needs to be processed to reveal C-terminal di-glycine motif. The mature form is then activated by SUMO-activating enzyme (E1) and transferred to the substrate by

Shu-Zhen Wang (ed.), *Retinal Development: Methods and Protocols*, Methods in Molecular Biology, vol. 884, DOI 10.1007/978-1-61779-848-1_25, © Springer Science+Business Media, LLC 2012

SUMO-conjugating enzyme (E2). Therefore, the C-terminal part of SUMO is covalently linked to the substrate protein by ε-amino group of lysine amino acid that is present within a short consensus sequence ψKX(E/D), where ψ is a hydrophobic residue. In contrast to ubiquitination, E3-ligase activity can augment but is not necessary in vitro for protein SUMOylation (9–11). In vertebrates, three SUMO proteins have been identified: SUMO1, SUMO2, and SUMO3. SUMO proteins show functional redundancy with the loss of SUMO1 being compensated by SUMO2/3 (12, 13). In general, only a small proportion of the substrate is SUMOylated, and this modification is transient and removed by the action of deSUMOylating enzymes explaining the relative recent discovery of this type of PTMs (14, 15). Thus, the development of reliable techniques is crucial for analyzing SUMOylation. For instance, site-directed mutagenesis (not elaborated here) is critical for defining lysine residues conjugated to SUMO.

NRL is a basic motif-leucine zipper transcription factor that determines rod photoreceptor cell fate and maintains homeostasis (16). We have used NRL protein as a model to describe the SUMOylation assays. The wild-type (WT) and mutant NRL expression constructs, previously employed for in vivo functional assays (17), are utilized here for in vitro conjugation of SUMO1 and immunoprecipitation. The protocols can be applied to any protein of interest (POI).

2. Materials

Prepare all solutions using ultrapure water and analytical grade reagents. All reagents can be prepared and stored at room temperature (RT), unless indicated otherwise.

2.1. GST-Protein Purification

1. Phosphate-buffered saline (PBS), pH 7.4.
2. Glutathione Sepharose column, 1 ml resin from GE Lifesciences.
3. Lysis buffer: 20 mM Tris–HCl, 150 mM NaCl, 1% Triton X-100, pH 7.4.
4. Elution buffer: PBS + 20 mM reduced glutathione.
5. Centricon spin concentrator with 10 kDa cutoff from Millipore.

2.2. In Vitro SUMOylation Assay

1. SUMOlink™ in vitro SUMOylation kits from Active Motif.
2. Microcentrifuge.
3. Waterbath/incubator setup at 30°C.
4. 2× SDS-PAGE loading buffer: 130 mM Tris–HCl, pH 6.8, 4% SDS, 0.02% Bromophenol Blue, 100 mM DTT, 20% glycerol.
5. Heating block setup at 90°C.

6. Equipment to run and transfer mini-gel SDS-PAGE.

2.3. Immunoprecipitation

1. HEK293T cells.
2. Dulbecco's modified Eagle's medium containing 10% fetal calf serum, 100 U/ml penicillin, and 100 μg/ml streptomycin.
3. Fugene 6 from Roche Diagnostics.
4. RIPA buffer: 150 mM NaCl, 1% NP-40, 0.5% DOC, 0.1% SDS, and 50 mM Tris–HCL, pH 7.5.
5. *N*-Ethylmaleimide (NEM) from Sigma-Aldrich.
6. Protease inhibitors from Roche Diagnostics.
7. Lysis buffer: RIPA buffer, 20 mM NEM, and protease inhibitors.
8. Monoclonal anti-SUMO1 from Invitrogen.
9. Monoclonal anti-FLAGM2 from Sigma-Aldrich.

3. Methods

3.1. Bacterial Expression Purification of GST-Tagged Proteins

1. Inoculate 5 ml of LB medium plus ampicillin (50 μg/ml) in a 15-ml tube with a bacterial colony from an agar plate containing BL21 cells transformed with pGEX4T2 plasmid containing the WT cDNA of interest and cDNA constructs mutated at putative SUMOylation sites (lysines).
2. Incubate the culture at 37°C overnight with shaking at 225 rpm.
3. Inoculate 500 ml of LB containing 50 μg/ml ampicillin in a conical flask with the overnight culture.
4. Incubate at 37°C in a shaker (225 rpm) for 3–4 h until the OD of the culture reaches 1.
5. Remove the flask from the shaker. Transfer 1 ml of the bacterial culture to a 1.5-ml tube for a quick SDS-PAGE analysis; step 7. To the remaining culture, add isopropyl β-D-1-thiogalactopyranoside (IPTG) to a final concentration of 0.3 mg/ml. Place the flask in a shaker incubator set at 25°C, 225 rpm. Keep overnight (8–10 h) for inducing expression of the recombinant GST protein.
6. Remove the flask from the incubator and chill on ice. Transfer 1 ml of the induced bacterial culture to a 1.5-ml tube for a quick SDS-PAGE analysis to check protein expression. Transfer the remaining culture to a 500-ml centrifuge bottle and spin at 10,000 × *g* for 30 min. Remove the bottle from the centrifuge, decant the supernatant, and wash the pellet by resuspending in 300 ml of PBS and spinning as described. Freeze the final pellet at –20°C until step 9.

7. Spin the 1-ml aliquot from steps 5 and 6 in a microcentrifuge (10,000 × *g*, 5 min), discard the supernatant, resuspend the pellet in 250 µl of 2× SDS-PAGE loading buffer, boil for 5 min, cool, and sonicate (5–6 pulses using a microtip) to shear the DNA. Spin the sample again (10,000 × *g*, 10 min) and collect the supernatant.
8. Analyze 20 µl of the supernatant on a 12% polyacrylamide mini gel (SDS-PAGE). Stain the gel with Coomassie colloidal blue and visualize the protein bands using a light box (see Note 1).
9. Thaw the pellet (from step 6) on ice; add 5 ml of the lysis buffer plus 2× (final concentration) of protease inhibitor cocktail. Lyse the bacterial cells using an ultra-sonicator, micro-tip, 50% power setting, 10–12 pulses (10 s each, with a 30-s cooling in between). Complete lysis is indicated by suspension becoming thinner and cloudy white in color. Keep the sample on ice throughout this procedure.
10. Spin the lysed sample at 10,000 × *g* for 30 min to remove the insoluble bacterial debris. Transfer the supernatant containing the soluble protein into another tube.
11. Equilibrate a Glutathione Sepharose column, with PBS (five-column volumes). This can be done by attaching a 10-ml syringe containing PBS to the column and pushing the plunger very slowly. Follow the manufacturer's instructions; never let the column dry at any of the steps.
12. Load the protein extract from step 10 onto the column very slowly using a 10-ml syringe. Alternately, the sample can be loaded onto the column by gravity flow using the syringe barrel connected to the column (as a reservoir) and letting the extract flow through very slowly (see Note 2).
13. Wash the column with eight- to ten-column volumes of PBS to remove any nonspecifically bound bacterial proteins. The washing step can be done using a syringe as described or by gravity flow. Collect the washes.
14. Elute the bound GST-POI with 10 ml of PBS containing 20 mM reduced glutathione (add fresh). Collect 1-ml fractions and determine the protein concentration. Pool the fractions containing the protein peak (usually fractions 1–5).
15. Concentrate the pooled fractions using Centricon spin concentrator; determine the protein concentration of the final solution and store in 50-µl aliquots at –80°C.

3.2. *In Vitro* SUMOylation Assay

1. The components listed in Table 1 are added in the indicated order.
2. Mix the tubes gently and incubate at 30°C for 3 h.
3. Stop the reaction by adding 20 µl of 2× SDS-PAGE loading buffer.

Table 1
Composition of SUMO1 conjugation reaction

Targeted protein	P53	P53	GST	POI WT	POI WT	POI mut1	POI mut2
SUMO1	+		+		+	+	+
mut SUMO1		+		+			
H_2O	11	11	11	11	11	11	11
Protein buffer	1	1	1	1	1	1	1
5× SUMOylation buffer	4	4	4	4	4	4	4
Targeted protein (0.5 μg/μl)	1	1	1	1	1	1	1
E1-activating enzyme	1	1	1	1	1	1	1
E2-conjugating enzyme	1	1	1	1	1	1	1
SUMO-1	1	1	1	1	1	1	1
Total volume	20	20	20	20	20	20	20

Protein of interest (POI). p53 is used as a positive control. mut SUMO1 is a mutated form of SUMO1 unable to conjugate to the target protein

4. Store at −20°C or proceed directly to the analysis of the reaction products by immunoblotting.
5. Heat samples at 90°C for 5 min and centrifuge briefly to collect all liquid at the bottom of tube.
6. Load 5–10 μl of each sample in parallel with appropriate molecular weight markers, on 10% polyacrylamide mini gel (SDS-PAGE), and run until migration front is out of the gel. Two identical gels are run in parallels; one will be used for probing with anti-POI and the other with anti-SUMO1 (see Note 3).
7. Transfer protein to nitrocellulose membrane.
8. Block membrane with PBS/5% skim milk for 1 h.
9. Dilute antibodies at the appropriate dilution in PBS/5% skim milk and incubate overnight at 4°C. One membrane is incubated with rabbit anti-SUMO1 (1/4,000) and one membrane is probed with rabbit anti-p53 (1/5,000) or antibody specific for the POI.
10. Rinse the membrane three times for 10 min in PBS/0.1% Tween-20.
11. Add appropriate secondary HRP-conjugated antibodies at the recommended dilution.
12. Rinse the membrane three times for 10 min in PBS/0.1% Tween-20.
13. Wash once with PBS for 5 min and use chemiluminescent substrate according to the manufacturer's instruction. The

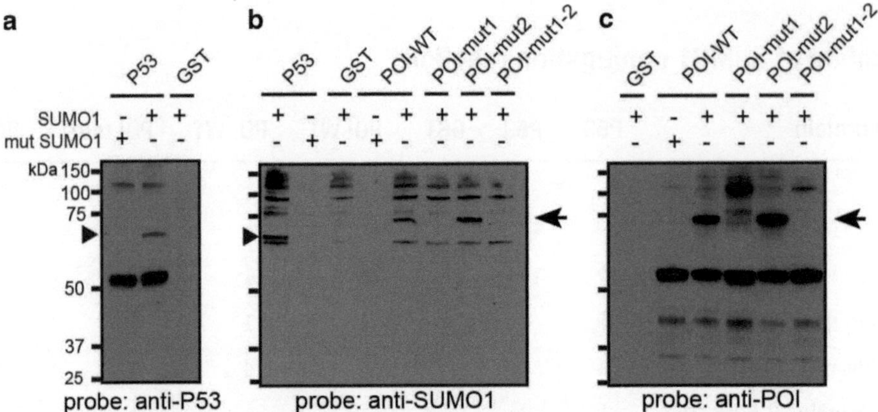

Fig. 1. SUMOylation of p53 and protein of interest (POI) with SUMO1. (**a**) Immunoblot probed with anti-p53 corresponding to the control experiments with p53 and Glutathione S-transferase (GST) in the presence of WT or mutant SUMO1. p53 was used as a positive control and showed an extra-band in the presence of SUMO1, but not with the mutant form of SUMO1. GST, used as a negative control, was not recognized by p53 antibody. (**b**) Under our assay conditions, p53, POI-WT, and POI-mut2 are conjugated with SUMO1 and not with mutant SUMO1. GST, POI-mut1, and the double-mutant POI-mut1–2 were not conjugated with SUMO1. (**c**) The POI tagged with GST showed a molecular weight of about 50 kDa. The extra-band observed with POI-WT and POI-mut2 corresponded to the conjugated forms with SUMO1. (**b**) The immunoblot probed with anti-SUMO1. (**c**) The immunoblot probed with anti-POI antibodies. *Arrowheads* show SUMOylated p53 control, and *arrows* indicate SUMOylated POI.

type of data obtained is illustrated in Fig. 1 with the WT form of the GST-tagged POI-WT, and mutant forms for two putative SUMOylation sites, POI-mut1, POI-mut2, and double-mutant POI-mut1–2. In this example, the in vitro SUMOylation assay revealed the absence of SUMOylation for POI-mut1 indicating that the PTMs occur on the amino acid mutated in POI-mut1 (see Note 4).

3.3. Immunoprecipitation

To test the conjugation of SUMO1 with the POI by immunoprecipitation, the following combinations of expression vectors are usually used for the initial transfection:

- Empty vector + SUMO1-FLAG (or any tagged SUMO1 construct): it will be the negative control
- POI alone
- POI-WT + SUMO1-FLAG
- POI-mut1 corresponding to the mutated amino acid, where the conjugation of SUMO1 occurs in WT
- POI-mut1 + SUMO1-FLAG

1. After trypsin treatment, determine the number of cells in suspension, plate 2.4×10^6 cells on a Petri dish, and incubate at 37°C.
2. The following day, mix 18 μl of Fugene + 182 μl of OptiMEM per transfection and incubate for 15 min.

3. Mix with 6 μg of total DNA (DNA: Fugene 6, ratio 1:3).
4. Tap 25 times. Keep at RT for 20–25 min.
5. Transfect the plates with the mix OptiMEM + Fugene 6/DNA. Disperse this mix by drops all over the plate. Do the same for the negative control (empty vector).
6. After 48 h, remove the medium, wash once with cold PBS, and add 800 μl of lysis buffer on each Petri dish (see Note 5). Incubate on ice for 10 min, use a cell scraper, and transfer the protein extract into a 1.5-ml tube (see Note 6).
7. Vortex vigorously, sonicate, and centrifuge at 16,000 × *g* for 10 min.
8. Transfer the supernatant and save 50 μl as an input to check the expression of the POI WT and mutant in transfected cells.
9. Wash 120 μl/sample of recombinant protein G agarose beads three times with PBS and resuspend in 120 μl of lysis buffer.
10. Clear the lysate by incubating at 4°C on a rotating wheel, and preclear the protein extract with 40 μl of beads to remove nonspecific binding.
11. After 1 h, centrifuge and transfer the supernatant.
12. Add 1 μg of monoclonal anti-FLAG (or the corresponding tag fused with SUMO1) and incubate overnight at 4°C on a rotating wheel.
13. The following day, add 80 μl of beads in each tube, and incubate at 4°C for 3 h on the rotating wheel.
14. Centrifuge briefly to pellet the beads–antibodies–proteins complex, remove the supernatant, and wash with PBS/0.5% Triton by tapping the tube.
15. Repeat step 14 five times.
16. Centrifuge, remove supernatant, and add 80 μl of 4× SDS-loading dye.
17. Vortex vigorously, boil the sample for 10 min, and centrifuge at 16,000 × *g* for 5 min.
18. Transfer the supernatant into a new tube.
19. Proceed to immunoblot analysis as described previously, with 40 μl from the immunoprecipitation reaction and 15 μl of the input.

4. Notes

1. A distinct band corresponding to the size of the fusion protein should be present in the sample that is taken after IPTG induction. A less intensity band of a similar size is often present in the

sample taken before IPTG induction; but the intensity of the band should increase five to ten times after IPTG induction.

2. Slow flow rate is important for efficient binding of GST to glutathione beads. Collect the flow through into another tube and save.
3. Five microliter of reaction is generally sufficient to give a strong signal.
4. In this case, the POI-WT has a molecular weight of 50 kDa when not conjugated to SUMO1. After SUMOylation, the non-conjugated form is still present but the molecular weight of the conjugated POI (70 kDa) indicates di-SUMOylation.
5. Add NEM in solution just before use to prevent hydrolysis of the maleimide group.
6. The IP protocol used for transfecting cells can be easily adapted for the analysis of tissue using monoclonal anti-SUMO1. For each tissue, the amount of starting material must be determined. After lysis of the tissue, initiate the protocol from step 7.

Acknowledgment

The authors are supported by intramural program of the National Eye Institute, National Institutes of Health, Bethesda, MD, USA.

References

1. Hunter T (2007) The age of crosstalk: phosphorylation, ubiquitination, and beyond. Mol Cell 28:730–738
2. Sims RJ 3rd, Reinberg D (2008) Is there a code embedded in proteins that is based on post-translational modifications? Nat Rev Mol Cell Biol 9:815–820
3. Ulrich HD (2008) The fast-growing business of SUMO chains. Mol Cell 32:301–305
4. Andreou AM, Tavernarakis N (2009) SUMOylation and cell signalling. Biotechnol J 4:1740–1752
5. Geiss-Friedlander R, Melchior F (2007) Concepts in sumoylation: a decade on. Nat Rev Mol Cell Biol 8:947–956
6. Zhao J (2007) Sumoylation regulates diverse biological processes. Cell Mol Life Sci 64:3017–3033
7. Wang Y, Dasso M (2009) SUMOylation and deSUMOylation at a glance. J Cell Sci 122:4249–4252
8. Gill G (2004) SUMO and ubiquitin in the nucleus: different functions, similar mechanisms? Genes Dev 18:2046–2059
9. Rytinki MM, Kaikkonen S, Pehkonen P, Jaaskelainen T, Palvimo JJ (2009) PIAS proteins: pleiotropic interactors associated with SUMO. Cell Mol Life Sci 66:3029–3041
10. Weissman AM (2001) Themes and variations on ubiquitylation. Nat Rev Mol Cell Biol 2:169–178
11. Wilkinson KA, Nishimune A, Henley JM (2008) Analysis of SUMO-1 modification of neuronal proteins containing consensus SUMOylation motifs. Neurosci Lett 436: 239–244
12. Evdokimov E, Sharma P, Lockett SJ, Lualdi M, Kuehn MR (2008) Loss of SUMO1 in mice affects RanGAP1 localization and formation of PML nuclear bodies, but is not lethal as it can be compensated by SUMO2 or SUMO3. J Cell Sci 121:4106–4113

13. Zhang FP, Mikkonen L, Toppari J, Palvimo JJ, Thesleff I, Janne OA (2008) SUMO-1 function is dispensable in normal mouse development. Mol Cell Biol 28:5381–5390
14. Dou H, Huang C, Van Nguyen T, Lu LS, Yeh ET (2011) SUMOylation and de-SUMOylation in response to DNA damage. FEBS Lett 585:2891–2896
15. Matunis MJ, Coutavas E, Blobel G (1996) A novel ubiquitin-like modification modulates the partitioning of the Ran-GTPase-activating protein RanGAP1 between the cytosol and the nuclear pore complex. J Cell Biol 135: 1457–1470
16. Swaroop A, Kim D, Forrest D (2010) Transcriptional regulation of photoreceptor development and homeostasis in the mammalian retina. Nat Rev Neurosci 11:563–576
17. Roger JE, Nellissery J, Kim DS, Swaroop A (2010) Sumoylation of bZIP transcription factor NRL modulates target gene expression during photoreceptor differentiation. J Biol Chem 285:25637–25644

Index

Shu-Zhen Wang (ed.), *Retinal Development: Methods and Protocols*, Methods in Molecular Biology, vol. 884,
DOI 10.1007/978-1-61779-848-1, © Springer Science+Business Media, LLC 2012

MIX
Papier aus verantwortungsvollen Quellen
Paper from responsible sources
FSC® C105338

If you have any concerns about our products,
you can contact us on
ProductSafety@springernature.com

In case Publisher is established outside the EU,
the EU authorized representative is:
Springer Nature Customer Service Center GmbH
Europaplatz 3, 69115 Heidelberg, Germany

Printed by Libri Plureos GmbH
in Hamburg, Germany